National Conference on Advanced Materials and Applications (NCAMA 2019)

IOP Conference Series: Materials Science and Engineering
Volume 798

Raipur, India
21 - 22 December 2019

ISBN: 978-1-7138-2798-6
ISSN: 1757-8981

Printed from e-media with permission by:

Curran Associates, Inc.
57 Morehouse Lane
Red Hook, NY 12571

CURRAN ASSOCIATES INC. proceedings.com

Some format issues inherent in the e-media version may also appear in this print version.

Institute of Physics
Dirac House, Temple Back
Bristol BS1 6BE UK

Phone: 44 1 17 929 7481
Fax: 44 1 17 920 0979

techtracking@iop.org

Additional copies of this publication are available from:

Curran Associates, Inc.
57 Morehouse Lane
Red Hook, NY 12571 USA
Phone: 845-758-0400
Fax: 845-758-2633
Email: curran@proceedings.com
Web: www.proceedings.com

TABLE OF CONTENTS

SEMICONDUCTOR PHYSICS

SOLAR CELL

NANOTECHNOLOGY

THEORETICAL PHYSICS

PREFACE

The National Conference on Advanced Materials and Applications (NCAMA-2019) has successfully taken place at National Institute of Technology (NIT), Raipur, India during December 21-22, 2019. NCAMA-2019 continues to be a forum mainly for the Physics, Chemistry, Material Science professional and academic communities engaged in research and development of smart luminescence materials, light emitting diodes, photovoltaic cells, thermoelectric device engineering, and related areas of engineering.

The major goal of NCAMA-2019 was to bring together scientists and practitioners, engineers and industry researchers to share their experience and innovations. The two-day conference provided a great opportunity to initiate discussions on topical issues of research & development on advanced materials testing and applications.

Scientists and practitioners, mostly from Chhattisgarh, Odisha, Madhya Pradesh and other nearby states of India submitted over 75 papers. All the papers were reviewed by the editors and invited experts. The most up-to-date and comprehensive manuscripts were selected for this volume. We sincerely hope that the NCAMA-2019 proceedings will show the readers a broad overview of the latest research results from the field of photovoltaic, luminescence, thermoelectrics and optoelectronics.

We thankfully acknowledge and express our sincere gratitude to Director, Dean (Research and Development) and TEQIP coordinator of NIT Raipur, India for financial support. At last, we would like to express our deep appreciation and gratitude to all the members of the NCAMA-2019 scientific committees for their strenuous efforts. Without their assistance and dedication, it would have not been possible to hold the discussion at such a high level and to collect quality papers for compiling the conference proceedings. Finally, we want to show appreciation of all the authors, participants and editors of the proceedings for their critical contribution to the work, holding the conference and to the preparation of the papers.

With kind regards,
Ayush Khare,
Bijay K Sahoo,
Sapan M Saini

IOP Conf. Series: Materials Science and Engineering **798** (2020) 011002 doi:10.1088/1757-899X/798/1/011002

Peer review statement

All papers published in this volume of *IOP Conference Series: Materials Science and Engineering* have been peer reviewed through processes administered by the proceedings Editors. Reviews were conducted by expert referees to the professional and scientific standards expected of a proceedings journal published by IOP Publishing.

IOP Conf. Series: Materials Science and Engineering **798** (2020) 012001 doi:10.1088/1757-899X/798/1/012001

Band energies in two-band model for Fe based superconductors in the coexistence state

B Pradhan[1], P K Parida[2,3] and S Sahoo[3]

[1]Department of Physics, B.J.B. College, Bhubaneswar-751014, Odisha, India
[2]Department of Physics, Dhenkanal (A) College, Dhenkanal-759001, Odisha, India
[3]Department of Physics, NIT, Durgapur-713209, West Bengal, India

*Corresponding author's e-mail address: brundaban73@gmail.com

Abstract. The superconductivity in iron based superconductors is hotly discussed since its discovery. The pairing mechanism and the gap structures for such superconductors are the core of attraction for debate. The superconductivity exists on the suppression of the pseudo gaps like antiferromagnetism and structural distortion. The Fermi surface band structure is studied by proposing a tight binding two-orbital model in $s^{\pm}$-wave pairing symmetry for the coexistence of Jahn-Teller distortion and antiferromagnetism with superconductivity.

1. Introduction

The iron-based superconductors (FeSCs) were discovered with chemical compositions [1, 2], such as 1111-material (LaFeAsO), 122-material ($BaFe_2As_2$), 111-material (NaFeAs) and 11-material (FeSe) have opened a new field to study the high temperature superconductivity. These superconductors motivated research on two-band superconductors where the pairing between electrons is produced by inter band electron-electron repulsion [3–6]. The development of sensible microscopic models is the first step towards the elucidation of the normal state and superconducting properties of the FeSCs. The effective low-energy model may be the starting point to study the main properties of the electronic states near the Fermi level as well as their interactions[7]. The models defined in the orbital basis provide qualitative insight into the role of the orbital degrees of freedom which analyse superconductivity and other instabilities without restricting to the Fermi surfaces. These models do not go beyond mean-field and do not distinguish between high and low energies. The models which are defined in the basis of bands, they use geometry of Fermi surfaces to capture low energy physics and allow one to go beyond mean-field and investigate the interplay between magnetism and superconductivity. The drawback of these models is that, they cannot capture orbital dependent features like spontaneous orbital order.

The Fermi surface is not confined to the gaps of different pairings though the pairing in the FeSCs is a strong coupling phenomenon [8]. The observed superconducting (SC) gap in any band is not able to pass through the Fermi surface which is being interpreted by the pairing mechanism. In the Fermi surface the SC gap for FeSCs may indicate the gap symmetry for different materials [3, 9–13]. The gap symmetry in FeSCs is like $s^{\pm}$-wave which have both electron and hole pockets [4, 5, 14]. The long range antiferromagnetic order has been observed along with the tetragonal to orthorhombic lattice distortion in the parent compounds of FeSCs [1, 15, 16]. This phenomenon prompted us to propose such a model for the coexistence of the SC, AFM and Jahn-Teller (JT) distortion. The rest of the paper is presented as the theoretical model in section 2, section 3 contains the necessary calculations for the three order parameters, the results so obtained are discussed in section 4 and finally concluded in section 5.

2. Theoretical Model

A tight-binding model Hamiltonian is considered for the coexistence of the SC, AFM and JT distortion with nearest neighbour hopping in the $s^{\pm}$-wave symmetry. The Fe lattice is divided into two sub lattices due to AFM spin alignment and the hopping of conduction electrons takes place between two degenerate Fe^{+}-orbitals. this is described by the Hamiltonian H_0as

$$H_0 = \sum_{\alpha,k\sigma} \varepsilon_k \left[\left(c^{\dagger}_{i,\alpha,k,\sigma} c_{j,\alpha,k,\sigma} + c^{\dagger}_{i,\alpha,k,\sigma} c_{j,\alpha',k,\sigma}\right) + H.c.\right] \qquad ...(1)$$

Here i and j represent two neighbouring sites, α= 1 and 2 are the two JT distorted orbitals with momentum k and spin σ. The conduction electron creation (annihilation) operators are $c^{\dagger}_{i,\alpha,k,\sigma}$ ($c_{i,\alpha,k,\sigma}$) and $c^{\dagger}_{j,\alpha,k,\sigma}$ ($c_{j,\alpha,k,\sigma}$) with the square lattice nearest neighbour tight-binding hopping in the form $\varepsilon_k = -2t_0$ ($\cos k_x + \cos k_y$) where t_0 is the hopping integral.

The sub-lattice magnetisation is produced in the neighbouring sites due to the Heisenberg exchange interaction between the magnetic moments. In this scenario the mean field AFM Hamiltonian for the sub-lattice magnetisation can be written as

$$H_{AFM} = \frac{h}{2} \sum_{\alpha,k,\sigma} s\left[c^{\dagger}_{i,\alpha,k,\sigma} c_{i,\alpha,k,\sigma} - c^{\dagger}_{j,\alpha,k,\sigma} c_{j,\alpha,k,\sigma}\right] \qquad ...(2)$$

Here, s = ±1 is the spin index and it is +1 for up and -1 for down spins respectively. The AFM order parameter h is described as

$$h = -\frac{1}{2} g_L \mu_B \sum_{\alpha,k,\sigma} s\left[\left\langle c^{\dagger}_{i,\alpha,k,\sigma} c_{i,\alpha,k,\sigma}\right\rangle - \left\langle c^{\dagger}_{j,\alpha,k,\sigma} c_{j,\alpha,k,\sigma}\right\rangle\right] \qquad ...(3)$$

Here, g_L and μ_B are the Lande-g factor and Bohr magnetron respectively.

The t_{2g} degenerate orbitals $3d_{xz}$ and $3d_{yz}$ are responsible for the JT distortion in the lattice. The single degenerate band splits into two bands as $\varepsilon_{1,2}(k) = \varepsilon_k \pm Ge$. We have considered here the static lattice strain by e and the electron-lattice interaction strength as G. The mean field JT Hamiltonian is defined as

$$H_{JT} = -Ge \sum_{\alpha,k\sigma} (-1)^{\alpha} \left[\left(c^{\dagger}_{i,\alpha,k,\sigma} c_{j,\alpha,k,\sigma}\right) + H.c. + \frac{1}{2} C_0 e^2\right], \qquad ...(4)$$

and the corresponding lattice strain e is defined as

$$e = \left(\frac{G}{C_0}\right) \sum_{\alpha,k\sigma} (-1)^{\alpha} \left[\left\langle c^{\dagger}_{i,\alpha,k,\sigma} c_{j,\alpha,k,\sigma}\right\rangle + \left\langle c^{\dagger}_{j,\alpha,k,\sigma} c_{i,\alpha,k,\sigma}\right\rangle\right]. \qquad ...(5)$$

In this expression the elastic energy is represented as $1/2C_0^2$with the elastic constantC_0.
The mean-field SC pairing Hamiltonian is written as

$$H_{SC} = -\sum_{\alpha,k} \Delta_k \left[\left(c^{\dagger}_{i,\alpha,k,\uparrow} c^{\dagger}_{i,\alpha,-k,\downarrow} + c^{\dagger}_{j,\alpha,k,\uparrow} c^{\dagger}_{j,\alpha,-k,\downarrow}\right) + H.c.\right] \qquad ...(6)$$

The SC gap parameter for $s^{\pm}$-wave symmetry will be $\Delta_k = \Delta(T) cosk_x \times cosk_y$. This SC gap parameter Δ_k is momentum dependent and is defined as

$$\Delta_k = -\sum_{\alpha,k,k'} \tilde{V}(k-k')\left(\left\langle c^{\dagger}_{i,\alpha,k',\uparrow} c^{\dagger}_{i,\alpha,-k',\downarrow}\right\rangle + \left\langle c^{\dagger}_{j,\alpha,k',\uparrow} c^{\dagger}_{j,\alpha,-k',\downarrow}\right\rangle\right), \qquad ...(7)$$

here the SC exchange interaction is $\tilde{V}$ and is expressed as $\tilde{V}(k-k') = -V_0 f_k f_{k'}$ with $f_k = (cosk_x \times cosk_y)$. The pairing form factor for $s^{\pm}$-wave symmetry is considered as $cosk_x \times cosk_y$. However, it is important to realize that the $s^{\pm}$-wave gap function is not synonymous to $\cos k_x \times \cos k_y$. The system is described by the total Hamiltonian as

$$H = H_0 + H_{AFM} + H_{JT} + H_{SC}. \qquad ...(8)$$

3. Calculation of order parameters

We have adopted the double-time single particle Green's function method for the calculation of order parameters [17] involved in the Hamiltonian in eqn.(8). The Green's functions are defined separately for *i* and *j* sites as follows:

$$A_1(k,\omega) = \langle\langle c_{i,\alpha,k,\sigma}; c^{\dagger}_{i,\alpha,k,\sigma} \rangle\rangle_{\omega} \quad ; \quad A_2(k,\omega) = \langle\langle c^{\dagger}_{i,\alpha,-k,\sigma'}; c^{\dagger}_{i,\alpha,k,\sigma} \rangle\rangle_{\omega}$$

$$A_3(k,\omega) = \langle\langle c_{j,\alpha,k,\sigma}; c^{\dagger}_{i,\alpha,k,\sigma} \rangle\rangle_{\omega} \quad ; \quad A_4(k,\omega) = \langle\langle c^{\dagger}_{j,\alpha,-k,\sigma'}; c^{\dagger}_{i,\alpha,k,\sigma} \rangle\rangle_{\omega}$$

$$B_1(k,\omega) = \langle\langle c_{j,\alpha,k,\sigma}; c^{\dagger}_{j,\alpha,k,\sigma} \rangle\rangle_{\omega} \quad : \quad B_2(k,\omega) = \langle\langle c^{\dagger}_{j,\alpha,-k,\sigma'}; c^{\dagger}_{j,\alpha,k,\sigma} \rangle\rangle_{\omega}$$

$$B_3(k,\omega) = \langle\langle c_{i,\alpha,k,\sigma}; c^{\dagger}_{j,\alpha,k,\sigma} \rangle\rangle_{\omega} \quad ; \quad B_4(k,\omega) = \langle\langle c^{\dagger}_{i,\alpha,-k,\sigma'}; c^{\dagger}_{j,\alpha,k,\sigma} \rangle\rangle_{\omega}$$

These coupled Green's functions are solved for quasi particle energy bands ω_i $(i = 1 - 4)$ given by

$$\omega_1 = \pm\sqrt{\frac{S_1 + \sqrt{S_1^2 - 4T_1}}{2}} \quad ; \quad \omega_2 = \pm\sqrt{\frac{S_1 - \sqrt{S_1^2 - 4T_1}}{2}}$$

$$\omega_3 = \pm\sqrt{\frac{S_2 + \sqrt{S_2^2 - 4T_2}}{2}} \quad ; \quad \omega_4 = \pm\sqrt{\frac{S_2 - \sqrt{S_2^2 - 4T_2}}{2}} \qquad ...(9)$$

with $S_1=2(E_1^2+2\varepsilon_k^2)$, $S_2=2(E_2^2+2\varepsilon_k^2)$, $T_1=E_1^4+4\varepsilon_k^2(h/2+\Delta_k)$ and $T_2=E_2^4+4\varepsilon_k^2(h/2-\Delta_k)$. Here, $E^2_{1,2}=(h/2\pm\Delta_k)^2+(Ge)^2$ with $\omega_i=\omega_i(Ge, h, \Delta_k)$, which shows that all the gap parameters interfere with each other.

The antiferromagnetic gap parameter defined in eqn.(3) is calculated to be

$$h = \frac{g_L \mu_B}{4} \sum_k \left[\frac{1}{\omega_1^2 - \omega_2^2} \{F_1(k,T) - F_2(k,T)\} + \frac{1}{\omega_3^2 - \omega_4^2} \{F_3(k,T) - F_4(k,T)\} \right] \qquad ...(10)$$

Where

$$F_{1,2}(k,T) = \frac{1}{\omega_{1,2}} \left(h/2 + \Delta_k\right)\left(\omega_{1,2}^2 - E_1^2 - 2\varepsilon_k^2\right) \tanh\left(\frac{\beta\omega_{1,2}}{2}\right),$$

$$F_{3,4}(k,T) = \frac{1}{\omega_{3,4}} \left(h/2 - \Delta_k\right)\left(\omega_{3,4}^2 - E_2^2 - 2\varepsilon_k^2\right) \tanh\left(\frac{\beta\omega_{3,4}}{2}\right).$$

The free energy expression in its standard form is written as

$$F = -k_B T \sum_{k,i} \ln[1 + \exp\{-\beta\omega_i(k)\}], \qquad ...(11)$$

with $\beta=1/k_B T$, k_B and T mentioned in the above expression are the Boltzmann constant and the absolute temperature respectively. We have considered $\hbar=k_B=1$ throughout the calculation. The self-consistent expression for lattice strain e can be obtained from the minimisation of the above equation. Hence the eqn.(5) now can written as

$$e = -\frac{Ge}{2C_0} \sum_k \left[\frac{1}{\omega_1^2 - \omega_2^2} \{F_5(k,T) - F_6(k,T)\} + \frac{1}{\omega_3^2 - \omega_4^2} \{F_7(k,T) - F_8(k,T)\} \right] \qquad ...(12)$$

Where

$$F_{5,6}(k,T) = \frac{1}{\omega_{1,2}} \left(\omega_{1,2}^2 - E_1^2\right) \tanh\left(\frac{\beta\omega_{1,2}}{2}\right),$$

$$F_{7,8}(k,T) = \frac{1}{\omega_{1,2}}\left(\omega_{3,4}^2 - E_2^2\right)\tanh\left(\frac{\beta\omega_{3,4}}{2}\right).$$

The SC gap parameter defined in eqn.(7) can be obtained from the Green's functions $A_2(k, \omega)$ and $B_2(k, \omega)$ as

$$\Delta_k = -\frac{1}{2}\sum_k \tilde{V}(k-k')\left[\frac{1}{\omega_1^2-\omega_2^2}\{F_9(k,T)-F_{10}(k,T)\} - \frac{1}{\omega_3^2-\omega_4^2}\{F_{11}(k,T)-F_{12}(k,T)\}\right] \quad ...(13)$$

where

$$F_{9,10}(k,T) = \frac{1}{\omega_{1,2}}\left(h/2 + \Delta_k\right)\left(\omega_{1,2}^2 - E_1^2 - 2\varepsilon_k^2\right)\tanh\left(\frac{\beta\omega_{1,2}}{2}\right),$$

$$F_{11,12}(k,T) = \frac{1}{\omega_{3,4}}\left(h/2 - \Delta_k\right)\left(\omega_{3,4}^2 - E_2^2 - 2\varepsilon_k^2\right)\tanh\left(\frac{\beta\omega_{3,4}}{2}\right).$$

In the above eqns.(10), (12) and (13) the *k*-sum takes the form of double integration in the a−b plane for the variables k_x and k_y. The summation $\sum_k = N(0)\frac{S}{(2\pi)^2}\iint dk_x dk_y$ in FeAs plane where *S* is the area of the square lattice and N(0) is the conduction electron density of states around Fermi surface.

4. Results and Discussion

4.1. Coexistence of order parameters

The coupled equations (10), (12) and (13) for the AFM gap (h), the JT gap $(e'=g_1\times e)$ and the SC gap (z) respectively are solved self-consistently numerically. We have considered the half-filling band situation with the Fermi level as zero ($\varepsilon_F = 0$) to be lying at the middle of the AFM band gap and the width of the conduction band $W= 8t_0$ ($\sim 1eV$). All the parameters involved in the gap equations are scaled by the conduction band width W. So, the dimensionless parameters are the SC gap$z=\Delta(T)/W$, the AFM gap $h_1=h/W$, the lattice strain $e=e/W$, the reduced temperature $t=k_BT/W$, the SC coupling constant $g=N(0)V_0$, the JT coupling constant $g_1=GN(0)/W$and the AFM coupling constant $g_2=g_L\mu_BN(0)/W$. Figure 1 shows the plots of the three order parameters in the independent and coexistence states. In the coexistence state the critical temperatures of the order parameters are in the order of$t_d>t_N>t_c$, where t_d is the JT distortion temperature, t_N is the Ne´el temperature and t_c is the SC transition temperature. In the coexistence state the SC order parameter is suppressed throughout the temperature range with respect to its independent state with the decrease of the transition temperature t_c from 0.0195 to 0.0125. Due to this the reduced SC gap value $2\Delta(0)/k_BT_c$=5.53 in the independent state is decreased to 2.72 in the coexistence state. For the multi-orbital FeSCs the reduced SC gap value $2\Delta(0)/k_BT_c$ is very high as compared to the universal BCS value of 3.52 which is an essential property of the FeSCs. The AFM order parameter is suppressed in the coexistence of the order parameters with enhancement of its Ne´el temperature t_N. But the JT transition temperature t_d is not changed due to coexistence; however, the gap value is suppressed for the temperature range less than t_c and shows mean-field behavior beyond t_c.

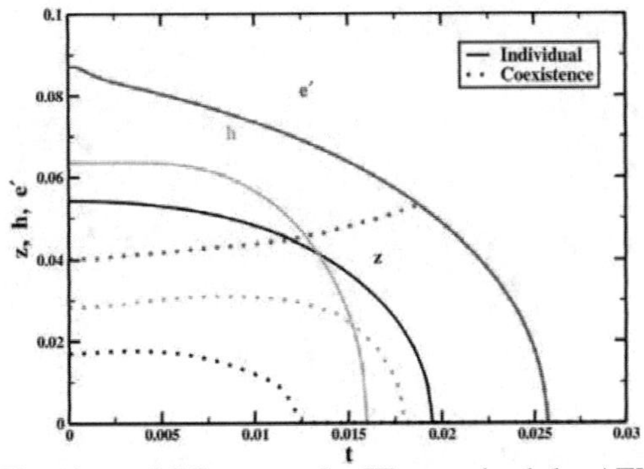

Figure 1: (Colour online) The plots of SC gap *z*, the JT gap *e'*and the AFM gap *h* in the independent and coexistence states *vs* reduced temperature *t* for fixed values of SC coupling $g=0.3$, the JT coupling $g_1=1.2$ and the AFM coupling $g_2= 1.0$.

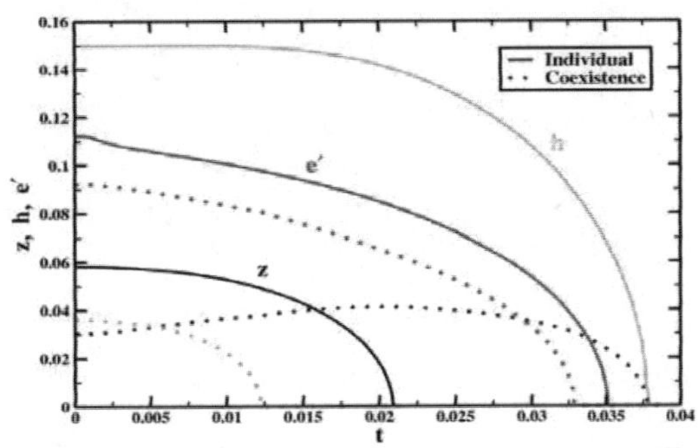

Figure 2: (Colour online) The plots of SC gap *z*, the JT gap *e'* and the AFM gap *h* in the independent and coexistence states *vs* reduced temperature *t* for fixed values of SC coupling $g = 0.32$, the JT coupling $g_1= 1.4$ and the AFM coupling $g_2= 2.2$.

Figure 2 shows the plots of the three order parameters for the situation in which their critical temperatures are in the order of $t_N>t_d>t_c$ in the independent state. Both the AFM and JT order parameters suppressed throughout the temperature range in the coexistence state. The rate of suppression of the AFM order parameter is so high that, the Ne´el temperature t_N goes so smaller to the JT transition temperature t_d. The SC gap parameter is suppressed towards the lower temperatures with an enhancement of the t_c from 0.021 to 0.038. So, the reduced SC gap value $2\Delta(0)/k_BT_c= 5.52$ in the independent state is also decreased in the coexistence state to 1.59.

4.2. Energy bands

The electronic origin of pairing interaction in FeSCs are suggested by their electronic structure and high transition temperature properties. The superconductivity in iron based superconductors is associated with the density of states near Fermi level which gets its maximum contribution from *Fe3d*-orbitals and *Fe* layers. The energy bands for the system in the proposed model is described in equation (9). Figures 3 and 4 shows the plots these energy bands and the dispersion $\varepsilon_k$vs k_x for $k_y= \pi$ in both the cases $td>t_N>t_c$ and $tc>t_d>t_N$ respectively for the coexistence state with the corresponding SC gap, JT gap and the AFM gap values at the zero temperature i.e., $t = 0$. These figures show that there is a strong interference of all the three gap parameters. At the saddle point ($k_x= 0$) the energy bands are pushed away from it and the gap value from the Fermi level is equivalent to the interference value of these gap parameters.

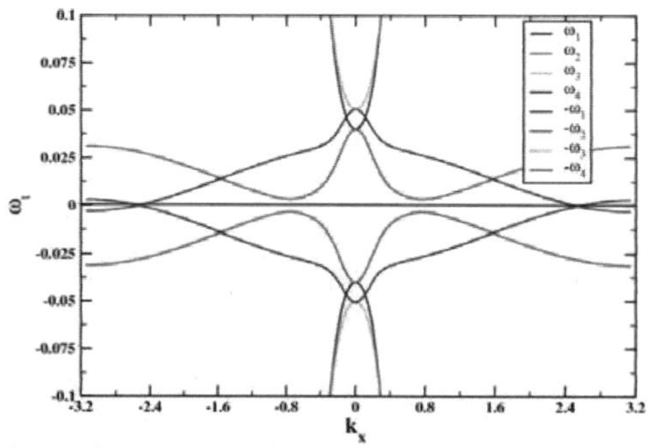

Figure 3: (Colour online) Plots of ω_i *vs* k_x for $k_y = \pi$ using parameters from figure 1 for $z = 0.017$, $e' = 0.040$ and $h_l = 0.028$.

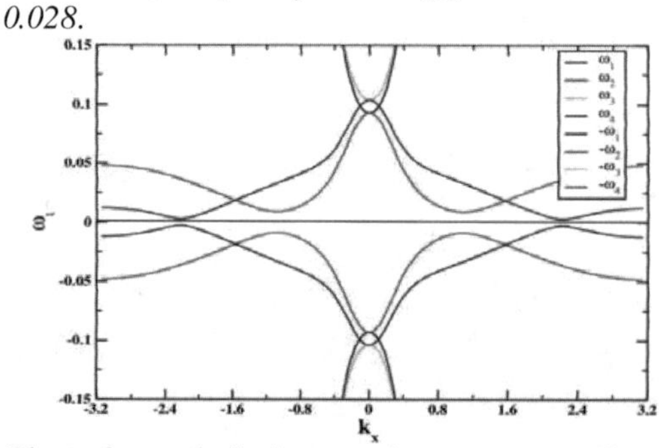

Figure 4: (Colour online) Plots of ω_i *vs* k_x for $k_y = \pi$ using parameters from figure 2 for z = 0.030, e′ = 0.092 and h_l = 0.036.

5. Conclusions

The interplay of three long range orders i.e., superconductivity, Jahn-Teller distortion and antiferromagnetic orders, present in FeSCs are discussed in this communication. In the phase diagram of iron pnictide superconductors display the coexistence of the interplay of SC, AFM and structural orders. Many people have been studied the coexistence of other order parameters with superconductivity for these materials. In the coexistence phase the interplay shows strong dependence of one another in this study. The superconductivity along with its transition temperature is suppressed in the interplay region for the case of $t_d > t_N > t_c$ and the transition temperature for superconductivity is enhanced with a suppression of the SC gap value for the case of $t_c > t_d > t_N$. The AFM order (*h*) is suppressed in the first case with increase of the t_N and in the second case it is suppressed throughout the temperature region. The distortion temperature t_d is not changed for its higher value than the others with a suppression in the coexistence region, however it is suppressed throughout the temperature range for the second case. The lattice strain energy is being arrested at t_c, as observed for different high-Tc superconductors, which can be observed here in presence of the AFM order. The coexistence of superconductivity with JT distortion and AFM order has been studies separately for FeSCs in presence of external magnetic field and reported [18, 19].

Acknowledgments

B. Pradhan acknowledges computational and reference facilities offered by the Institute of Physics, Bhubaneswar, India and Dr. B. K. Panda of Ravenshaw University, Cuttack, India for timely discussions.

References

[1] Kamihara Y, Watanabe T, Hirano M and Hosono H 2008 *J. American Chem. Soc.* **130** 3296
[2] Rotter M, Tegel M and Johrendt D 2008 *Phys. Rev. Lett.* **101** 107006
[3] Hirschfeld P J, Korshunov M M and Mazin I I 2011 *Rev. Prog. Phys.* **74** 124508
[4] Chubukov A V, Efremov D and Eremin I 2008 *Phys. Rev. B* **78** 134512
[5] Mazin I I, Singh D J, Johannes M D and Du M H 2008 *Phys. Rev. Lett.***101** 057003
[6] Garaud J, Corticelli A, Silaev M and Babaev E 2018 *Phys. Rev. B* **98** 014520
[7] Fernandes R M and ChubukovA V 2017 *Rep. Prog. Phys.* **80** 014503
[8] Miao H, Qian T, Shi X, Richard P, Kim T K, Hoesch M, Xing L Y, Wang X -C, Jin C -Q, Hu J –P and Ding H 2015 *Nature Commun.* **6** 6056
[9] Fernandes R M, Chubukov A V and Schmalian J 2014 *Nature Phys.* **10** 97
[10] Chubukov A V 2012 *Annu. Rev. Cond. Matt. Phys.* **3** 57
[11] Basov D N and ChubukovA V 2011 *Nature Phys.* **7** 241
[12] Wen H H and Li S 2011 *Annu. Rev. Condens. Matter Phys.* **2** 121
[13] Paglione J -P and Greene R L 2010 *Nature Phys.* **6** 645
[14] Kuroki K, Onari S, Arita R, Usui H, Tanaka Y, Kontani H and Aoki H 2008 *Phys.Rev. Lett.***101** 087004
[15] Dai P C 2015 *Rev. Mod. Phys.* **87** 855
[16] Stewart G R 2011 *Rev. Mod. Phys.* **83** 1589
[17] Zubarev D N 1960 *Sov. Phys. Usp.* **3** 320
[18] Pradhan B, Goi S K and Mishra R N 2016 *Solid State Comm.* **248** 110
[19] Pradhan B, Goi S K and Mishra R N 2016 *Solid State Comm.* **254** 6

IOP Conf. Series: Materials Science and Engineering **798** (2020) 012002 doi:10.1088/1757-899X/798/1/012002

Improving the Mechanical, Water Vapor Permeability, Antimicrobial properties of Corn-Starch /Poly Vinyl Alcoholfilm (PVA): Effect of Rice husk fiber (RH) & Alovera gel(AV)

Nirlipta Mallick[1], A.B Soni[2] and Dharm Pal[3*]

[1]Research Scholar in the Department of Chemical Engineering, National Institute of Technology, Raipur-492010, CG, India.
[2]Professor in the Department of Chemical Engineering, National Institute of Technology, Raipur- 492010,CG,India.
[3] Assistant Professor in the Department of Chemical Engineering, National Institute of Technology,Raipur-492010,CG,India.

*Corresponding author's e-mail address: dpsingh.che@nitrr.ac.in

Abstract: Essentially because of ecological points, petroleum-based plastics are being supplanted by natural polymers. In the most recent decades, the starch idea is outstanding as a green material. Because of the developing interest for Starch biodegradable materials and the film-forming capacity it has been significantly utilized in the food packaging area.Polyvinyl alcohol (PVA) is one of the synthetic biodegradablematerials that has the upsides of good film formation and high thermal stability. It has been utilized as coatings, packaging material, and in drug delivery applications. *Aloe vera(AV)* gel, an individual is a promising material in food preservation and has great antimicrobial and anti-inflammatory properties. Because of its film-forming capacity, it has been utilized in the edible film and coatings. Naturalfibers from agricultural waste are finding their significance in the food packaging sector because of the numerous favourable circumstances, for example, their lightness, minimal effort and being earth amicable. The rice husk (RH) is acquired from agro-industrial waste, can be utilized as filler in different bio compositesmaterials. Themain purpose of this examination was to investigate the impact of Rice husk fiber and Alovera gel on the Mechanical, Water Vapor Permeability, Antimicrobial properties of Corn-Starch based film.

1. Introduction

Extensive ecological awareness has prompted the formation of various biopolymers in ecological materials capable of supplanting petroleum-based synthetic materials in the packaging sectors.As one of the biodegradable polymers, starch has become a promising applicant attributable to its accessible as carbohydrate assets from sustainable stock feeds in grains and tubers, edible, nontoxic biopolymers, nonpolluting, low cost and thermoplastic biopolymer[1].It has the superb film-forming ability yet its application is incredibly influenced by the intrinsic hydrophilicity and non-resistance from microorganisms [2].PVA is a biocompatible, water-soluble polymer, which is a sort of thin-film material with fantastic packaging performance and high thermal stability[3]. Be that as it may, It has low dimensional stability because of high moisture absorption. Moreover, it has a reasonably huge cost appeared differently in relation to other commercial polymers. Therefore, mixed with renewable and abundant agro-active ingredients, forexample,polysaccharides,in particular starch, can be used to reduce cost. [4]. Mixing with starch produced greater resistance to moisture and accelerated degradation. Aloe vera is a stemless cactus-like plant that has a place with the Liliaceae family [5]. The restorative employments of the gel juice (orally) are against blockage, skin illnesses, and cardiovascular issues [6].As of late other significant property of Aloe vera has been accounted for, for example, mitigating and anti-infection activities. It has antibacterial property against gram-positive and gram-negative bacteria [7]. Aloe vera can be applied as consumable coatings for organic products as its biological activities prevent loss of moisture, control respiration rate and reduce microorganism production. In present investigation effect of AV antimicrobial properties on Starch/PVA biocomposite was studid and observed that by increasing concentration of AV antimicrobial properties of Starch/PVA biocomposite film was increased. The composite based on natural fibers has various tremendous points of interest over composites with engineered fibers, for instance, ease, lighter weight, available as plants or waste, is not toxic, and doesn't cause skin irritations [8]. The comfort of these composites lies in the manner that the ingredients are gained viably from normal or agricultural waste and consequently the composites can be made generally effectively. In this investigation RH utilized as natural fiber. During rice processing rice husk is produced as a by product and around 18.3 million tons of rice husk is produced each year. It was seen that increase of rice husk concentration as filler caused improvement in mechanical properties but decraesd water vapor permeability.

2.Experimental

2.1.Materials

The starch that was utilized in the preparation of the film was corn starch and was acquired from the local market. Polyvinyl Alcohol (PVA, Mol.wt.1, 20,000) was purchased from Acrylamide India. Glycerol (Mol.wt:92.09) was purchased from Estelle Chemicals, Maharashtra, India. Rice Husk was obtained from a neighbourhood rice factory forreinforcement in Starch/PVA biocomposites at an ostensible price, Maharastra, India. Completely developed aloe vera leaves were gathered from a solitary nursery plant.

2.2 Preparation Methods

Two biocomposite films were prepared. Such as husk reinforced Starch/PVA bio composite and AV gel based Starch/PVA bio composite.

2.2.1. Preparation of Starch/PVA/RH biocomposite

The Starch/Polyvinyl alcohol/Rice husk biocomposite for food packaging film was prepared with the solvent casting method. In this method, the film solution was prepared by mixing (40/60) composition of Starch/PVA in distilled water to form a homogenous gel solution. The blend was persistentlymixed usingan electric stirrer for around 40 min at 90^0C .Then the plasticizers, for example Glycerol (0.1%) and cross linkingagent Citric acid (20%) were added to the above blend.We have included Rice husk fiber for the reinforcementof the material. To expel the impurities from the fibers, they were washed with a mild detergent required during the extraction of fibers. The fibers were dried in a hot air oven maintained at 40◦C for12 h. Subsequent to drying the fibers were granulated to change over into fine particles. These particles of different concentration (10%,20%,30%,40%) have been added to the previous matrix. The suspension in this way formed was placed on a Teflon plate to prepare the biocomposite film. The biocomposite film was dried at room temperature for 72 h.

IOP Conf. Series: Materials Science and Engineering **798** (2020) 012002 doi:10.1088/1757-899X/798/1/012002

2.2.2 Preparation of Starch/PVA/AV biocomposite

Develop fresh leaves of Aloe Vera were washed with water. At that point, their thick epidermis was evacuated and the strong mucilaginous gel was gathered in a sterile container. Gel of various concentration (10%,20%,30%,40%) was blended in with (40/60) composition of Starch/PVA.The development of Starch/PVA suspension was depicted above. In a similar route here Starch/PVA suspension was additionally framed. The prepared suspension was poured onto a Teflon plate to form the biocompositefilm. The biocomposite film was dried at room temperature for 24 h.

2.3. Mechanical properties of Rice husk(RH) fibre reinforced St/PVA biocomposite film.

The tensile strength (TS) and percentage of elongation (%E) of the St/PVA bicomposite filmwith RH fiber as reinforcing material were performed with the computerized Universal Testing Machine based on the ASTMD638 standard. Theaverage sample thickness was around 1 mm and was operated at a cross-head speed of 20 mm/min at 30◦C. The test was performed with five samples and averaged.

2.4.Water vapor Permeability of Rice husk(RH) fibre reinforced St/PVA biocomposite film.

The water vapor transmission rate (WVTR) of the Starch/PVA/RH biocomposite films with a surface of 50 cm2, was determined with a water vapor permeation meter according to ISO 15106-1 test method. The WVTR was determined at 35 °C and 90% relative humidity. As the water solubilized the membrane and penetrated through the sample films, the nitrogen gas swept and transported the water vapor molecules in to a calibrated infrared sensor.

2.5. Antimicrobial activity of Alovera gel based St/PVA biocomposite film.

The Antimicrobial activities of Starch/PVA/AV films dependent on various concentration of AV(10%,20%,30%,40%) against *E.coli and Klebsilla pneumonia*utilized as model of Gram-negative bacteria and *Bacillus cereus* as a Gram-positive model. The antimicrobial activity was performed using the disc diffusion method, the Starch/PVA/AV biocomposites were cut into circles with a diameter of 1.5 cm, sterilized in an autoclave for 20 min at 130 °C, and were put on a surface of supplement agar media previously incubated by bacterial suspensions target of *E.coli, Klebsilla peneumonia and Bacillus cereus*, and afterward incubated for 27 h at 40 °C and evaluated the inhibition zones.

3.Results & Discussions

3.1. Mechanical strength

The outcomes acquired on rice husk reinforced Starch/PVA/RH biocomposite at various weight percentage for example 10wt. %, 20wt. % ,30wt. % and 20wt. % indicated that the mechanical properties improved with the addition of fiber weight percentage.

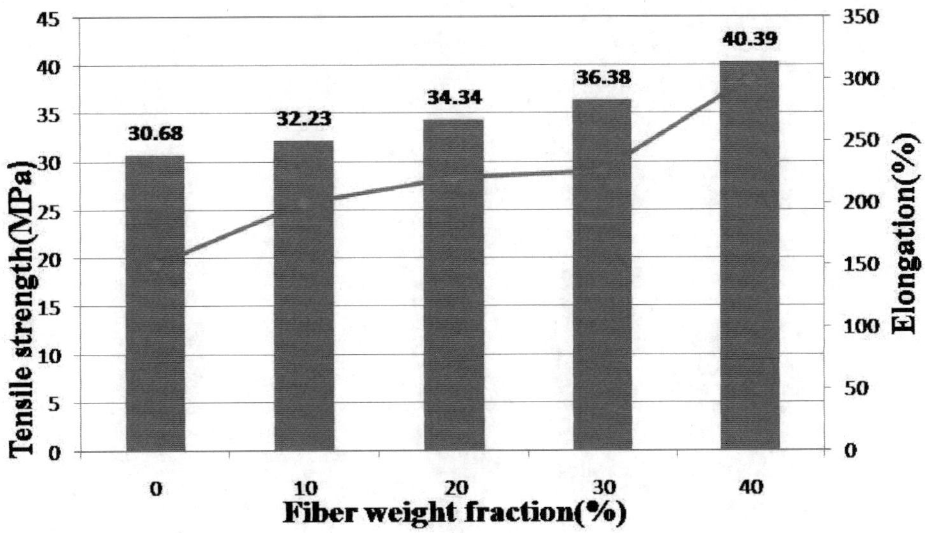

Figure 1. Tensile strength as a function of fiber weight fraction

The tensile strength and elongation at break of Starch/PVA/RH biocomosite are shown in figure 1.With an increase in the Rice husk fiber content, the tensile strength of the Starch/PVA/RH biocomposite film increased. The tensile strength of biocomposite film increased from 30.68Mpa to 40.39 Mpa with the increase in the concentration of Rice husk fiber .The mechanical impact of Rice husk fiber on Starch/PVA bio composite film was remarkable. This outcome recommends that Rice husk fiber was all around dispersed in to the starch/PVA matrix.Furthermore, there was a strong interface between the Starch/PVA matrix and the Rice husk fiber, which permitted appropriate stress among matrix and fiber, bringing about the increase of tensile strength [9].

The elongation at break arrived at a maximum value of 300.02% with a fiber content of 40%.With the increase in the concentration of Rice husk fiber, the elongation at break of the biocomposite film has increased from 150.03% to 300.02%,which could be due to the great compatibility between the Rice husk fiber and Starch/PVA matrix. Thepresence of the – OH group in the Rice husk fiber and its hydrophilic nature are compatible with Starch/PVA.Another resonance due to a greater elongation in Starch/PVA/RH biocomposite due to the increase in the crystalinity of the cellulose present in Rice husk fiber.

3.2.Water Vapor Permeability

Figure 2 showing the water vapor permeability of the starch/PVA/RH biocomposite with respect to different concentration of Rice husk fiber after conditioning at room temperature.

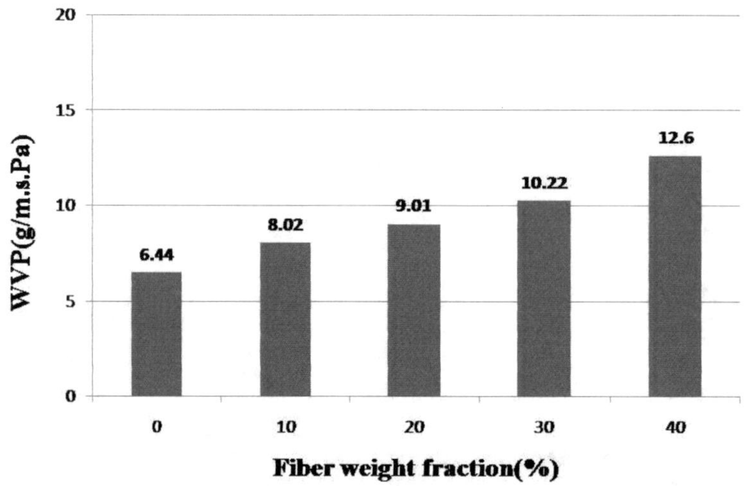

Figure 2. Water vapor permeability as a function of fiber weight fraction

As indicated by the findings appeared above, the water vapor transmission rate of Starch/PVA/RH biocomposite with various Rice husk concentrations has increased from 6.44 g.mm/day.m2 to 12.6 g.mm/day.m2. As per the above information, the water vapour transmission rate expanded with the increment of Rice husk fiber weight fraction from 0 to 40%. It has beenreported that fibers, for example, rice husk, kenaf, and wood compare an excessive amount of more waterabsorption when the fiber loading is expanded therefore water vapor permeability additionally expanded [10,11]. The water absorbing property of rice husk causes an increment of water vapor permeability in biocomposites. In general, the bio composites reinforced with natural fiber show enormous water vapor permeability because of the presence of micro voids at the interface of reinforcement/matrix.

3.3. Antimicrobial assessment

In the present examination, the antibacterial activity of the biocomposite film of Starch/PVA based on Alovera gel at various concentrations against Gram positive bacteria and Gram negative bacteria was assessed with the disc diffusion technique. The four concentrations of Alovera gel-based Starch/PVA biocomposite was examined in the present examination showed differing level of inhibitory effect against the choosen bacterial pathogens as shown in figure 3(a),(b),(c).The choosen microorganisms present a significant hazard to human health[12]. They were utilized to test the antimicrobial activity of the examined biocomposite films. *S. aureus* was an significant food adulterant that produces bacterial contagion. *E.coli, Klebsilla peneumonia and Bacillus cereus* were used to evaluate the antibacterial activity of the prepared biocomposite films.

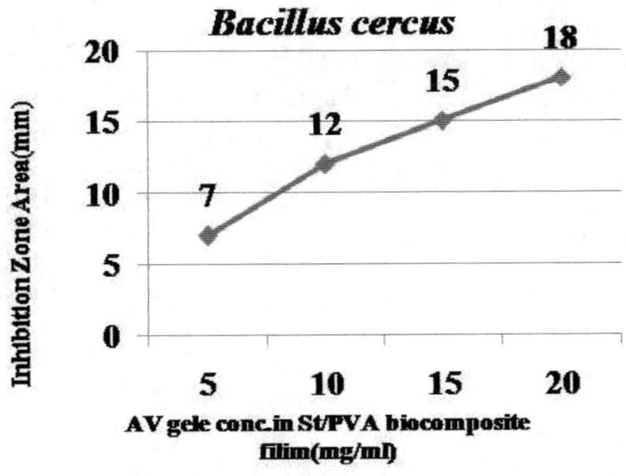

Figure 3. (a)

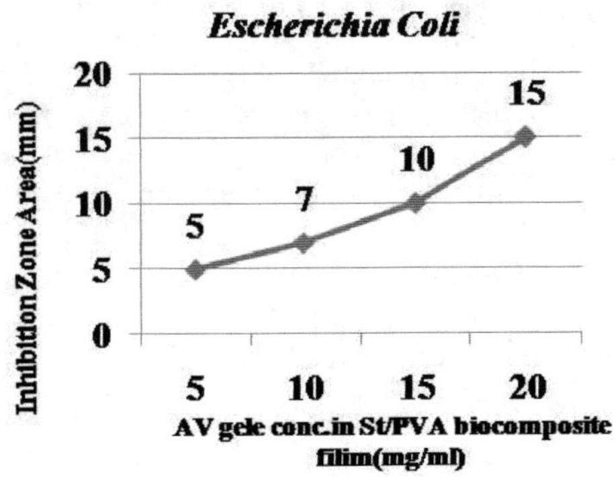

Figure 3. (b)

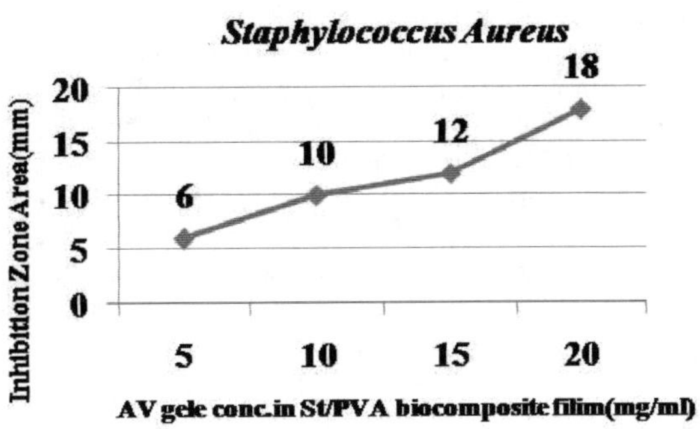

Figure 3. (c)

Figure 3. (a), (b), (c) Minimum inhibition concentration (MIC) of the prepared biocomposite films against gram positive and gram negative bacteria.

The present examination demonstrated the anti-bacterial property of the Starch/PVA biocomposites based on Alovera gel against the choosen strains of human pathogenic bacteria, the level of inhibition varied according to the concentration of the alovera extract.Highest concentration of Alovera gel based Starch/PVA biocomposites showed most extreme zone of inhibition. The antibacterial activity of the Alovera gel based Starch/PVA biocomposites was seen as progressively compelling as indicated by the increase of Alovera concentration in Starch/PVA biocomposite. This could be ascribed to presence of more amount of the anthraquinones and phenolic antioxidants in the Alovera gel [13].

4. Conclusion

In this investigation, it was discovered that the content of RH fiber influenced the structural integrity, tensile strength and water vapor permeability of the biocomposites. According to the observation, the increase in the quantity of rice husk in the Starch/PVA biocomposite has led to an increase in tensile values. The water vapor permeability of the Starch/PVA biocomposite filled with RH was estimated and it was seen that 40 wt. % rice husk has a higher water vapor transmission rate.Rice husk with a minimum content should be consideredan alternative additive to diminish the water vapor transmission rate. The antimicrobial activity of Alovera gel-based Starch/PVA biocomposite film against *E.coli, Klebsilla peneumonia and Bacillus cereus* were assessed. It came about that bacterial inhibition increases with the increase of Alovera gel concentration in Starch/PVA biocomposite film. These characteristics makethe biocomposite a polymer in perspective for food packaging applications.

Acknowledgement

The authors are thankful to the ELCA LAB for providing facilities to perform this study.

References

1. Bhanu P, Gupta V, Pathania D and SinghaA (2014)Synthesis,characterization and antibacterial activity of biodegradablestarch/PVA composite films reinforced with cellulosic fibre*Carbohydrate Polymers 109 171–179.*
2. ALAY S and Meireles M (2015)Physicochemical properties, modifications and applications of starches from different botanical sources*Food Sci. Technol Campinas* 35(2): 215-236 Abr.-Jun.
3. Singha A, Bhanu P and PathaniD (2015) *Cornstarch/Poly (vinyl alcohol)* Biocomposite Blend Films: Mechanical Properties, Thermal Behavior, FireRetardancy, and Antibacterial Activity*International Journal of Polymer Analy. and Charact*20: 357–366.
4. Tan B, Ching Y,Poh S ,Abdullah L and Gan S(2015)A Review of Natural Fiber Reinforced Poly(Vinyl Alcohol) Based Composites: Application and Opportunity Polymers 7 2205–2222 doi:10.3390/polym7111509.
5. Jain S,Rathod N, Nagi R, Sur J,laheji A,Gupta N, Agrawal P and Prasad S(2016)Antibacterial Effect of Aloe Vera Gel against Oral Pathogens: An In-vitro StudyJournal of Clinical and Diagnostic ResearchVol-10(11): ZC41-ZC44.
6. Chauhan S, Gupta K.C and Agrawal M(2014) Application of Biodegradable Aloe vera gel to control post harvest decay and longer the shelf life of Grapes*Int.J.Curr.Microbiol App Sci*3(3): 632-642.
7. Yehuala and EmireS(2013)Antimicrobial Activity, Physicochemical andMechanical Properties of Aloe (Aloe debrana)Based Packaging FilmsBritish *Journal of Applied Science & Technology3*(4): 1257-1275, 2013.
8. K. Oksman, M. Skrifvars, and J. F. Selin (2003) Natural fiber asreinforcement in Polylactic Acid (PLA) composites," J. Composite Sci.and Tech*vol.**63**,* pp. 1317-1324.
9. K Rassiah, T W Sin and M Z Ismail(2016)A study on flexural and water absorption of surface modified rice husk flour/E-glass/polypropylene hybrid compositeIOP Conference Series: *Materials Science and Engineering*152 012061.
10. Nosbi N, kil H, IshakZ.A, and BakarA (2011) Behavior of Kenaf Fibers after Immersion in Several Water Conditions BioResourses 6 950-60.
11. Andrews JM (2001) Determination of minimum inhibitory concentrationsJ Antimicrob Chemother 48:5.
12. Sun Y, Sun G. Novel regenerable N-halamine polymeric biocidesI(2001)Synthesis, characterization and antimicrobial activity of hydantoin-containing polymers*J Appl Polym Sci.*;80:2460.
13. Formela, K.; Zedler, Ł.; Hejna, A.; Tercjak(2018)A. Reactive extrusion of bio-based polymer blends and composites—Current trends and future developments*Express Polym. Lett,* 12, 24–57.

Thermal insulation ceramic material by using coal combustion waste fly ash and waste glass with medicinal phytochemical repellent

Saroja Kumar Sahoo[1], Umakant Shrivastava[1]and Mohammad Ziyauddin Khan[2]

[1]Research Scholar, Department of Physics, Dr. C. V. Raman University, Kota Bilaspur (C.G.) INDIA.
[2]Associate Professor, Department of Physics, Dr. C. V. Raman University, Kota Bilaspur (C.G.) INDIA.

Corresponding Author :E-mail: sarojkumar7690@gmail.com, mzkhan077@gmail.com

Abstract:

Toreduce in environmental load from pollution after primary using of various products (waste product), which is supporting to secondary utilization in preparation of thermal insulation materials using coal fly ash (40-60%), siliceous glass material(30-55%) ,some phytochemical medicinal additives and (2-8%) cellulose. We know that thermal energy is effective for nature. By the scientific ideas and methods 'energy' has been presenting a greatest continuous chainthat frequently more economical to transforms the various form of ceramic materials. The capability of ceramic material such as "tiles", cooking utensil, crucible etc chosen the better quality for treatment of heat insulation materials based on waste coal combustion of fly ash with waste glass and regular active form of adding medicinal organic adsorbates.In heretheorganic medicinal additives acts as reducing over materials from harmful diseases of microbial. It is favor for developing to high response to health care purposing. Over the materials, thermal activitydepends on some factors such as chemical composition of substances. From this typical point, the Fly Ash and Waste Glass reduce the flow of heat. To testing after preparation of materials as water absorption in boiling and cold water acts with (4-5) days .By the thermal treatment, the material individually acts as (700-1000)°C. The mechanical durability with other physical and chemical properties for this material is supporting with environmental parameters.

Keywords:
Ceramic materials, Waste FA, Siliceous Materials, Cellulose, thermal insulation, medicinal Phytochemical.

1. Introduction

Ceramic materials are the abundance materials that used in various purposes. Its requisitionby environ respondingnature. For robust environment, we put the antibacterial ceramic materials, with thermally insulate,which is absolutely necessary in quality to dominate and annihilate the harmful microorganisms. In case of food ,drink ,water and other multifunctional concept of ceramic materials , the antibacterial material tends to obtaining the vast solutions to reduce the environmental harmful disease from microorganisms. In case of building materials of ceramic tiles, the antibacterial effect is the best concept for reducing harmful microbes over tiles surfaces.[1]. By sol-gel technique, the surface cleansing propertyof Titanium dioxide (TiO_2) in nanocomposites can be used[2]. By this application, it would be remove the harmful bacteria in daily useful material.

Fig-1(a) Ceramic materials in various purposes.

First of all thinking about, pathogenic microorganism attacking over several kinds of ceramic materials and damaging in ceramic materials(as the cause of diseases and pollution load in environment)[3].

1.1 About Pathogenic microorganism over damage ceramic materials:

The skin contact with the pathogenic microorganisms over daily use of ceramic materials invarious area are considerable for increasing the rate of harmful diseases in environment. It is necessary to treat the materials in clinics, food industries and domestic spaces the vast utilization in antibacterial ceramic materials.

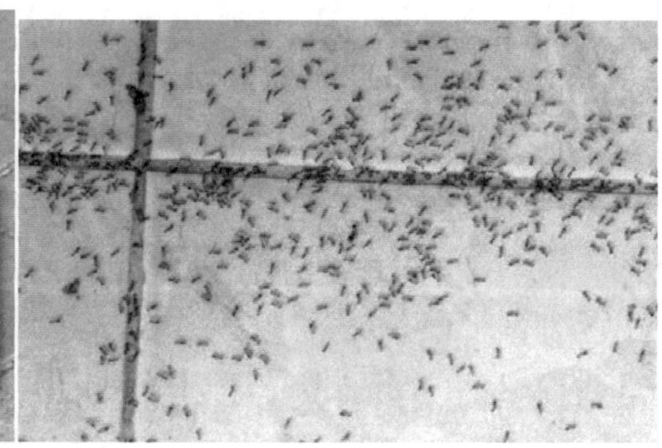

Fig-1.1(a)Damage part of ceramic tile materialsFig-1.1(b) Microorganism & Insects

1.2 Phytochemicals:

Various types of phytochemical that reducing the harmful microorganism over ceramic materials. In India the medicinal plants as traditional medicine as date back to approx.60,000yearsago[4].In this paper, the medicinal plants could be used as well as harmful microorganism repellent.

Fig-1.2 Medicinal plant for microorganism repellent

1.3 Waste materials

The defective consumption material is known as waste materials. The production of waste materials has grown highly in amount with a lot of problem faces by environment. The higher harmful chemical and radiations corrupt our biodiversity. Now a days, the seriousness for attend to control the matter of waste material management is a global issue that remarkable for each and every humankind in our world. The heavy metaldamage to the organ part of humanthat mixing with water from any chemicalindustries. The issues are avoiding by utilize the waste material to various purposes that includes designing & producing from waste materials to make up the good and healthy environment.

IOP Conf. Series: Materials Science and Engineering **798** (2020) 012003 doi:10.1088/1757-899X/798/1/012003

The paper is finding a solution for dealing with waste materials such as coal combust FA and WG. A large amount of fly ash produced from the raw material as coal in thermal Power plant using electrostatic precipitators (ESP) device that filter the fine particles. Both fly ash (FA)&waste glass (WG)are environmental meaningless materials but by scientific ideasit has to be possible and showing the capacity which may be developed to lead the future success or usefulness resource material. Nowadays the FA is used in cement, concrete and other cement based applications in world. The interdisciplinary name of the garbage product ascribed to combustion of coal or lignite in the boiler of a thermal power plant is turned to dust as fine particles that carried in air, since the FA at bottom ash, pond ash or, mound ash. From the bottom of boilers by any acceptable procedure is termed as Bottom Ash. In here the terminology Pond Ash is used when FA or bottom ash or both mixed in any portion is transfer in the form of insoluble particles that suspension in water as deposited in pond. When the mixture of these in any portion is transferred or carried and deposited in dry from, then it is called as Mound Ash.

First of all, Scientist discovered the materials using waste FA i.e, cement specially requires for building bricks, concrete etc. The essential quality of FA in India uses in cement and concrete, which is managed by distribution of framing for fly ash to use of pozzolana (siliceous and aluminous compounds divided in the presence of alkali/ water) and admixture. The major valuabledifference in fly ash in India and elsewhere is in the glass(SiO_2) content which described as a fine dust of mostly spherical glass particles having pozzolanic properties.

Nowadays the various use of fly ash that is gaining move faster scientific development for world[5]. It concerns to put fly ash for making disposal, which is increasing the list of research areas of application. In world, fly ash is not only being used in civil (building and roadconstruction)but also in, material science (ceramics, metallurgy), agriculture and environmental-related areas. There are common areas for using are cement, concrete, ready-mixed concrete, cement or lime-based fly ash bricks and blocks for walling prefabricated building elements, land reclamation, soil stabilization, road constructions, embankments, land fills etc and in case of Non-engineering applications such as in agriculture, plant nutrients, ceramics, neutralizing soil acidity, metal extraction etc.

The coal is mostly carbon which is formed by dead plants fossils with heat and pressure under soil. It is of four types mainly Peat, Lignite, Bituminousand Anthracite. And others are semi types of coal.

Peat - less than 60% carbon.

Lignite - (60 – 70) % carbon.

Sub-bituminouscoal -(35 – 45)%carbon

Bituminous coal -85% carbon

Anthracite -87% carbon

Graphite - (95 - 99)% carbon

Cannel coal - lower than in fixed amount of carbon than Bituminous

Coal Fly Ash is classified in two types such as,

1) FA,which are obtained from bituminous coal with fractions ofSiO_2+ Al_2O_3+ Fe_2O_3greater than 70%.
2) FA, which are obtained from lignite coal with fractions of of SiO_2+ Al_2O_3+ Fe_2O_3greater than 50%. The two kinds of Class C and Class F depending on the type of coal and the effective chemical analysis.

IOP Conf. Series: Materials Science and Engineering **798** (2020) 012003 doi:10.1088/1757-899X/798/1/012003

Class C FA is normally produced from the burning of lignite or sub bituminous coals, contains CaO higher than 10% and possesses cementations properties in addition to pozzolanic properties.

Class F FA is normally produced from the burning of bituminous or an anthracite coal contains CaO below 10% and possesses pozzolanic properties.

The classification, based on the boiler operations, which is classified with two distinct identities:

1. Low temperature(LT) fly ash, marked out of combustion temperature below 900°C.

2. High temperature(HT) fly ash, marked out of combustion temperature below 1000°C.

Fig-1.3 Fly Ash (FA)

Characteristics of FA are given as follows

1.3.1 Physical Characteristics

In general, the physical characteristics of fly ashes differ over a notable range correlate to their source.The fitness is likely effective more by features such as coal combustion and ash collection with classification than by the nature of the coal itself. Similarly, the type of fly ash exhibited no apparent influence on the specific surface as measured by the Blaine technique. Moreover, except in some cases, there was very little relationship between the specific surface as measured by the Blaine and the fineness as determined by percentage retained on a 45μm filter. It is a fine grained material consisting mostly of spherical, glassy particles containing irregular or angular particles.The Size and Shape of Fly Ashparticle varies depending on the sources. It consists of silt sized particles as spherical and typically ranging in size between 10 to 100 micron.

1.3.2 The Color of FA

The color is grey, depending on its chemical and mineral constituents. Yellowish - Brown and light colors are typically associated with high lime content. A brownish color is commonly associated with the iron content. A dark grey to black color is typically assigned to upraised unburnt content. FA color is usually very consistent for each and individually power plant and coal source. The fineness of FA. Generally, FA is dry and wet filtering. Hence it is used to measure the fineness of fly ashes.

1.3.3 The specific surface of fly ash

The surface area of FA is the area of a unit of mass, can be measured by various techniques. There is a common skill as Blaine specific-surface method, which measures the resistance of compressed particles to air flow.

1.3.4 The Specific Gravity of FA

Specific heat capacity for different composition of fly ash is found as 0.753 J/g-°C. This is in the range of pure Cement Concrete. It can be observed that specific heat carrying capacity of composite is gradually increasing with increase in fly ash percentage. It is mainly because the fly ash particle exhibits higher specific heat capacity. The variation of specific heat with respect to fly ash grain size is higher for small sized particles. It is varied over a wide range.

Table 1.3 The Chemical Composition of Fly Ash

COMPONENTS	PERCENTAGE %
SiO_2	54.20
Al_2O_3	25.13
Fe_2O_3	11.72
CaO	3.57
MgO	1.2
K_2O	0.76
Na_2O	0.31
TiO_2	1.1
P_2O_5	0.6
Mn_2O_3	0.2
SO_3	0.35
LOI	0.86

It depends on the sources of coal and also on operating parameters of boilers that the various quality from source to source and within the source also. With the use of pulverized coal and systematic combustion system, LOI(Loss on ignition) is a measurement of unburned carbon remaining in the ash. Generally, the mineralogical Characteristic of FA of coal as a source and types of both influence on its mineralogical composition. It depends on the system of combustion and some unburnt coal may be collected with ash particles. In addition, the amount of glassy material, each fly ash may contain one or more of the four major crystalline phases: quartz, mullite, magnetite and hematite. In sub-bituminous fly ashes, the crystalline phases may include calcium sulphate and alkali sulphates. The reactivityfly ashes are related to the noncrystalline phase or glass for the high reactivity of high-calcium fly ashes may partially lie in the chemical composition of the glass.

Experiments are conducted to determine the thermal properties such as thermal conductivity, coefficient of thermal expansion, specific heat and thermal diffusivity. There is a decrease in thermal conductivity, increase in specific heat capacity and decrease in thermal expansion [4].

Many of the experiments which are conducted on fly ash based is mainly on the thermal conductivity and resistivity properties with WG is very less research has been conducted on its thermal properties. Thermal properties are very important factors of the fly ash based materials, because in real scenarios the structures undergo many physical and chemical changes due to the temperature effects. These temperature effects cause the thermal failure of the other materials and discharging the ion, protect the surface etc formation in the structure takes place. When there is a high heat of temperature, it puts in danger the ordinary materials like electronics appliances that the structures fails to withstand the high temperature and the structure collapses due to the overheating. The work presented in this paper reports an investigation on the behavior of best heat resistive materials produced from Fly Ash and WG. Attempts are made to study the effect of thermal properties on different composition of Fly ash and grain sizes of fly ash. Some of the earlier investigations showed that the thermal properties are enhanced with increase in fly ash percentage. With thermal property, thermal properties are also varying. In order to explore an interesting, useful engineering material and to check whether the Fly ash is able to withstand the change in temperature effects, with increase in percentage of Fly ash this study is carried out.

1.3.5Thermal conductivity

It defined as the ability of material to conduct heat.The TC of pure concrete is around 1.38 W/m-K. and the range 1.21.4 W/m-K of the Pure Cement Concrete . The greater are the porosity and pore size, the smaller is TC. That is for, if the porosity and pore size increased, the pores in concrete would be filled with more air, and the thermal conductivity of the gas is much smaller than the solid, compared to pure cement concrete, fly ash concrete has lower thermal conductivity.

1.3.6 Coefficient of thermal expansion

The ratio of change in length to product of temperature change and original length. For coefficient of thermal expansion by using digital Vernier caliper, the initial specimen length is measured, and then the specimen is kept inside the Oven for heating to get desired temperature. After the specimen attains the desired temperature the length of specimen is measured, this data used to find the Co efficient of thermal expansion.

1.3.7 Thermal diffusivity of the material

It is defined as the thermal conductivity divided by density and specific heat capacity at constant pressure. Thermal Diffusivity for different composition of fly ash .The thermal diffusivity is directly dependent on specific heat and thermal conductivity, the thermal diffusivity decreases with increase in the fly ash percentage this is because of the porosity and pore size of the fly ash. Also we can see that the diffusivity of the higher grain size is lesser than that of the smaller grain size because of the fine powder.

1.4About waste glass (WG)

Glass is aproduct of supercooling melted liquid mixture that consists of sand, soda ash in which there is no crystallization and retains the organization for internal structure of melted liquid. After crushed the waste glass, it exhibits the particle size like sands.

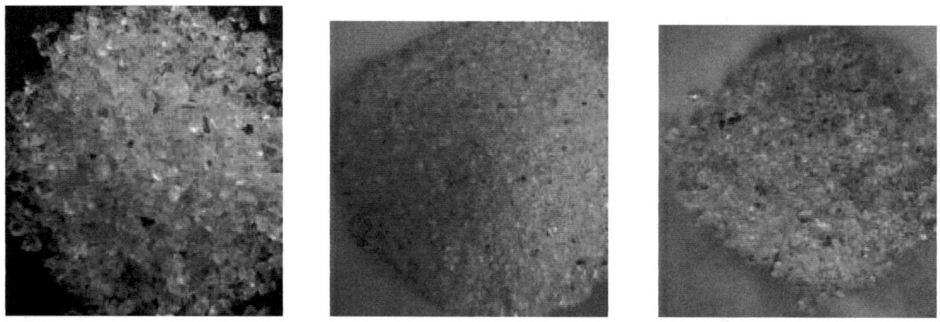

Fig-1.4 Waste glass in different size

The properties of waste glass (WG) materials are given as follows.

1.4.1 Physical propertiesof WG

WG with crushed particle are generally angular shape and contains some flat and elongated particles depend on degree of crushing. In here small particles exhibits less angularity and reduced the quantities of flat.

1.4.2 Chemical properties of WG

The elementary compounds of glass as Silicon when combined with Oxygen, silicon dioxide in the form of sand which is the most common glass former. It contains 68% silicon dioxide. As soda ash (anhydrous sodium carbonate) acts as the fluxing agent in melt. The limestone & dolomite are also used in lieu of soda ash. Types of glasses, such as made from soda lime glass, Lead alkali silicate glasses are used in manufacturing the light bulb, neon sign, optical fiber etc. Borosilicate glass having extra ordinary chemical resistances with high temperatures that used in research in laboratories.

1.4The chemical compositions of glasses.

Components	**Value Percentage (%)**
SiO_2	71.10 %
Al_2O_3	01 %
CaO	9.2 %
Fe_2O_3	0.29 %
MgO	4.4 %
Na_2O	12.6 %
SO_3	0.5 %
K_2O	0.81 %
Loss of ignition	0.04 %

The glass is not chemically resisted to hydrofluoric acid and alkali. It is amorphous in nature.

1.4.3Mechanical properties.

It is a brittle material that fractures from tensile stress. The internal friction angle or shear strength and bearing capacity of crushed blended aggregates high, its compatibility insensitive to moisture content. It's having good soundness with frictional properties.

1.4.4Thermal properties of WG

The glass having heat retention as low heat conductivity.Actually themixing of chemical composition in glass can help to decrease the depth of penetration.

The thermal conductivity of glass is not high. It is about k = 1 W/ m-K.Waste FA and WG are selecting for construct the heat resistive and conductive materials. As we know that heat resistive means thermal insulation and heat conductive means thermal conduction.

1.4.5 Thermal Insulation materials

It may be defined as a combination of materials that retard the flow of heat depending surface shape and size.

2. Advantages:

- To control the processing of temperature and protect the electronic wares.
- It prevents on cold surfaces and resulting corrosion.
- It provides the fire protection and absorbs vibration.

The materials which have basic requirement to dealing with heat transfer that the rate of heat flows in order to minimize the change of temperature[6]. It is very important to select right type of insulation materials considering temperature of system for the mode of heat transfer involved.

The thermal insulation materials like fibrous materials, building walls (paints, bricks wall), aluminum foil insulation etc. The metal foil which used to increase the insulating value of air spaces by reducing heat transfer by radiation. A film of paint is relatively thin that offers resistance to flow of heat conduction. The color of external paint the effect on absorption of heat from the sun. since the light-colored paint absorbs less heat from the sun which is the standpoint of insulation. The coating like paint contains metallic flakes i.e, aluminum paint.

The high temperature of heat resistant lightweight materials having high application in electronic appliances(battery electrode material, capacitors), furnaces, tunnel layers, anticorrosive materials (paint) etc[7]. The resistive material fully effective for high temperature that protect the devices and building structures.

For Environmental control; Focus is being continued on:

i)Development of new / incremental technologies,ii) Adaptation/large scale application of technologies, iii) Preparation of standards / specifications,iv) Support & facilitation to other Government Ministries / Department

The dumping of waste FA is one of the major concerns and challenge to the society, as the products are precarious as well as proper method is to be come after for the dumping of waste. By scientific ways, the waste product from the factory is one of the best solutions for dumping problems.

3.Research Methodology:

3.1 Materials

The actuality of which physical objects are unperturbed, in here the term is used notional. The word established some specific kind, quality, or quantity of matter, primarily as propose for use (cotton material,volatile materials), some materials are used for making this new thermal conductive&resistive materialpreparations are given below.

3.2Waste glass

Waste glass which contains silica. The proportion of waste glass is 40-60%.

3.3 Fly ash

It is a waste material which is collected from NTPC SIPAT thermal power plant in Bilaspur of Chhattisgarh. The proportion of this material is (30-55)%.

3.4 Phytochemical repellent

A small quantity of percentage in Neem, Tulsi etc. pointing to medicinal phytochemicals for reducing the viral diseases from harmful diseases by microbes. It is added with preparation of ceramic materials that support to develop the intermolecular bonding character in substances. These are functioning for health care and response with ceramic materials.

3.5 Cellulose

The organic polymeric compound as Cellulose such as is used as binder,to bind the pattern of molecules by statistical thermodynamically arrangement. The proportion of cellulose is (2-8)%.

4. Manufacturing process:

4.1 Heat treating method

The material includes of the fine structure called "grains". The nature of the grains (i.e. grain size and composition) is one of the most constructive factors. Hot plate provides an efficient way to manipulate the properties the rate of diffusion and the rate of cooling within the fine structure.

The structure consists of atoms that are grouped in a very specific arrangement, called a lattice. In most elements, this order will rearrange itself, depending on bonding factors like temperature and pressure[8]. This rearrangement may occur several times, at many different temperatures. In the soluble state, the process of diffusion causes the atoms of the dissolved element to spread out, attempting to form a homogenous distribution within the silica

The heat resistive and conductive materials were made from waste fly ash, waste glass, organic additives and definite quantity of HF which prepared as following procedure.

In order to shaping the mixture of FA, WG finely by polymeric cellulose as binder with suitable compositions.The bonding between molecules depends upon reacting species for substances[9]. By the help of laboratory apparatus, the substance of definite amount of mixture initiating to pelletization by hydraulic pressure machine. And the pellets would be dry with temperature at 100°C for 5 hr by calibrator. After all, it is fired at muffle furnace at (1000-1300)°C.

4.2Testing of material

4.2.1 Hot plate method

There are a number of chances to compute heat conductivity. It is verified in comparison with the thermal conductivity of nanofluid [10], each of them acceptable for restricted range of materials, depending on the thermal premises and the medium temperature.In general there are two basic mechanism of computation:

5. Observation & Analysis:

Table 5.1 Raw Material composition

Serial No.	Components	Sample No.		
		Sample No.1	Sample No.2	Sample No.3
1.	WG (waste glass)	(45 – 50)%	(50 – 55)%	(55 – 60)%
2.	Coal FA (flyash)	(35 - 39)%	(35 - 39)%	(28 - 32)%
3.	Binder (Organic cellulose)	(0 - 04)%	(02 - 06)%	(03 - 08)%

Mathematically, the thermal Conductivity of a material can be formulized.

$$\lambda = \frac{q \times d}{T_1 - T_2}$$

Whereas, 'q' quantity of heat passing through a unit area of the sample in unit time

$$q = \frac{Q}{A} w/m^2$$

Where Q = Quantity of heat passing through a base area of sample in w.

A = Base area of the sample.

d = Distance between the two sides of sample.

T_1= Temperature on warmer side of the sample.

T_2 = Temperature on colder side of the sample.

IOP Conf. Series: Materials Science and Engineering **798** (2020) 012003 doi:10.1088/1757-899X/798/1/012003

Table 5.2 Observation table for thermal conductivity and resistivity of materials

Sample No.	Q in w	A in m^2	$q = \frac{Q}{A}$ in w/m^2	. T_1 in K	T_2 in K	$T_1 - T_2$ in K	d in m	λ
S_1	10	1.5	6.67	1373	1273	100	0.75	0.05
S_2	12	2	6	1423	1323	100	1	0.06
S_3	11.5	0.5	23	1473	1373	100	0.25	0.0575

The average thermal conductivity of this ceramic material is 0.0558.

6. Result

The thermal conductivity of this new thermal ceramic antibacterial insulation material is manufactured from waste glass and fly ashwith medicinal phytochemical repellent is approximately 0.0558W/mk.

7. Conclusion

Thermal insulation depends on preparation of particle size.And we know that the thermal insulation is the constant rate of reducing fluxion of heat from high to low temperature. In material te interatomic interactions develop the insulation rate in materials. In this paper, the coal FA , waste glass WG, organic medicinal additives are main component to prepare the materials. At (700-1000)°C, the pysical parameters such as porosity, compressive strength, molecular size etc. to detect the advancing new materials.

To manage the waste materials which contributes in investigation of utillization of waste material.

The main conclusion marked as,.

a) The specimen that utilize mainly in building materials such as ceramic tiles materials, cooking utensil etc .

b) Thermal insulation depends upon particles size, absorbtion of water is mainly relative and supportive for environment.

The waste material utilising of FA & WG is the low cost effective material Production.

8. Suggestion

we suggest that the source of waste products represents increasingly the problem for human kind.As waste combustion coal fly ash and waste glass are the two basic ingredients and challenges for control the environmental pollution.The FlyAsh of oxide materials and WG as waste glass having siliceous compounds are the main selective components for making thermal insulation materials that contribute to minimize the negative impact of FA and WG for environment. In other hand , the production of heat insulation ceramic materials are selective usage for building materials such as tiles, cooking utensil etc. to protect the environment. Mainly the harmful radiation and diseaseswill reduce by thisceramic material by the presence of oxide materials in FA and WGand phytochemical microorganism repellentfrom harmful microbial. It is balancing the both matter in case of FA & WG to protect the electronic and building materials. There is so cost effective by using in building materials.

9. References:

[1] Selcuk OZCAN, Nurcan CALIS ACIKBAS, Gokhan ACIKBAS (2017), "Formation of Antibacterial Effect on ceramic tiles surfaces", Anadolu university Journal of Science technology A- Applied Sciences & Engineering, Vol. 18, 122-130.

[2] Hasmaliza Mohamad, Foo hui Shian (2015), "Titania (TiO2) as Antibacterial material in ceramic tiles: Effect of composition", Material Science Forum, Vol. 840, 14-18.

[3] M. Hasmaliza, H.S.foo& K. Mohd (2016), "Anatase as Antibacterial material in ceramic tiles", Procedia chemistry, Vol. 828-834.

[4] DA Dhale (2013), " Plants used for insect & pest control in north maharahtra India" , The journal of Ethrobiology Medicine, Vol. 118, 379-388.

[5] Sett Rupnarayan (2017), "Fly Ash: characteristics, problems and possible utilization ", Pelagia Research Library, Vol. 8, 32-50.

[6] S. I. Abu-Eishah (2002), "Correlation for the thermal conductivity of metals as a function of temperature", International Journal of Thermophysics, Vol. 22, 1855-1868.

[7] Romil Patel, Chintan Patel, Pratik Patel (2016), "A Review paper on Measure Thermal Conductivity", Journal of Emerging Technologies and Innovative Research, Vol. 3, 51-53.

[8] Caminati, W., Melandri,S (2006), " Relative strength of Hydrogen bonding for synthesize the substances", Chemical Int.Ed, Vol 45 ,2438.

[9] Wiberg, K.B. (1996), "To control the Thermal energy in molecular mechanics to bonding between different molecules", Accts Chem. Res., Vol 29, 229.

[10] Ravi Sankar B., Nageswara Rao D., Srinivasa Rao Ch. (2015), "Nanofluid thermal conductivity – A Review", International Journal of Advances in Engineering & Technology, Vol. 5, 13-28

A study on components of fly ash of different locations and its utilization for production of improved quality materials: A Review

Umakant Shrivastava[1], Mohammad Ziyauddin Khan[2], and Vikas Mishra[1]
[1] Research Scholar, Department of Physics, Dr. C. V. Raman University, Kota Bilaspur (C.G.) India.

[2] Associate Professor, Department of Physics, Dr. C. V. Raman University, Kota Bilaspur (C.G.) India.

***Corresponding author's e-mail address:ukscvru@rediffmail.com, mzkhan077@gmail.com**

Abstract: During the combustion of coal for energy production an industrial by-product is generated known as Fly Ash, is recognized as a hazardous environmental pollutant. Globally, a lot of considerable research has been undertaken by the researchers and scientists on this subject because of various environmental problems caused by Fly Ash. In recent past it is only being treated as waste and a source of air and water pollution but is in fact a resource material and has also proven its worth over a period of time. The present paper reviews on the potential applications for coal Fly Ash as a eco-friendly raw material: in construction of building materials like bricks and Portland cement, in painting application, application in TL Dosimetries, as a soil amelioration agent in agriculture, in the synthesis of zeolite, in the synthesis of geo-polymers, as an adsorbent for flue gases, as a removal agent for toxic metals, dyes, organic & inorganic compounds from waste water and for the extraction of the metals. Since at different locations the property of Fly Ash varies; thus we required analyzing its variability parameter for utilizing it effectively. Hence, before the utilization of Fly Ash as a eco-friendly resource material, it is necessary to study properties of Fly Ash from different sources, so that it can be used beneficially.
Thus Fly Ash management is a cause of concern for the future. This article attempts to highlight the management of waste Fly Ash to make use of this solid waste, in order to save our environment.

Keywords: Eco-friendly, Fly Ash, TLD badges, Construction Materials, Waste Management.

1. Introduction

In present era fly ashes are generated from combustion of coal at thermal power generating plants. The production of fly ash in rapidly increased due to vastly growing industrialization. In the power plants before the combustion of coal it is been grind and converted into fine powder to utilize most from it. The safe disposal and waste management is the most important threat in the whole process of fly ash production. In order to maintain the ecological balance of the nature we need to properly dispose the toxic and complex characterized waste produced from various industries so that it will not impose any sort of harmful effect on our environment and social life. [1].

The amount of coal waste (fly ash), released by factories and thermal power plants has been increasing throughout the world, and its disposal has become a serious environmental problem. The present day utilization of ash on worldwide basis varied widely from a minimum of 3% to a maximum of 57%, yet the world average only amounts to 16% of the total ash [2]. (Table 1) simply shows that India is the world largest producer of fly ash but only 38 % of produced ash is being utilized in various applications as shown in (Table 2) from year 1993 to 2012 and futuristic utilization during 2031-32. On the other hand countries like Denmark, Italy, and Netherlands were utilizing all produced fly ash from their power plants.

Table 1: Country wise production and utilization of FA as on 2014 [3].

Countries	Production of FA (MT/year)	Utilization % of produced
INDIA	112	38
CHINA	100	45
USA	75	65
GERMANY	40	85
UK	15	50
AUSTRALIA	10	85
CANADA	6	75
FRANCE	3	85
DENMARK	2	100
ITALY	2	100
NETHERLANDS	2	100

Table 2: Fly Ash Generation and Utilization in India [5].

Year	Generation (Mt)	Utilization	
		Mt	Percent of generation
1993-94	40	1.2	3
2004-05	112	42	38
2006-07	130	60	46
2011-12	170	170	100% utilization mandated.
2031-32	600	---	Not yet planned.

Now, we will look after a comparison of 18 Indian states during the year 2016-17 of their power plant installation capacity, fly ash generation and fly ash utilization in million tons.

Graph 1: State wise Fly Ash generation and its utilization during year 2016-17 [4].

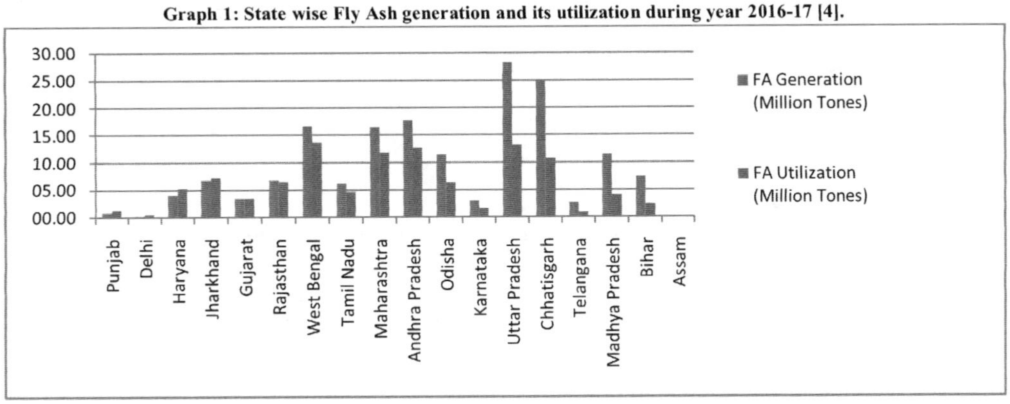

About 75 % of power generation capacity of India is based on thermal energy and it is mainly dependent on coal based power generation. It is around 90% of whole thermal power generation which is based on combustion of coal and rest others are merely 10%. The main reason behind the large amount of fly ash generation is the presence of high ash content in the Indian coal of about 30-50 percent. Fly ash management would remain an important area of national concern just because from past to present and also in future coal will be a major source of energy generation.

The most shocking stats for the two states of Uttar Pradesh and Chhattisgarh since these two were the largest producers of fly ash while both of these states were below 50 % line in utilizing the fly ash in various fields. In the state of Chhattisgarh there are some major thermal power plants are situated at Korba district which generates about 24000 metric tons per day or 8.7 million tons of fly ash annually that is approximately $9/10^{th}$ part of the total fly ash produced in the Chhattisgarh State.

2. Types of fly ash and its classification

Fly ash is the residue of coal being burnt and obtained from combustion process involved in the thermal power plants. Due to the combustion of coal in thermal power plants some different varieties of ashes are produced. They are:

i. Fly Ash: This kind of ash is extracted from flue gases through electrostatic precipitator in dry from. This ash is fine materials and possesses good pozzolanic property.
ii. Bottom Ash: This kind of ash is collected in the bottom of the boiler furnace. It is comparatively coarse materials and contains higher unburned carbon. It possesses zero or little pozzolanic property.
iii. Pond Ash: Both fly ash and bottom ash with a large quantity of water to make it in slurry form and deposited in ponds wherein water gets drained away.
iv. Mound ash: - Fly ash and bottom ash or both mixed in any proportion and deposited in dry form in the shape of a mound is termed as mound ash.

$1/5^{th}$ of the total fly ash generated is bottom fly ash while $4/5^{th}$ part is the fly ash generated from power plants. Both bottom and fly ash are together dumped into the ash ponds.

Table 3: Chemical compositions of Indian coal ashes [6].

Compounds	Fly Ash %
SiO_2	38–63
Al_2O_3	27–44
TiO_2	0.4–1.8
Fe_2O_3	3.3–6.4
MnO	0–0.5
MgO	0.01–0.5
CaO	0.2–8
K_2O	0.04–0.9
Na_2O	0.07–0.43
LOI	0.2–3.4

Table 4: Classification of fly ash components as per the types of coal [2].

Component (wt.%)	Bituminous	Sub-bituminous	Lignite
SiO_2	20–60	40–60	15–45
Al_2O_3	5–35	20–30	10–25
Fe_2O_3	10–40	4–10	4–15
CaO	1–12	5–30	15–40
MgO	0–5	1–6	3–10
SO_3	0–4	0–2	0–10
Na_2O	0–4	0–2	0–6
K_2O	0–3	0–4	0–4
LOI	0–15	0–3	0–5

(Table 3) and (Table 4) shows the percentage of components in fly ash, percentage of components as per the type of coal. Depending on the concentration of calcium oxide, silica, alumina and iron oxide there are two grades of Fly Ash.

a. Grade I fly ash, having fractions SiO_2+Al_2O_3+Fe_2O_3 greater than 70 % and which are derived from bituminous coal.

b. Grade II Fly ash, having fractions SiO_2+Al_2O_3+Fe_2O_3 greater than 50 % and which are derived from lignite coal.

ASTM C618 specified two categories of fly ash, Class C and Class F depending on the type of coal and the resultant chemical analysis [7].

The Fly Ash mainly consists of micro sized particles of silica, alumina and iron. It is easy to form mixture of fly ash with any other component as the particle of fly ash is spherical in nature. The capillarity is one of the best properties for fly ashes to be utilized during the preparation of concrete [1]. The Quality requirements for fly ash varied with respect to the kind of use made out of it. Fly ash quality is mainly affected by types of coal and combustion process. The four most relevant characteristics of fly ash which affects the quality for use in various applications are loss on ignition (LOI), fineness, chemical composition and uniformity.

Fly ash produced from the burning of anthracite and bituminous coal produces Class F fly ash whereas combustion of lignite or sub-bituminous coal produces Class C fly ash. Class C fly ash generally contains more than 20% lime (CaO) and has both pozzolanic and self cementing characteristics whereas class F contains less than 10% lime (CaO) [3]. In Class C fly ash the amount of alkali and sulphate (SO_4) present is higher than Class F fly ash.

It is found that all the Indian coal ashes possess pozzolanic activity. ASTM defines a pozzolan as "a siliceous and aluminous materials which possess very little or no cementious property within it but in finely divisible form and when the moisture is present, react chemically with calcium hydroxides at room temperatures to form compounds possessing cementious properties" [4]. For this reason, it has been chosen for preparing building materials and materials for some other constructive works.

3. Composition of fly ash at different locations

Now we will look after some statistics through graphs regarding the composition of fly ash and the percentage comparison of different locations in India and outside India. In this statistics power plants in 11 different locations in India were shown with the composition of fly ash different at each location. Similarly, 4 locations of Russia (Outside India) were shown in the statistics for composition of fly ash.

Graph 2: Composition of Fly Ash in 15 locations in India and outside India (Russia) [8-9].

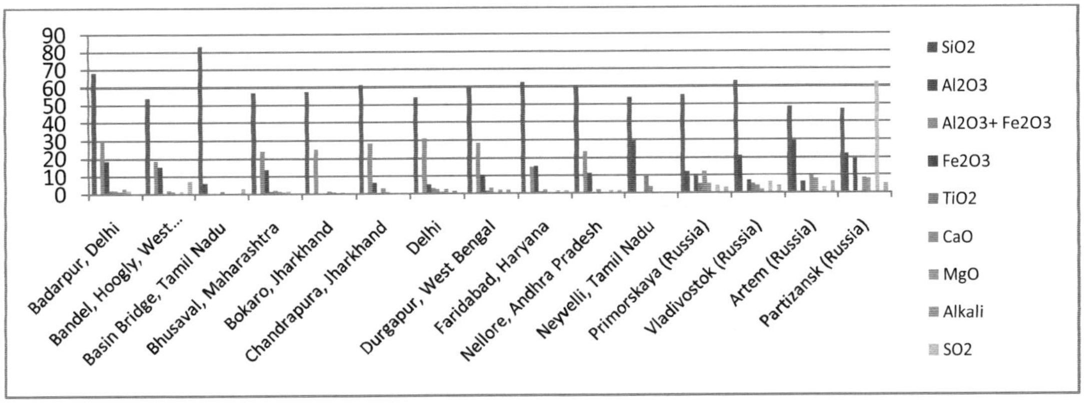

IOP Conf. Series: Materials Science and Engineering **798** (2020) 012004 doi:10.1088/1757-899X/798/1/012004

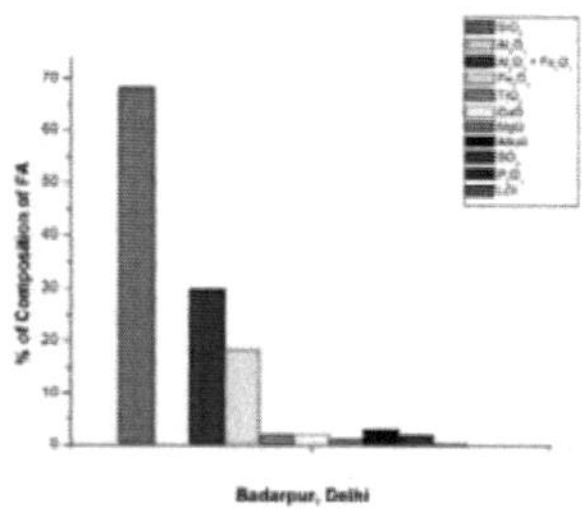

Graph 3

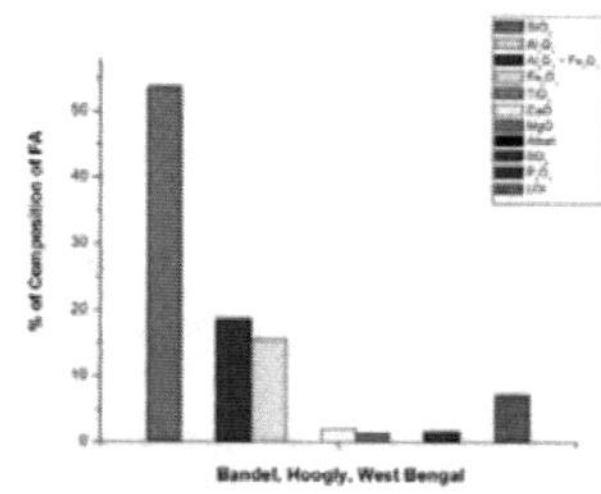

Graph 4

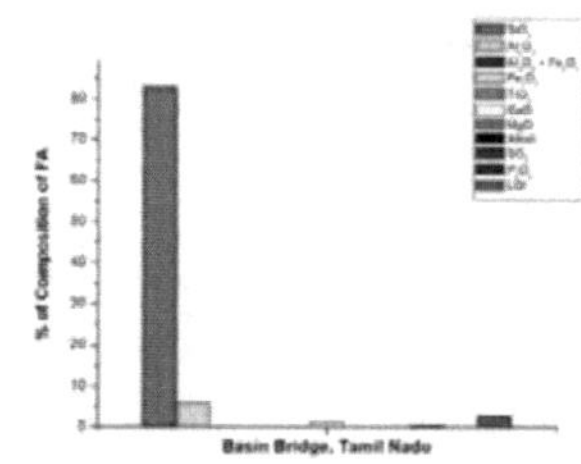

Graph 5

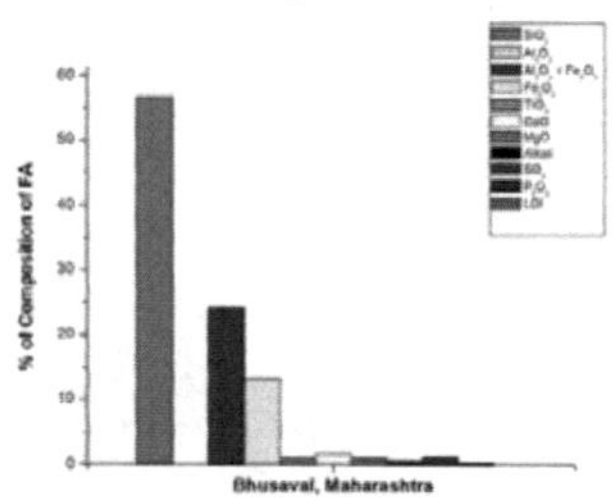

Graph 6

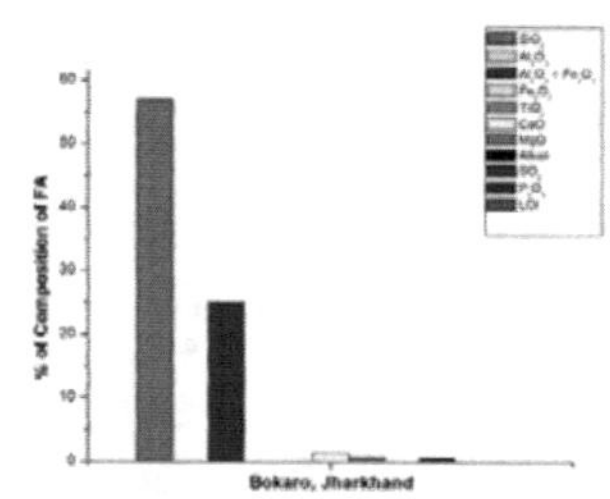

Graph 7

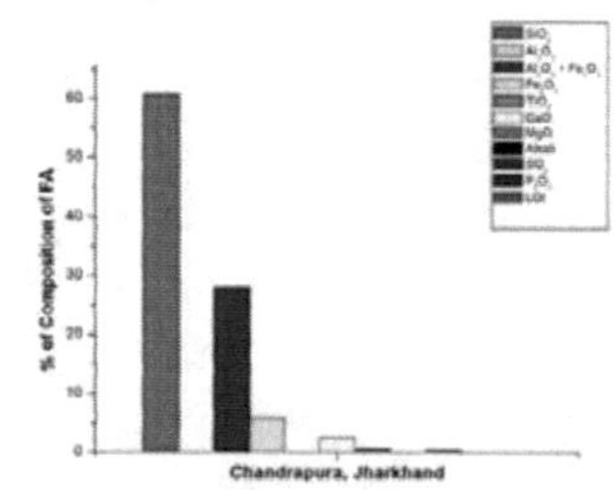

Graph 8

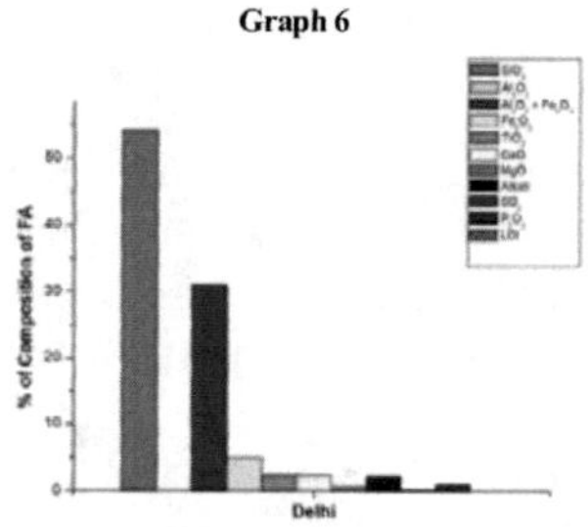

Graph 9

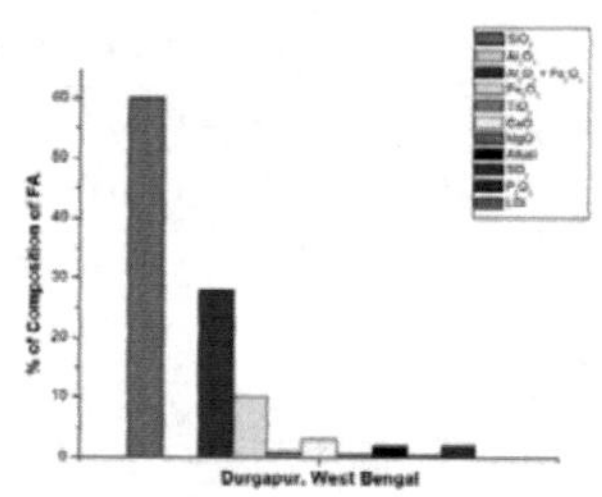

Graph 10

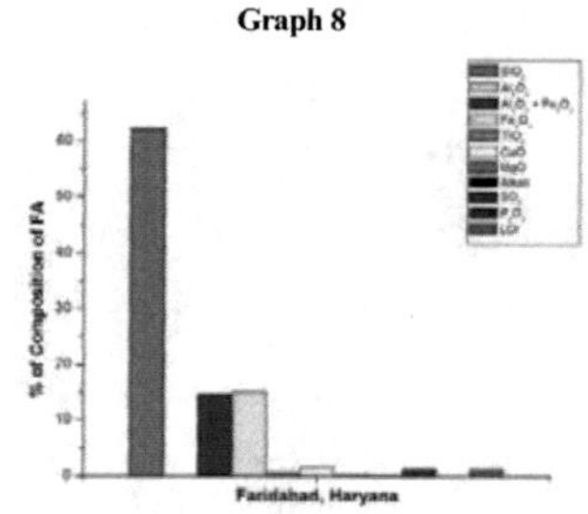

Graph 11

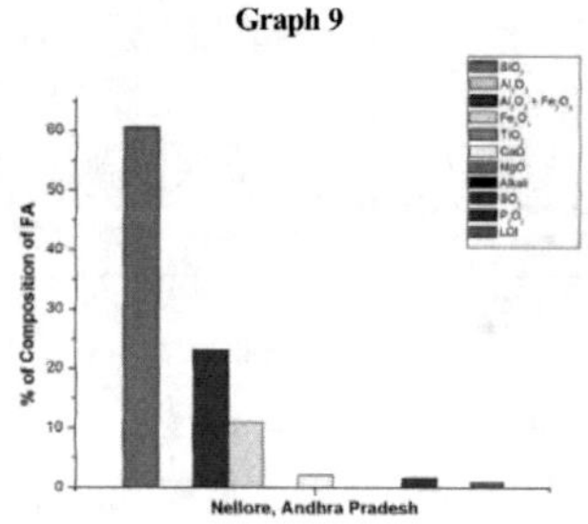

Graph 12

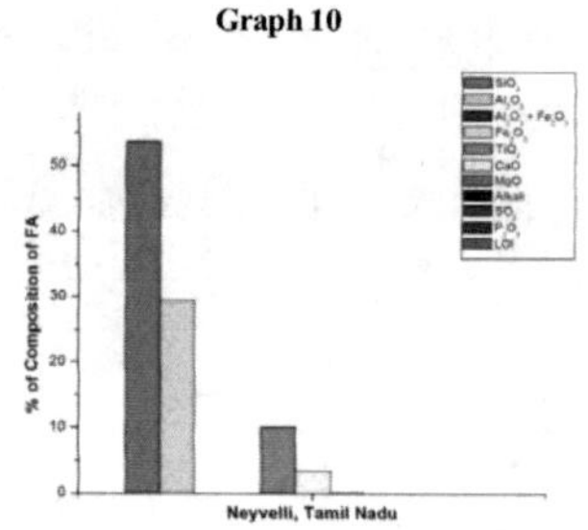

Graph 13

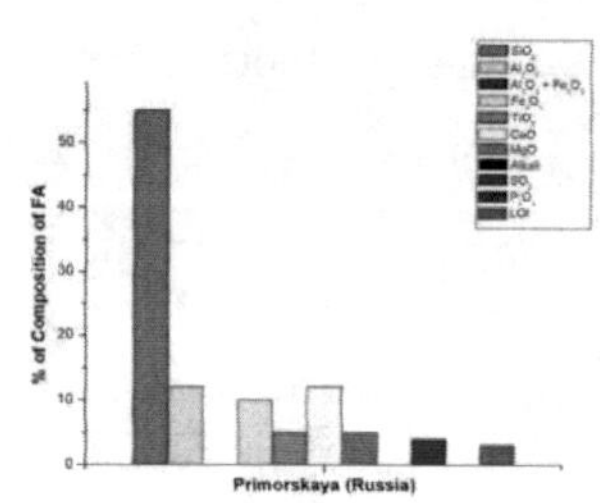

Graph 14

IOP Conf. Series: Materials Science and Engineering **798** (2020) 012004 doi:10.1088/1757-899X/798/1/012004

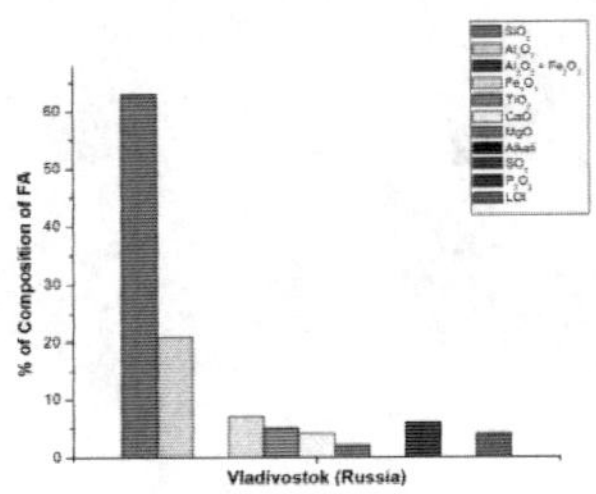

Graph 15

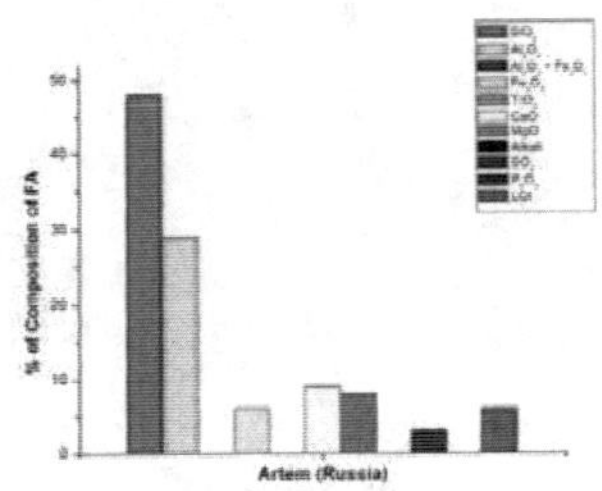

Graph 16

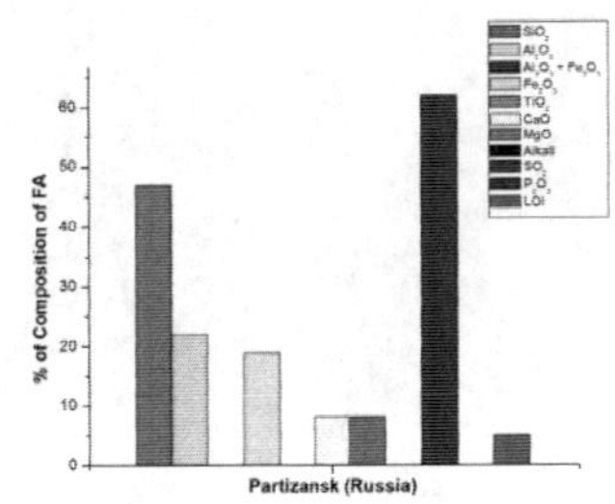

Graph 17

From Graph 2 to Graph 17 shows the percentage composition of fly ash at different locations in India and outside India (Russia) comparatively and individually in different graphs [8-9].

4. Utilization of fly ash as eco-friendly source material

Due to large scale production of fly ash and increasing environmental issues originated from it, vital steps toward the management of fly ash is to be taken. Most importantly fly ash has tremendous qualities and potential to be utilized as initial material. Keeping in view this versatility of fly ash, several Government and Non Government Organizations are involved in fly ash utilization and its safe disposal. It is now being used as a source material in various applications and is been replacing various natural materials in order to save it. Some of the applications are discussed below:

4.1 Use of Fly Ash in manufacturing of eco-friendly Bricks

A better replacement of fired clay bricks is fly ash based bricks as a construction material as it can be utilized as an eco-friendly source. It is eco-friendly in the sense that it will stop the depletion of top layered soil as clay bricks required a lot of soil mass where as fly ash is produced using ash produced during the combustion of coal and some other components like sand, cement and many more. Apart from that we need very high temperature in order to produce clay bricks where as different compositions of fly ash bricks can be produced at room or normal temperature. The different compositions of fly ash brick sets and gain a shape in the presence of moisture. With the time passes this fly ash based bricks became more durable due to its moisture absorbing capacity and thus gain strength.

According to some statistics only 5 million tones fly ash is utilized in preparation of fly ash bricks which saves around 20 crores INR out of 5770 crores INR which is quite less in comparison to other applications. So the use of fly ash in bricks preparation needs to increase for the betterment of environment and to maintain socio-economic condition [10].

In such a way we can have many different mixtures for fly ash bricks to make it eco-friendly and low cost. Now a day various attempts were made by researchers and scholars to study new properties of bricks prepared by inclusion of waste materials such as fly ash, lime and gypsum. So that these bricks can be used as a replacement to burnt clay bricks. Fly ash bricks can be economical and having good compressive strength compared to fired clay bricks [3].

IOP Conf. Series: Materials Science and Engineering **798** (2020) 012004 doi:10.1088/1757-899X/798/1/012004

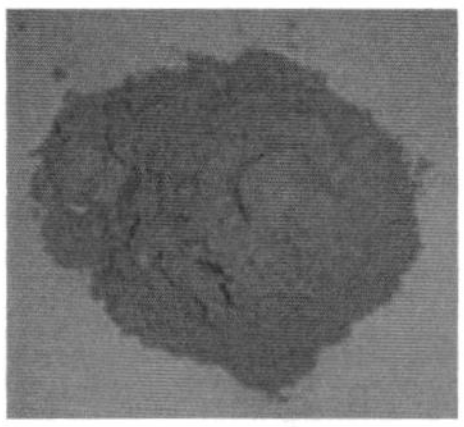

Figure 1: Fly Ash

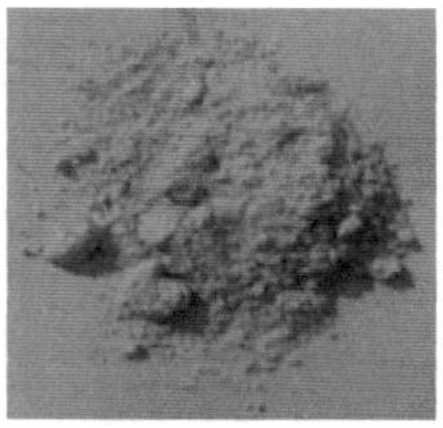

Figure 2: Lime

Figure 3: Kalmegh

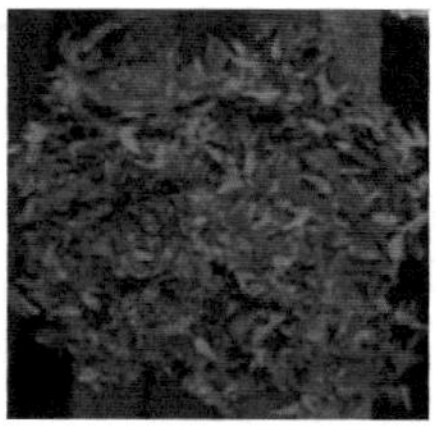

Figure 4: Neem

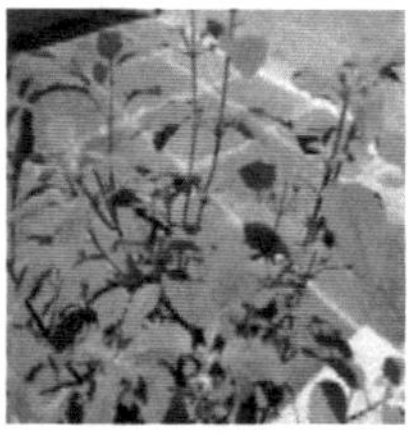

Figure 5: Tulsi

Since we know that fly ash bricks does not require soil but the inclusion of some medicinal plants content can make it more eco-friendly. The dry leafs of neem, tulsi, bhuineem (Kalmegh) etc. can be mixed with fly ash in a proper ratio in powder form for maintaining the ecological balance. Along with that bio-waste like rice husk and egg shell can be mixed with fly ash as rice husk contains silica where as egg shell can be a source of CaO which are constituents of fly ash chemically.

4.2 Use of Fly Ash in manufacturing of eco-friendly concrete

On economic grounds the utilization of fly ash is quite unique as it replaces cement in preparation of concrete mixtures and it also possess some positive benefits as it demands less water for similar workability, reduces the bleeding mechanism and does not require large amount of heat. Its application lies particularly in mass concrete production [2]. Since mix concrete is very popular in western countries but in countries like India merely 5 % of cement is replaced by fly ash (Cement like material) in concrete production. In ready mix concrete we need to control the ration of ingredients and quality parameters but in concrete prepare at sites it impossible to attain exact accuracy which consumes higher amount of fly ash and quality factor may go down [11]. It provides natural protective sheet against natural weathering action. To understand the complex aggregate action and sulphate resistance fly ash is utilized in concrete.

An agricultural residue Rice husk which accounts for 20% of the 649.7 million tons of rice produced in the world per annum. When this residue is burnt partially in the rice milling plants and when used as fuel produces pollution up to great extent. One more natural waste Chicken eggshell is a by-product in food processing and several tons of eggshells are sent to landfills on daily basis without further processing. So it is generous to apply eggshell bio-waste which is of very low cost as a commercial product for increasing its value. Thus (Table 5) gives a comparative idea of Rice Husk and Egg Shell.

Table 5:Comparision of Chemical properties of Rice Husk Ash (RHA) and Egg Shell [12, 19].

Composition (% by mass)	RHA	Egg Shell
SiO_2	88.32	0.09
Al_2O_3	0.46	0.03
Fe_2O_3	0.67	0.02
CaO	0.67	50.7
MgO	0.44	0.01
Na_2O_3	0.12	-
K_2O	2.91	-
Na_2O	-	0.19
P_2O_5	-	0.24
SrO	-	0.13
NiO	-	0.001
Cl	-	0.08
SO_3	-	0.57
LOI	5.81	47.8
Specifc gravity	2.11	-

So utilizing this agricultural residue with fly ash as a supplementary cementing material gives a remedy for pollution generated by its burning and also it provides an eco-friendly and economical option for concrete production. The compressive strength of RHA was investigated with replacing the percentage of cement in the mixture of concrete and variable grinding period for different period; 1, 3, 7, 28 days [12].

Graph 18: Compressive strength of Concrete mixtures with RHA with variable grinding period [12].

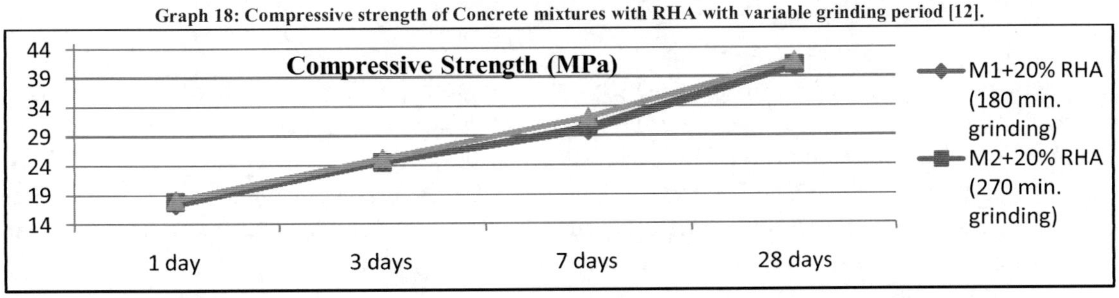

It basically possess 10 % overall weight of hen egg and contains 94 % weight of calcium carbonate as calcite along with magnesium carbonate, calcium phosphate and organic. It was used for adsorbing heavy metal ions such as Cu, Cr, Cg, Ag and Zn from solution [13-17] and removing dry effluents. It can also be used as fertilizers, soil conditioner and additives for animal feeding as it contains nutrients like calcium, magnesium and phosphorous [18].

4.3 In Thermo Luminescence Dosimeter applications

For the identification of radiation level in any body or material Thermoluminesence Dosimetry are used. It is helpful in determination of dose absorbed by the material irradiated by ionizing radiation source. Since some elements of the procedures have broader application, the specific area of concern is radiation level testing of electronic devices using phosphor based Dosimetry with the inclusion of waste fly ash as a source material [20]. By using acid evaporation in open system Dysprosium doped Calcium Sulfate ($CaSO_4$:Dy) has been prepared by *Yamashita* in 1968 which faces some problems related to health issues, corrosion and pollution to the environment due to sulfuric acid vapor. A linear response obtained at 0.25 mol % with higher sensitivity level from 5μGy to 10 μGy and at 75 μm grain size optimum sensitivity is achieved [21]. The TL response is studied when exposed to gamma radiation through a curve between temperature (T) and TL intensity in arbitrary units.

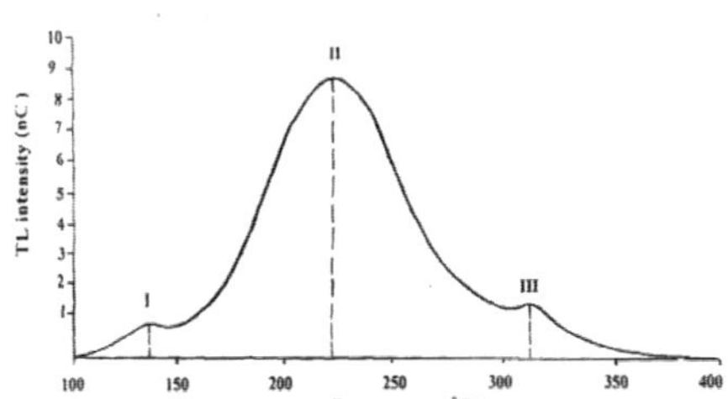

Figure 6: Glow curve for temperature (T) in °C vs TL Intensity in arbitrary units when exposed to gamma dose for ($CaSO_4$:Dy) [21].

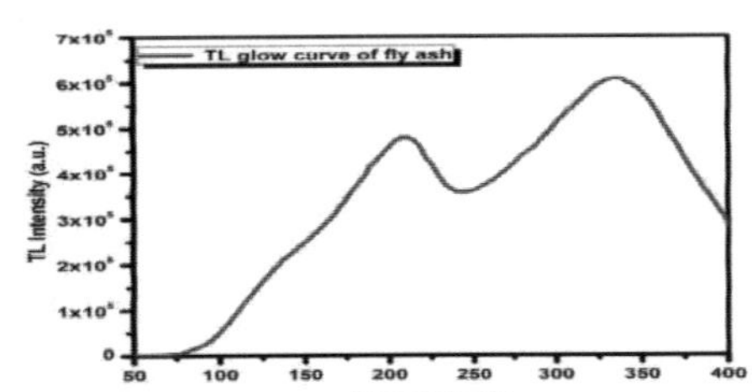

Figure 7: Glow curve of T (°C) vs TL Intensity for Fly Ash when exposed to gamma radiation [22]

In a ^{60}Co gamma field the behavior of fly ash in powder was tested whether it can be used as a material for dosimeter. Fly ash is irradiated to different doses of gamma rays varying from 10-60 kGy and linear response was obtained at 40 kGy of irradiation. T_m-T_{stop} method is used to estimate the possible number of TL peaks present in the glow curve and the TL response of fly ash is found to be satisfactory in high dose Dosimetry applications [22].

4.4 As a Soil amelioration agent in agriculture

Recent agricultural investigations suggests that fly ash when combined with organic materials like cow dung, mud, paper factory sludge, farmyard manure, sewage, residue of crops and with degraded soil gives very positive applications toward eco-friendly and balancing ecosystem.

Its utilizes 'slash' which is a mixture of sewage, sludge and lime as a soil ameliorating agent which also shows positive impact on soil pH and elemental contents in it. Thus we can say that amendment of fly ash in the agriculture field will be beneficial for the crop production as shown in (Graph 19).

Graph 19: pH value and Specific Conductivity of Mixture of soil and FA in different incubation periods [23].

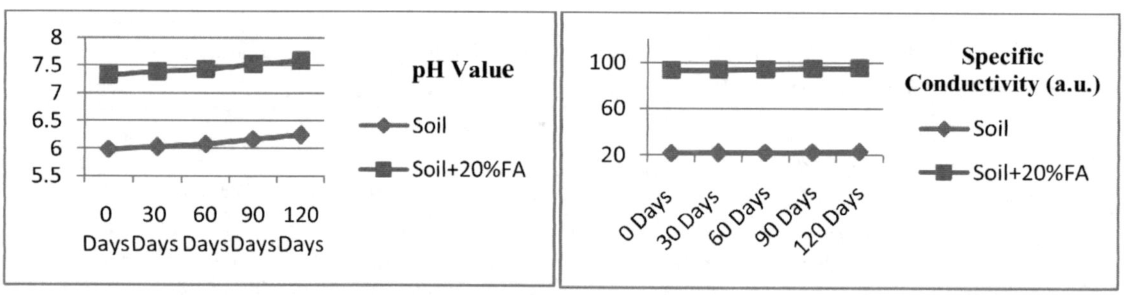

Sao *et.*al; investigated the effect of varying levels of fly ash and growth hormones on the determination of chlorophylls. For fresh leaf, chlorophylls content varies in the plain soil from 0.29 to 0.64 mg g-1, which is less for photosynthetic activities [24].

The inclusion of fly ash in soil also improves permeability status of soil, fertility of crop yield, improves porosity, reduces crust formation, provides micro nutrients like Fe, Zn, Cu, Mo, B, Mn, etc. and macro nutrients as K, P, Ca, Mg, S etc., as a substitute of gypsum for reclamation of saline alkali soil and lime for reclamation of acidic soils, provides ash pond as a place for the nutritional growth of plants, and in improving the socio-economic condition of the inhabitants around the area [14].

5. Conclusions

In this paper we studied about the different locations of fly ash generation and the composition difference at various locations some in India and some outside India. Overall we can see that the percentage of SiO_2, Al_2O_3, Fe_2O_3 and CaO is quite higher in comparison to other components of fly ash in all the locations examined in the paper. The utilization of fly ash as brick and concrete manufacturing material is been done earlier but now we are trying to make it eco-friendly and low cost by inclusion of some part of medicinal plants leafs and also waste materials like egg shell, rice husk, banana bark, etc. From several studies in the agriculture field we came to know that it is perfect candidate for enhancing the fertility of soil. Fly ash also can be utilized as source material for the preparation of radiation absorbing discs as it mixed with phosphor material. Thermoluminesence Dosimetry is useful for radiation workers to take care of exposure from radiation in radiation areas.

Finally, we can conclude that fly ash is a resource which can be utilized in such a way that it will definitely replace many high cost ingredients in various applications discussed above. Its utilization itself solves the problem of environment pollution either it is air, soil or land pollution. Fly ash consumption allows us to get free landmass for agriculture purpose.

References

[1] Updated on: May 2013, http://ethesis.nitrkl.ac.in/4744/1/211CE1230.pdf

[2] M. Ahmaruzzaman, 2010, A review on the utilization of fly ash, Progress in Energy and Combustion Science, 36: 327–363.

[3] Aakash Dwivedi and Manish Kumar Jain, 2014, Fly ash – waste management and overview: A Review, Recent Research in Science and Technology, 6(1): 30-35.

[4] Report on Fly Ash generation at Coal/Lignite based Thermal Power Stations and its utilization in the country for the year 2016-17, Central Electricity Authority New Delhi December 2017.

[5] Anjan K. Chatterjee, 2010, Indian Fly Ashes, Their Characteristics, and Potential for Mechano-Chemical Activation for Enhanced Usability, International conference on sustainable construction materials and Technologies, http://www.claisse.info/proceedings.htm

[6] Gorai Soma, 2018, J. Mater. Environ. Sci., 9 (2): 385-393.

[7] ASTM C618 – 08a. , 2005, Annual Book of ASTM Standards, Concrete and Aggregate, American Society for Testing Materials, 04, 02.

[8] Techno Economic Feasibility report on Fly Ash bricks, Building Materials & Technology Promotion Council Ministry of Housing & Urban Poverty Alleviation Government of India, New Delhi.

[9] R S Fediuk and A M Yushin, 2015, The use of fly ash the thermal power plants in the construction, IOP Conf. Series: Materials Science and Engineering 93: 012070.

[10] Price GC, 1961, Peterson R., Performance of concrete, Toronto, ON: University of Toronto Press.

[11] J.Alam1 and M.N Akhtar, 2011, Fly Ash utilization in different sectors in Indian scenario, International journal of emerging trends in Engineering and Development, 1(1).

[12] Ghassan Abood Habeeb, Hilmi Bin Mahmud, 2010, Study on Properties of Rice Husk Ash and Its Use as Cement Replacement Material, Materials Research. 13(2): 185-190.

[13] Baláž M, Ficeriová J, Briančin J., 2016, Influence of milling on the adsorption ability of eggshell waste, Chemosphere, 146: 458-471.

[14] Chojnacka Katarzyna, 2005, Biosorption of Cr(III) ions by eggshells, Journal of Hazardous Materials, 121(13): 167-173.

[15] Park HJ, Jeong SW, Yang JK, Kim BG, Lee SM., 2007, Removal of heavy metals using waste eggshell, Journal of Environmental Sciences, 19(12): 1436-1441.

[16] Vijayaraghavan K, Jegan J, Palanivelu K, Velan M., 2005, Removal and recovery of copper from aqueous solution by eggshell in a packed column. Minerals Engineering, 18(5): 545-547.

[17] Zheng W, Li X-m, Yang Q, Zeng G-m, Shen X-x, Zhang Y, 2007, Adsorption of Cd(II) and Cu(II) from aqueous solution by carbonate hydroxylapatite derived from eggshell waste, Journal of Hazardous Materials,147(12): 534-539.

[18] Phonphan Kamkum, Wanwilai Vittayakorn, Panpailin Seeharaj, Somsak Woramongkolchai, Rangson Muanghlua and Naratip Vittayakorn, 2018, Utilization of Eggshell as a Low-Cost Precursor for Synthesizing Calcium Niobate Ceramic, doi: 10.1680/jgrma.18.00049

[19] M. N. Freire, J. N. F. Holanda, Characterization of avian eggshell waste aiming its use in a ceramic wall tile paste, 2004, Cerâmica, 52 :240-244.
[20] J. C. Humphreys and S. E. Chappell, 1979, Standard Procedure for Use of Thermoluminesence Dosimetry in Radiation-Hardness Testing of Electronic Devices.
[21] Salah M.Kamal, Awad S. Gerges and Mohamed A. Al-, 2003, Thermoluminescence Properties of Home made $CaSO_4$:Dy For Dosimetry Purposes, Conference on Nuclear and Particle Physics, Fayoum , Egypt.
[22] Swati Joshi , Karan Kumar Gupta, S.K. Ubale, S.J. Dhoble, 2017, High dose TL response of fly ash collected from coal fired thermal power plant, Radiation Measurements, 30: 1-7.
[23] Manorama Gupta, S. P. Singh & K. Kaur, 2015, Impact of Neem Leaf on Physical and Chemical Profile of Fly Ash Incorpoted Soil, Asian Journal of Advanced Basic Science, 4(1), 28-32.
[24] Shweta Sao, Pankaj K. Sahu, 2013, Influence of Fly Ash and Growth Regulator with Soil for Determination of Chlorophyll in Arachis hypogaea L., American Journal of Plant Sciences, 4: 1744-1749.

IOP Conf. Series: Materials Science and Engineering **798** (2020) 012005 doi:10.1088/1757-899X/798/1/012005

Preparation and characterization of amphiphilic polymer coating for marine biofouling control

Lokeshwar Mahto[1] and Dharm Pal[1]

[1]Department of Chemical Engineering, National Institute of Technology Raipur, Raipur, Chhattisgarh, 492010, India

Abstract. Biofouling is a major issue prevalent in marine industry resulting in losses in billions of dollars for repairing and maintenance. Coatings developed to tackle marine biofouling are either toxic towards environment or have shorter lifespan. Amphiphilic block copolymer coatings have gained wide attention recently to mitigate biofouling. In this work, performance of synthesized coating towards biofouling was studied. Hydroxyl terminated polydimethylsiloxane as hydrophobic and chitosan as hydrophilic moiety was used to synthesize the amphiphilic coating. Characterization of the prepared coating using Scanning Electron Microscopy (SEM) showed heterogeneous microstructure. Antifouling performance using egg white as the protein testing probe was done as it simulates the adhesives used by biofouling organism. Resistance to protein adhesion was seen on the surface of the coating. So, the developed environmental-friendly coating can be used effectively for the mitigation of marine biofouling.

1. Introduction

Marine biofouling is the growth of undesired marine microorganisms like bacteria, algae, sponges, etc. along with macro-organisms like barnacles, mussels, balanus etc. over the surfaces immersed in seawater[1,2]. This results in increased weight and skin-friction which increases the frictional resistance and deteriorates the surface[3]. This leads to a reduction in the speed and manoeuvrability of the ships and vessels and increases the fuel consumption for the same distance travelled without fouling[4]. The more fuel burnt releases more harmful emissions from ships which are also a monetary loss in billions for the marine industry. The deterioration of ship hull and its coating generate other problems as well like corrosion, deposition of the released coating in the sea bed leading to chemical waste[2]. Therefore, over the years, scientists have researched to find a novel technique to protect the marine industry from fouling by developing various protective antifouling coatings.

Amphiphilic copolymer composite coating is the dominating fouling control strategy. They have low interfacial energy and low surface energy because of hydrophobic moiety which makes the adhesion bond weak and easy to detach. Also, they possess low modulus of elasticity which deforms the organisms attached to the surface by peeling off[5]. They have hydrophilic moiety which forms a strong hydration layer. Because the interfacial energy of strong hydration layer with the surface is high enough, the fouling organisms are not able to make a bond with the surface which makes the surface protein resistant and biofouling settlement[6,7]. Polydimethylsiloxane (PDMS) is the most generally used hydrophobic polymer used for marine applications. Hydroxyl terminated PDMS has more functional groups make the crosslinking easier by blending. Many hydrophilic polymer coatings like polyethylene glycol (PEG), zwitterionic, grafted and blended coatings have been used to control

IOP Conf. Series: Materials Science and Engineering **798** (2020) 012005 doi:10.1088/1757-899X/798/1/012005

marine fouling[8–10].They have hydrophobic as well as hydrophilic elements at the surface rendering both detachment of biofoulants and prevention of attachment mechanism into a single polymer by presenting an ambiguous surface to the protein adhesive to have a better antifouling coating than the standalone[11–13]. Currently used amphiphilic polymers have some issues with the synthesis and applicability. Instability of hydrophilic polymers is an issue like oxidation of PEG occurs which deters the surface[14]. Chemistries involved in the fabrication are quite complex in harsh environment which make it difficult to commercialize[11,15]. The blending technique can provide an easy step to synthesize an amphiphilic copolymer[16].

Recently, chitosan has gained wide application due to biocompatibility, anti-microbial activity, non-toxicity, hydrophilicity[17]. It is derived from chitin which is a by-product of sea-food industry[18]. It has polycationic nature which imparts more functionalities to the surface[19].

In this work, we developed a novel and stable amphiphilic coating for marine biofouling control. A direct crosslinked network of hydroxyl terminated PDMS and chitosan was formed at room temperature. The prepared coating showed excellent protein resistance which can infer good antifouling performance indicating its potential to use as a non-toxic marine biofouling control coating.

2. Experimental

2.1. Materials
Epoxy resin consisting of Bisphenol A diglycidyl ether (BADGE) andsilicone elastomer hydroxyl terminated polydimethylsiloxane (HTPDMS) was purchased from Sigma Aldrich, Inc.(3-Aminopropyl)triethoxysilane (APTES)was purchased from Sisco Research Laboratories Pvt. Ltd. and was used as an elastomer coupling agent. Epoxy hardener triethylenetetramine (TETA), chitosan flakes and acetic acid were purchased from Loba Chemie Ltd. All the reagents were used as received without purification.

2.2. Preparation of substrate and chitosan solution
Mild steelplates of size 7cm×5cm×0.5cm were used as primary substrate. The top side of plates are roughened to make the surface rough enough so that the coating is well adhered onto it. The surface is cleaned with acetone and distilled water to remove remaining grease and impurities. Chitosan is insoluble in water but dissolves in weak acids like acetic acid. So, 2g of 75% deacetylated chitosan flakes were dissolved in 200ml of 0.1M acetic acid at 80°C under constant stirring at 200rpm with magnetic stirrer. All the undissolved large molecules and impurities present in the solution was filtered using syringe filter and a homogeneous chitosan solution is obtained.

2.3. Preparation of epoxy primer coating
10g of BADGE and a stoichiometric amount 1.43g of TETA curing agent is mixed thoroughly. The desired stoichiometric amount of curing agent required is calculated using the following equation:

$$Amount\ of\ curing\ agent = \frac{Equivalentweightof\,amine}{Equivalentweightof\,epoxyresin} \times Amountof\,epoxy \qquad (1)$$

In present work, molecular weight of TETA is 146g/mol, number of active hydrogen atoms of TETA is 6, molecular weight of BADGE is 340g/mol and the number of epoxy groups of epoxy resin is 2. In accordance to the above equation, the stoichiometric amount of TETA required to cure 10g of epoxy resin is 1.43g. After homogeneous mixing, the solution is applied on the substrate using hand bar coater and allowed to cure. Curing of epoxy mixture was done at room temperature for 12 hours.

2.4. Synthesis of HTPDMS-chitosan coating

10ml of HTPDMS silicone elastomer and 1ml of APTES coupling agent are added into a beaker. 1ml of prepared chitosan solution is added to it. The solution is mixed thoroughly in magnetic stirrer at 800 rpm for 3 hour at room temperature. The prepared solution is coated over the epoxy coat using a hand bar coater. Curing of this solution mixture wascarried out for 7 days and at room temperature.

2.5. Characterization of the prepared sample
The surface morphology of the synthesized coating was revealed using a scanning electron microscopy (ZIESS EVO 18 Special Edition) at an accelerating voltage of 15kV. The sample was coated with gold to make it conductive before taking the images. The dispersion studies were carried out. Scratch resistant test was done to check the durability of the applied coating using a knife and applying moderate hand pressure to it. Water contact angle analysis and critical surface tension was also calculated using Zisman plot with pure liquids of known surface tension like water, glycerol and methanol.

2.6. Antifouling performance testing
The initial step of biofouling is conditioning film formation which is mainly protein. The antifouling performance assessment were tested using a pseudo-model of protein mixture contained in egg white simulating the adhesives secreted by the biofouling organisms to adhere. 1ml of egg white was applied on the coating surface and left to dry for 24 hours. Analyzing the dried egg white visually gives a better understanding of the adhesion strength. The dried egg white fell off the coating surface by simply shaking.

3. Results and discussion
Chitosan has very weak solubility in water and when it is allowed to react with acetic acid, amine group present in chitosan gets protonated which forms polycation chitosan[18]. This results in dissolution of the solid chitosan. The epoxy primer applied to the pristine metal plate adheres strongly to it. The heterogeneous amphiphilic microstructure can be confirmed from the SEM images shown in figure 1. The chitosan size ranges from hundreds of micrometre to nanometres. This gives the topographic surface an irregular heterogeneity. This is beneficial for the antifouling performance because fouling organisms ranges from few nanometres to millimetres and they prefer to attach to surfaces non-familiar to their behaviour[20]. The formation of blend can also be seen evidently from the image. Scratch resistant test was done using a knife and applying moderate hand pressure. The surface resisted scratches from moderate pressure but applying high pressure scratched the surface.

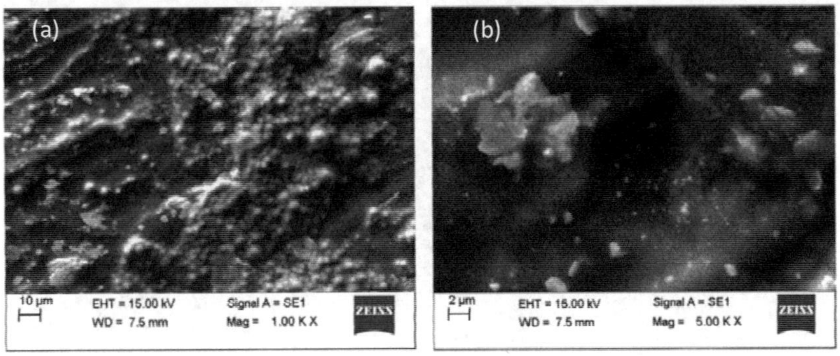

Figure 1. SEM image of the prepared sample showing amphiphilic heterogeneous structure

Average water contact angle was found to be 72.44° implying that surface became hydrophilic. This confirms the formation of amphiphilic copolymer network. Because, the water contact angle of polydimethylsiloxane is around 116.7° rendering hydrophobicity[21]. Decreasing of contact angle shows incorporation of hydrophilic chitosan in the polymer matrix. The critical surface tension calculated using Zisman plot was found to be 19.34 mN/m. So to completely wet the surface, a liquid with critical surface tension of 19.34 mN/m is required.

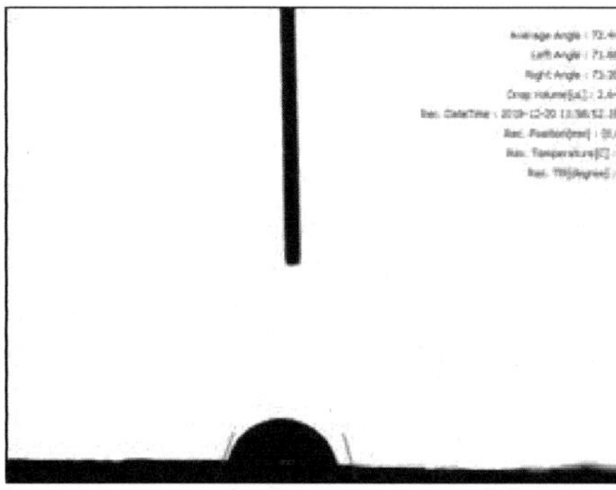

Figure 2. Water contact angle measurement of the coating surface

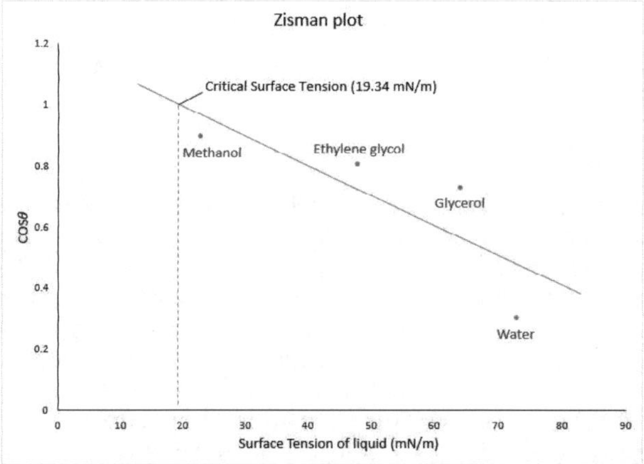

Figure 3. Zisman plot for calculating critical surface tension

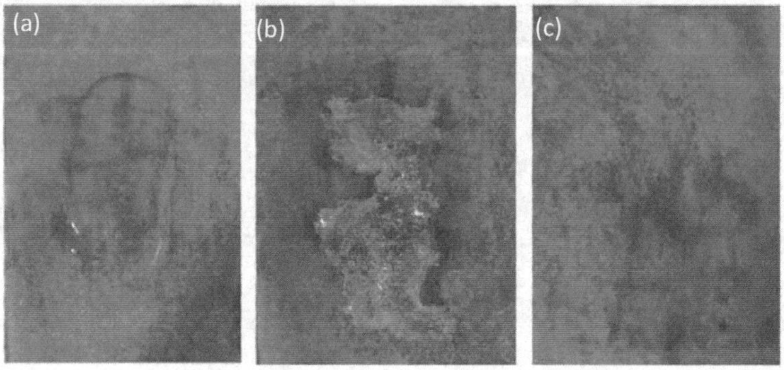

Figure 4. Egg white protein adhesion test simulating protein adhesives secreted by biofouling organisms showing image (a) applied egg white (b) egg white after 24 hours (c) egg white shaken off

The antifouling test performed using egg white confirms the applicability of the prepared sample[22]. 1ml of egg white when allowed to adhere on the coating didn't attach at all and was easily dried after 24 hours and shaken off without any effort leaving the surface clean as before. The pseudo model assumed here showed excellent antifouling performance towards protein adherence.

4. Conclusion

In this study, a novel amphiphilic marine antifouling coating was developed which shows excellent antifouling property and fair mechanical strength. The surface was heterogeneous with variable blend size domains. This will make the surface effective against most of the biofouling organisms when real time field tests will be performed.

References

[1] Callow J A and Callow M E 2011 Trends in the development of environmentally friendly fouling-resistant marine coatings *Nat. Commun.* **2** 244

[2] Schultz M P, Bendick J A, Holm E R and Hertel W M 2011 Economic impact of biofouling on a naval surface ship *Biofouling* **27** 87–98

[3] Schultz M P, Walker J M, Steppe C N and Flack K A 2015 Impact of diatomaceous biofilms on the frictional drag of fouling-release coatings *Biofouling* **31** 759–73

[4] Lejars M, Margaillan A and Bressy C 2012 Fouling Release Coatings: A Nontoxic Alternative to Biocidal Antifouling Coatings *Chem. Rev.* **112** 4347–90

[5] Detty M R, Ciriminna R, Bright F V and Pagliaro M 2014 Environmentally Benign Sol–Gel Antifouling and Foul-Releasing Coatings *Acc. Chem. Res.* **47** 678–87

[6] Acevedo M S, Puentes C, Carreño K, León J G, Stupak M, García M, Pérez M and Blustein G 2013 Antifouling paints based on marine natural products from Colombian Caribbean *Int. Biodeterior. Biodegrad.* **83** 97–104

[7] Srinivasan M and Swain G W 2007 Managing the Use of Copper-Based Antifouling Paints *Environ. Manage.* **39** 423–41

[8] Banerjee I, Pangule R C and Kane R S 2011 Antifouling coatings: Recent developments in the design of surfaces that prevent fouling by proteins, bacteria, and marine organisms *Adv. Mater.* **23** 690–718

[9] Martinelli E, Suffredini M, Galli G, Glisenti A, Pettitt M E, Callow M E, Callow J A, Williams D and Lyall G 2011 Amphiphilic block copolymer/poly(dimethylsiloxane) (PDMS) blends and nanocomposites for improved fouling-release *Biofouling* **27** 529–41

[10] Pollack K A, Imbesi P M, Raymond J E and Wooley K L 2014 Hyperbranched fluoropolymer-polydimethylsiloxane-poly(ethylene glycol) cross-linked terpolymer networks designed for marine and biomedical applications: Heterogeneous nontoxic antibiofouling surfaces *ACS Appl. Mater. Interfaces* **6** 19265–74

[11] Yilgör E and Yilgör I 2014 Silicone containing copolymers: Synthesis, properties and applications *Prog. Polym. Sci.* **39** 1165–95

[12] Brady R F 1999 Properties which influence marine fouling resistance in polymers containing silicon and fluorine *Prog. Org. Coatings* **35** 31–5

[13] Kaffashi A, Jannesari A and Ranjbar Z 2012 Silicone fouling-release coatings: effects of the molecular weight of poly(dimethylsiloxane) and tetraethyl orthosilicate on the magnitude of pseudobarnacle adhesion strength *Biofouling* **28** 729–41

[14] Bernhard M, Eubeler J P, Zok S and Knepper T P 2008 Aerobic biodegradation of polyethylene glycols of different molecular weights in wastewater and seawater *Water Res.* **42** 4791–801

[15] Leng C, Hung H-C, Sun S, Wang D, Li Y, Jiang S and Chen Z 2015 Probing the Surface Hydration of Nonfouling Zwitterionic and PEG Materials in Contact with Proteins *ACS Appl. Mater. Interfaces* **7** 16881–8

[16] Guo H, Yang J, Zhao W, Xu T, Lin C, Zhang J and Zhang L 2019 Direct formation of amphiphilic crosslinked networks based on PVP as a marine anti-biofouling coating *Chem. Eng. J.* **374** 1353–63

[17] Rutnakornpituk M, Ngamdee P and Phinyocheep P 2005 Synthesis, characterization and

properties of chitosan modified with poly(ethylene glycol)–polydimethylsiloxane amphiphilic block copolymers *Polymer (Guildf).***46** 9742–52

[18] Park S-I, Daeschel M A and Zhao Y 2004 Functional Properties of Antimicrobial Lysozyme-Chitosan Composite Films *J. Food Sci.***69** M215–21

[19] Arkoun M, Ardila N, Heuzey M and Ajji A 2018 Chitosan-Based Structures/Coatings With Antibacterial Properties *Handbook of Antimicrobial Coatings* (Elsevier Inc.) pp 357–89

[20] Hoipkemeier-Wilson L, Schumacher J F, Carman M L, Gibson A L, Feinberg A W, Callow M E, Finlay J A, Callow J A and Brennan A B 2004 Antifouling potential of lubricious, micro-engineered, PDMS elastomers against zoospores of the green fouling alga Ulva (Enteromorpha) *Biofouling***20** 53–63

[21] Chuah Y J, Koh Y T, Lim K, Menon N V., Wu Y and Kang Y 2016 Simple surface engineering of polydimethylsiloxane with polydopamine for stabilized mesenchymal stem cell adhesion and multipotency *Sci. Rep.***5** 18162

[22] Barletta M, Aversa C, Pizzi E, Puopolo M and Vesco S 2018 Design, manufacturing and testing of anti-fouling/foul-release (AF/FR) amphiphilic coatings *Prog. Org. Coatings***123** 267–81

Oxygen vacancy related conduction behavior in $BaZr_{0.05}Ti_{0.95}O_3$ ceramic

G. Nag Bhargavi[1], Tanmaya Badapanda[2], Ayush Khare[3], M. Shahid Anwar[4] and Nameeta Brahme[5]

[1]*Department of Physics, Govt. Pt. Shyamacharan Shukla College Dharsiwa Raipur- 493221, India*
[2]*Nanophotonics Laboratory, Department of Physics, C.V. Raman College of Engineering, Bhubaneswar-752054, India*
[3]*Department of Physics, National Institute of Technology, Raipur-492010, India,*
[4]*Colloids & Materials Chemistry, Institute of Minerals and Materials Technology, Bhubaneswar-751013, India*
[5]*School of Studies in Physics and Astrophysics, Pt. Ravishankar Shukla University, Raipur-492010, India*

Abstract. For this study, the microcrystalline powder of $BaZr_{0.05}Ti_{0.95}O_3$ (BZT) was prepared by the conventional solid state reaction method. The sample was calcined at 1200 °C for 4 hours and sintered at 1300 °C. The calcined powder was structurally characterized by X-ray diffraction (XRD), which showed that the specimen has a Perovskite structure having orthorhombic structure. On analyzing the scanning electron microscope (SEM) the calculated crystal size was observed to range between 20- 30 μm. The dielectric study of BZT showed normal phase transition behavior. The conductivity studies as a function of temperature and frequency has been performed to study the role of oxygen vacancies. The results of the frequency dependence of the conductivity suggest that oxygen vacancy hopping processes, due to relaxations in oxygen vacancy-related dipoles, being mainly responsible for the conduction behavior in the studied system.

Keywords: Solid state reaction method, XRD, SEM, Dielectric study, Conductivity

1. Introduction

Barium titanate (BT) is a well known Perovskite structured material, which is known for its high dielectric constant, spontaneous polarization, ferroelectric properties and non linear optical properties. It is given in the open literature that the properties of BT can be enhanced by the substitution at Barium and/or Titanium sites by homovalent or aliovalent ions [1-6]. This lead (Pb) free compound has numerous applications in DRAMs, MLCCs, ultra sonic transducers, optical data storage at high density, sensors, actuators and optoelectronic devices [7-9]. Among the homovalent substituents Zr^{4+} in place of Ti^{4+} has been reported to improve the dielectric properties of BT [10-12]. This ceramic with general formula $BaZr_xTi_{1-x}O3$ shows promising piezoelectric and electrostrictive properties [13, 14]. Better thermal and chemical stability of Zr^{4+} than Ti^{4+} makes Zr^{4+} a best choice as isovalent substituent for substitution at Ti sites [15]. Also, the ionic radius of Zr^{4+} (0.087 nm) is larger than Ti^{4+} (0.068 nm) which expand the unit cell and the possibility of electron hopping between Ti^{4+} ions and Ti^{3+} ions reduces to a large extent [15].

In case of all the oxide Perovskites the oxygen vacancies are the fundamental intrinsic defects that play a very critical impact on the electrical properties. The oxygen vacancies do not lost their identity even after the process of sintering. In all the oxide ferroelectric materials the dielectric relaxation is usually related to the vacancies produced at oxygen sites. It is extensively accepted that the dielectric relaxation in oxide materials specifically at high temperatures are usually related to oxygen vacancies, apart from this dielectric relaxation appear in the low-frequency region is also attributed to oxygen-vacancies [16, 17]. So, it is very interesting to study the conduction behavior of BZT system in terms of oxygen vacancies.

The material performances are closely dependent on the methods of synthesis. In case of ferroelectric ceramics the method of synthesis is prominent in determining the various structural and electrical properties. There are various methods of synthesis like solid state reaction method, ball milling, wet chemical synthesis, combustion technique, Sol-gel technique etc. [18-20]. Among these some are chemical routes while some are physical (high temperature diffusion) routes. In the present work, we have chosen the high temperature solid state reaction methods. The chemical routes of processing require costly chemicals which are very sensitive to the environment and physical conditions like light heat and moisture. On the other hand the solid state reaction method uses carbonates and oxides which are easily available and also cost effective.

So, in this study the main objective of the authors is to synthesize $BaZr_{0.05}Ti_{0.95}O_3$ ceramic system by the conventional solid state reaction method and to study in detail the electrical conduction behavior in terms of oxygen vacancies.

2. Experimental Procedure

The sample ($BaZr_{0.05}Ti_{0.95}O_3$) was synthesized by the traditional solid state reaction method. The stoichiometric mixture of powders ($BaCO_3$, TiO_2, ZrO_2) was mixed and grinded in an agate mortar with a little amount of distilled water. The mixture was fired at 1200 °C for calcination for 4 hours in a high temperature furnace. The chemical reaction for the above process is shown in eq. (1). The Perovskite phase structure of the sample was confirmed under the X-ray diffraction analysis (X'pertPro Diffractometer). The analyzed 2θ range was 10°-90° with a step rate of 0.02° and counting time of 0.5 minute for each step. The microstructure of the sintered sample was observed by scanning electron microscope (Zeiss EVO 18). The fired powder was formed as a thick disc (1mm) by cold pressing and sintering was carried out at 1300 °C for 4 hours. The sintered disc was polished with silver paint to make the surface conducting to carry out the electric studies (N4L-NumeriQ LCR meter, Model PSM 1735).

$$BaCO_3 + 0.05ZrO_2 + 0.95TiO_2 \rightarrow BaZr_{0.05}Ti_{0.95}O_3 + CO_2 \uparrow \quad (1)$$

3. Results and Discussion

3.1 *Structural Characterization*

Fig. 1 is presenting the XRD pattern of $BaZr_{0.05}Ti_{0.95}O_3$ system which is calcined at 1200 °C for 4 hours. The X- ray diffraction pattern could be indexed as a pure Perovskite structure. No impurity peak corresponding to any deleterious phase was found which confirms the single phase nature of the compound. Moreover, the sharp intense and well defined diffraction peaks indicate that the ceramic material has a long range degree of crystallinity. The obtained phase was indexed using the orthorhombic space group *Amm2* which is in close agreement with the earlier reports [15, 21].

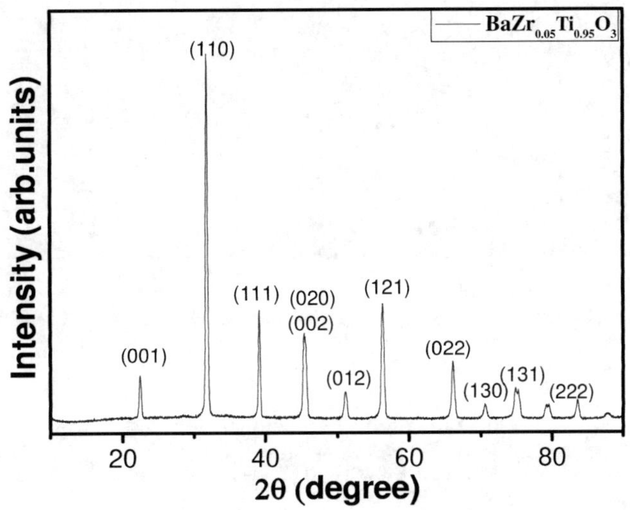

Fig. 1 XRD pattern of $BaZr_{0.05}Ti_{0.95}O_3$ ceramic

3.2 *Microstructural studies*

Fig. 2(a) shows the morphological image of $BaZr_{0.05}Ti_{0.95}O_3$ ceramic pellet sintered at 1300 °C taken from SEM. The SEM investigation has shown that the microstructures of the specimen exhibit a good growth of grains. The image showed that the grains are in irregular polygon shape. Also the grains developed with irregular morphology have multiple grain boundaries. This characteristic is mainly attributed to the matter transportation mechanism at high temperatures during the solid state reaction. Also, the average crystallite size is estimated by line intercept method which is observed to be 20- 30 μm. The compositional characterization of $BaZr_{0.05}Ti_{0.95}O_3$ ceramic sintered at 1300 °C was also done using Energy dispersive X-ray spectroscopy (EDX) which is shown in fig. 2(b). The measurements were taken at different locations and then averaged. Table 1 shows the calculated and EDX derived composition of $BaZr_{0.05}Ti_{0.95}O_3$ ceramic. There exist small variation in both the data which possibly due the deficiency of oxygen during sintering at high temperature. Moreover, in solid state reaction method it is always difficult to get the microscopic uniformity.

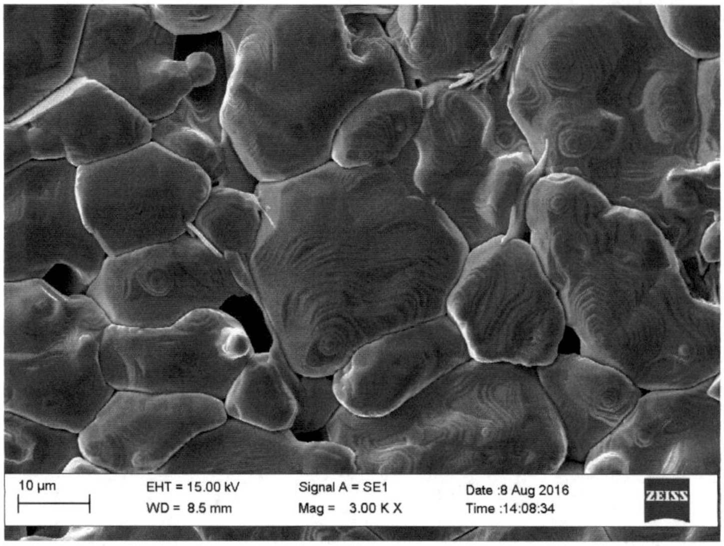

Fig. 2(a) SEM image of $BaZr_{0.05}Ti_{0.95}O_3$ ceramic sintered at 1300 °C

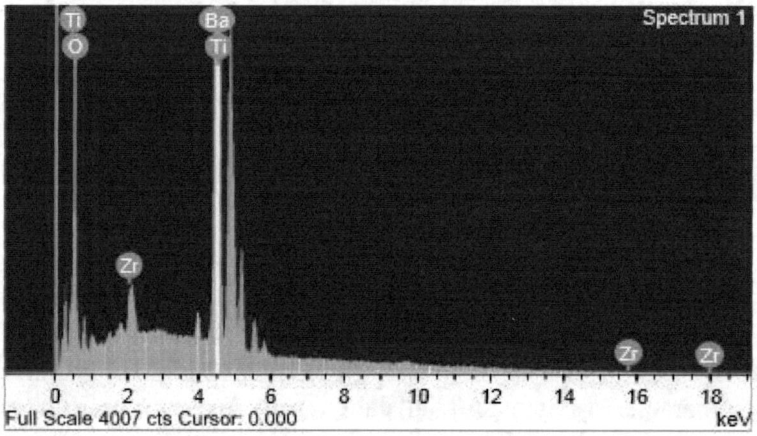

Fig. 2(b) EDX spectrum showing the elemental composition of $BaZr_{0.05}Ti_{0.95}O_3$ ceramic

Table 1 Calculated and EDX derived composition of $BaZr_{0.05}Ti_{0.95}O_3$ ceramic

Element	Wt.% calculated	Wt.% from EDX
Ba	57.59	57.65
Zr	1.48	1.59
Ti	18.52	18.12
O	22.41	22.64

3.3 The Ferroelectric – Paraelectric phase transition

Fig. 3 depicts the change in inverse of permittivity ($1/\varepsilon'$) as a function of temperature performed at 100 kHz. It is known that for normal ferroelectrics the ferro- to – paraelectric phase transition is known to be first order when $T_0 < T_c$ and of second order when $T_c = T_0$, here T_c is the Curie-Weiss temperature which is defined by the following relation:

$$\frac{1}{\varepsilon_r} = \frac{T - T_C}{C} \tag{2}$$

Curie - Weiss law is no longer valid in the case where relaxation and diffuseness are present. For all the relaxor ferroelectrics (having diffuse phase transition) a deviation from Curie – Weiss law is observed which is a typical behavior of relaxors. The parameter $\Delta T_m = T_{CW} - T_m$, where T_{CW} denotes that temperature from which the dielectric permittivity starts to deviate from the Curie - Weiss law while T_m is the temperature of dielectric maximum. In the present study for $BaZr_{0.05}Ti_{0.95}O_3$ the phase transition was observed to be first order i.e. no deviation from ideal Curie Weiss law is observed. Also, the phase transition temperature was measured to be 400 K.

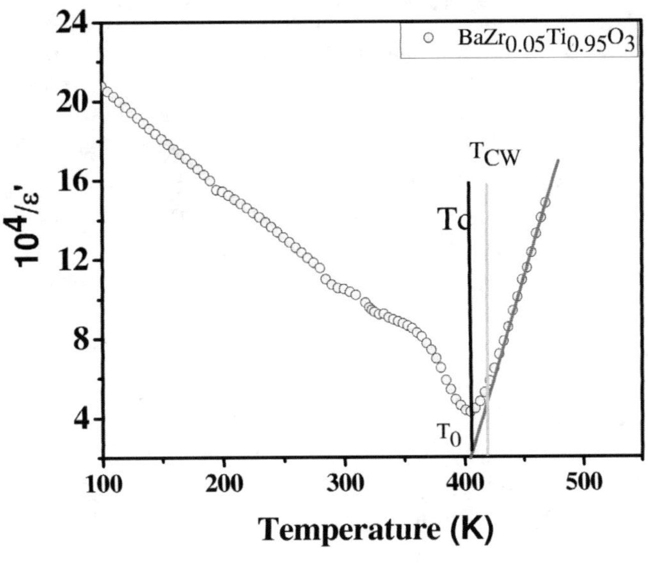

Fig. 3 Change in $1/\varepsilon'$ with respect to temperature measured at 1 kHz

3.4 AC Conductivity in ferroelectric phase

Fig. 4 shows the variation of AC conductivity of $BaZr_{0.05}Ti_{0.95}O_3$ with reciprocal of temperature ($10^3/T$) measured at 1 KHz. The observed variation in AC conductivity with temperature could be divided into two distinct regions in the scale namely the lower temperature region and the higher temperature region. Being activated by the various temperatures the oxygen vacancies behaves as trapping centres for the mobile charge carriers. Also an increase in the conductivity considerably in the high temperature region is seen probably this is due to the hopping of the charge carriers among the available oxygen vacancies. Another important fact we observed from the graph is the changing slopes that appear in different regions of temperature scale, which is an indication of involvement of multiple activation processes having different activation energies. The AC conductivity first increases with temperature up to 125 °C and then decreases with increasing in temperature, which also represents the transition temperature.

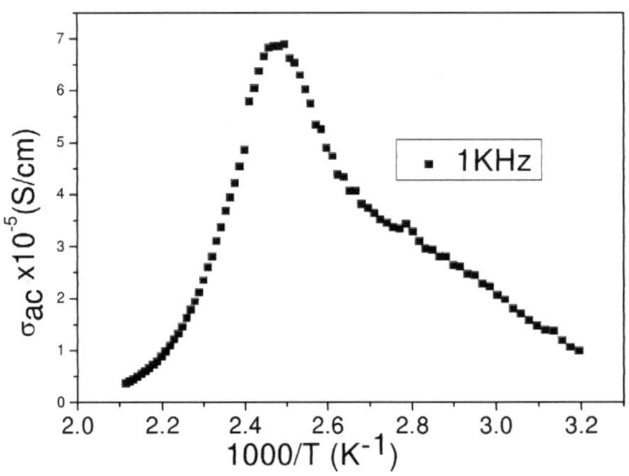

Fig. 4 AC conductivity of BZT ceramic plotted against 1000/T measured at 1 kHz

In all the dielectric materials there exist several polarization mechanisms; these mechanisms have some relationships with the relaxation behaviour like space charge polarization, interfacial polarization, dipolar polarization etc. The dielectric relaxation due to space charge polarization, interfacial polarization, domain wall motion, long range structural disorder and defect relaxation etc. usually occurs at low frequency region while the dipolar relaxation occurs at higher frequency region. It is given in the open literature that the dielectric relaxation in all the dielectric materials in the temperature above room temperature is usually related to the available oxygen vacancies and hopping motion of singly and doubly ionized oxygen vacancies leads to the relaxation process.
Thus, it can be considered that in the low frequency spectral region the relaxation belongs to space charge that is related to the oxygen vacancies and the defect sites. While sintering at high temperatures produces Barium vacancies which results in oxygen vacancies due to the constraints of charge neutrality. At high temperatures the ionization of oxygen vacancies takes place that create conducting electrons in the Perovskite structure as a result single ionized and double ionized oxygen vacancies are left where $V_O^{'}$ and $V_O^{''}$ are respectively the single ionised and double ionised states. Thus, as described by the Kroger -Vink notation:

$$V_O \leftrightarrow V_O^{'} + e^{'} \tag{3}$$
$$V_O^{'} \leftrightarrow V_O^{''} + e^{'} \tag{4}$$
$$Ba_{Ba}^{x} + O_O^{x} \rightarrow V_{Ba}^{''} + V_O^{''} + BaO \uparrow \tag{5}$$

Further to understand the transport mechanism in the samples the electrical conductivity behavior was investigated. The transport mechanism is basically due to the presence of space charge in the material. The origin of conductivity mechanism is related to the level and mobility of oxygen vacancies which is a common phenomenon in perovskites. As it is known the AC conductivity of the samples is directly related to the dielectric properties of the sample. To evaluate the bulk conductivity (σ_{ac}) of the samples the formula used is $\sigma_{ac} = t/Z'A$ where t is the thickness, A is the surface area of the samples and Z′ is the real part of the impedance. Fig. 5(a) indicates the ac conductivity of the $BaZr_{0.05}Ti_{0.95}O_3$ samples at various frequencies (from RT to T_c). Initially, with increasing frequency, σ is almost independent of frequency followed by dispersion at higher frequencies. By extrapolating the graph towards lower frequency side the value of σ_{dc} can be determined which is credited to the long range translational motion of charge carriers. It is also observed that with increasing temperature conductivity increases for all frequencies. Conduction in perovskites is basically due to the hopping mechanism of different species of ions present in the material. The excess electrons interact with the lattice and distort the surroundings as a result a potential well is created which raise the conduction. As the temperature increases the number of excess electrons increases and get sufficient energy to cross the well, thus with increasing temperature conductivity increases. The change in AC conductivity in the lower frequency region is mainly due to the polarization effect at electrode and electrolyte interface. At much lower frequencies the charge accumulates at the electrode and electrolyte interface and reduces the conductivity almost equal to the DC conductivity. At higher frequencies the characteristic of conductivity spectra can be explained as according to the Jonscher's power law. According to which $\sigma_{ac} = \sigma_{dc} + A\omega^n$ $(0 < n < 1)$ where, A is the pre-exponential factor and 'n' is the exponent of power law which gives the idea of interaction between mobile ions and lattice. 'A' gives the idea of strength of polarizability in the samples.

The variation of AC conductivity with respect to inverse of temperature by means of Arrhenius law is shown in fig. 5(b). The slope of this graph determines the value of activation energy involved in the conduction process. Conduction induced by oxygen vacancies becomes dominant with increasing temperature because the mobility of oxygen ions becomes higher in high temperature region. It has been reported in the literature that the when the activation energy lies in the range of 0.3-0.4eV, the oxygen vacancies lies in the single ionized state while for the doubly ionized oxygen vacancies the value lies between 0.6-1.2eV. The experimental value of activation energy associated with the DC conductivity for the studied sample is 1.1eV. The results are in good agreement with the values reported for the doubly ionized oxygen vacancies.

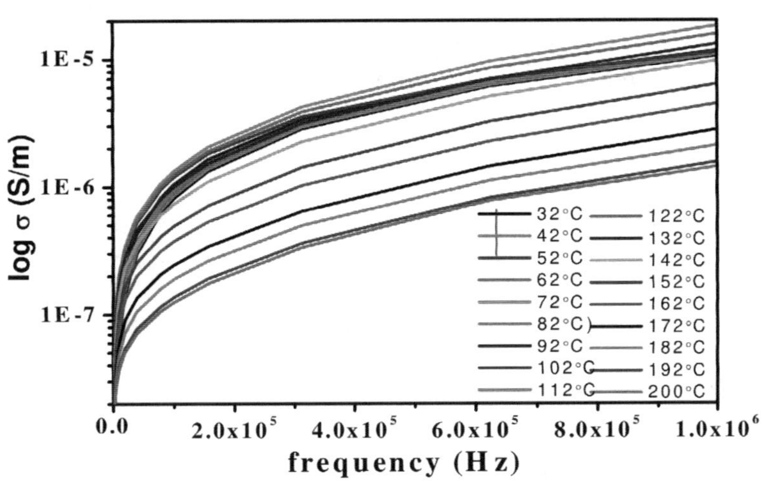

Fig. 5(a) AC conductivity as a function of frequency at various temperatures

IOP Conf. Series: Materials Science and Engineering **798** (2020) 012006 doi:10.1088/1757-899X/798/1/012006

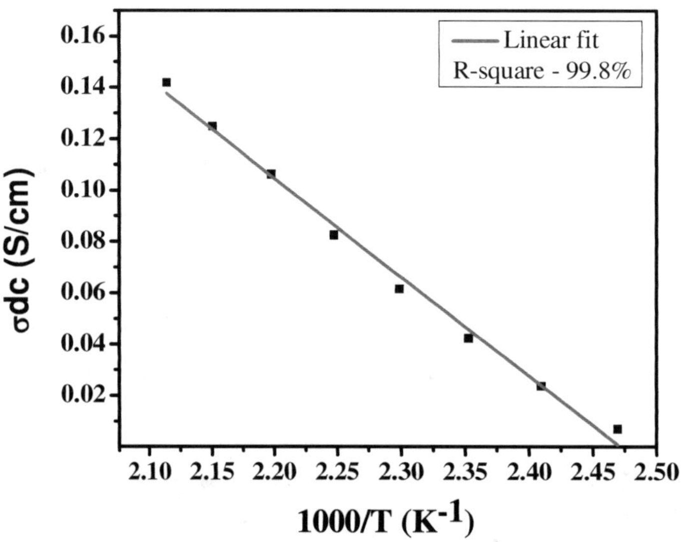

Fig. 5(b) Arrhenius dependence for dc conductivity as a function of 1000/T (K^{-1})

4 Conclusions

Thus, we conclude that polycrystalline sample of $BaZr_{0.05}Ti_{0.95}O_3$ ceramic has been synthesized by the high temperature solid state reaction method. The X-ray diffraction studies revealed the presence of orthorhombic phase. The dielectric measurements of $BaZr_{0.05}Ti_{0.95}O_3$ ceramic with respect to temperature at 1 kHz has been analyzed, no deviation from Curie – Weiss law has been seen indicating the normal ferroelectric phase transition in the sample. The studies show that the conduction was caused by the presence of oxygen vacancies. Also, the dielectric relaxation at high temperature was related to the presence of oxygen vacancies.

References

[1] Belous A G, V'yunov O and Khomenko B S 1998 *Inorg. Mater.* 34(6) 725-729

[2] Bąk W, Kajtoch C, Ptaszek S, Lisińska-Czekaj A, Czekaj D, Ziętek D, Glos T and Garbarz-Glos B 2016 *Arch. Metall. Mater.* 61(2) 905–908

[3] Guo Z G, Yang L, Qiu H, Zhan X, Yin J and Cao L 2012 *Mod. Phys. Lett. B* 26(9) 1250056

[4] Slipenyuk A M, Glinchuk M D, Bykov I P, Yurchenko L P, Mikheev V A, Frenkel O A, Tkachenko V D and Garmash E P 2003 *Cond. Mat. Phys.* 6 2(34) 237-244

[5] Wang Y, Pu Y and Zhang P 2015 *J. Alloys Compds.* 653(25) 596-603

[6] Lyu J, Fina I, Solanas R, Fontcuberta J and Sánchez F 2018 *Scientific Reports* 8 495 DOI:10.1038/s41598-017-18842-5.

[7] Ertuğ B 2013 *Amer. J. Eng. Resear* 02(08) 01-07

[8] Wu L, Chure M C, Wu K, Chang W C, Yang M J, Liu W K and Wu M J 2009 *Ceram. Int.* 35(3) 957-960

[9] Jiang B, Iocozzia J, Zhao L, Lin Z, Zhang H, Harn Y W and Chen Y 2019 *Chem. Soc. Rev.* 48(4) DOI: 10.1039/C8CS00583D.

[10] Kumar R, Asokan K, Patnaik S and Birajdar B 2015 *DAE Sol. Stat. Phys. Symposium AIP Conf. Proc.* 1731 030025-1–030025-3; DOI: 10.1063/1.4947630.

[11] Feng H, Hou J, Qu Y, Yao G and Shan D 2012 *J. Alloys Compds.* 512(1) 12–16

[12] Kheyrdan A, Abdizadeh H, Shakeri A and Golobostanfard M R 2018 *J. Sol-Gel Sci. Tech.* 86(1) 141–150

[13] Rehrig P W, Park S E, McKinstry S T, Messing G L, Jones B and Shrout T R 1999 *J, Appl.Phys.* 86(3) 1657

[14] Wieczorek K, Ziebiniska A, Ujma Z and Roleder K 2006 *Ferroelect.* 336(1) DOI: 10.1080/00150190600695743.

[15] Bhargavi G N, Khare A, Badapanda T, Anwar M S and Brahme N 2017 *J. Mater. Sci.: Mater. Electron.* 2816956–16964 DOI 10.1007/s10854-017-7617-8

[16] Barranco A P, Guerra J D S, Noda R L and Araujo E B 2008 *J. Phys. D: Appl. Phys.* 41 215503

[17] Zhang T F, Tang X G, Liu Q X, Lu S G, Jiang Y P, Huang X X and Zhou Q F 2014 *AIP Advances* 4 107141

[18] Badapanda T, Sarangi S, Behera B, Sahoo P K, Anwar S, Sinha T P, Luz Jr. G E, Longo E and Cavalcante L S 2014 *Curr. Appl. Phys.* 14 708-715

[19] Dixit A, Majumder S B, Savvinov A, Katiyar R S, Guo R and Bhalla A S 2002 Mater. Letts. 56 933 – 940

[20] Ahmad A and Razak K A 2017 *AIP Conf. Proc.* 1865 060001 DOI.org/10.1063/1.4993377

[21] Bhargavi G N, Khare A, Badapanda T, Ray P K and Brahme N 2018 *Ceram. Int.* 44 1817–1825

IOP Conf. Series: Materials Science and Engineering **798** (2020) 012007 doi:10.1088/1757-899X/798/1/012007

Synthesis and ionic conductivity measurement of a new potassium ion conducting solid polymer electrolytes

Angesh Chandra*[,1], Subhashis Basak [2], M.Z. Khan [2], Archana Chandra [1], R.S. Dhundhel [1] and Alok Bhatt [3]

[1] Shri Shankaracharya Institute of Professional Management & Technology, Raipur, Chhattisgarh- 492015, India
[2] Dr. C.V. Raman University, Kargi Road, Kota, Bilaspur, Chhattisgarh, India
[3] Bharti College of Engineering & Technology, Durg, Chhattisgarh, India
*E-mail: chandrassi@gmail.com

Abstract. Synthesis and ionic conductivity measurement of a new PEO-based K^+ ion conducting solid polymer electrolytes (SPEs): (1-x) PEO: x KCl, where $0 \leq x \leq 50$ in wt%., are reported. The present SPEs have been prepared by hot-press technique. Ionic conductivity measurements have been carried out with the help of LCR meter. The activation energy of newly synthesized SPE has also been calculated with the help of Arrhenius plot. Materials characterizations have been carried out with the help of DSC analysis. Ionic nature of the present SPE has been explained with the help of ionic transference number (t_{ion}) measurement.

1. Introduction

Amongst the known superionic systems, solid polymer electrolytes are attracted special attention due to their relatively higher room temperature conductivity as well as easier of material handling and synthesized. These electrolytes have been used to fabricate flexible, compact, laminated, batteries [1-4]. The SPE films are casted usually by traditional solution-cast/ sol-gel techniques. Recently, a hot-press method has been developed which can be potentially used for casting SPE films and this technique is more rapid, least expensive as well as solvent-free procedure to form completely dry polymer electrolyte films complexed with variety of ionic salts [5-7].

In the present investigation, author reports synthesis and ionic conductivity (σ) measurements of a new PEO-based K^+ ion conducting solid polymer electrolytes (SPEs): (1-x)PEO: xKCl, where $0 \leq x \leq 50$ in wt%. Temperature dependent ionic conductivity (σ), measurements have been carried out for the determination of activation energy (E_a). Ionic transference number (t_{ion}) has been determined at room temperature.

2. Experimental

AR grade precursor chemicals: poly (ethylene oxide) PEO (10^5 Mw, Aldrich, USA) and a potassium ion conducting salt KCl (purity > 98%, Merck, India) have been used for synthesis of present solid polymer electrolytes (SPEs): (1-x) PEO: xKCl, where $0 \leq x \leq 50$ in wt%. The dry powders of constituent chemicals PEO and KCl in appropriate wt.% ratios were homogeneously mixed for ~ 30 min at room temperature then heated separately ~ 70 ^{0}C (close to the melting point of PEO) for ~ 30 min with mixing continued. It resulted into a soft lump which was then pressed (~ 1.25 ton/ cm^2) between two

SS- cold blocks, giving rise to a uniform film of thickness ~ 0.015 cm. The details related to hot- press casting of other SPEs have also been given reported in the literature [8,9].
The ionic conductivity (σ) -measurements were carried out by using the following equation:

$$\sigma = \frac{l}{R_b . A} \qquad [S.cm^{-1}] \qquad (1)$$

where R_b is the bulk resistance, l is the thickness and A is the cross sectional area of the polymeric sample material. The ionic transference number (t_{ion}) was also determined with the help of following equation [10-12]:

$$t_{ion} = 1 - \frac{I_e}{I_T} \qquad (2)$$

where I_e is the electronic current and I_T is the total current of the cell: [SS // SPE OCC // SS].

3. Results and discussions

The salt concentration dependent conductivity (σ) for hot-pressed newly synthesized solid polymer electrolytes (SPEs): (1-x) PEO: x KCl, is shown in figure 1. Ionic conductivity increased abruptly (~10^2 times) after the addition of salt in the host polymer. However, a σ-maxima appeared at x = 30 wt.% and then decreases on further addition of salts. SPE film: (70PEO:30KCl), with room temperature conductivity (σ ~5.6×10^{-7} S/cm), exhibited maximum conductivity and this has been referred to as Optimum Conducting Composition (OCC). The conductivity enhancement in the present SPE OCC is due to the increase in degree of amorphicity and/or available of more number of conducting K^+ ions. This has been confirmed by the DSC analysis.

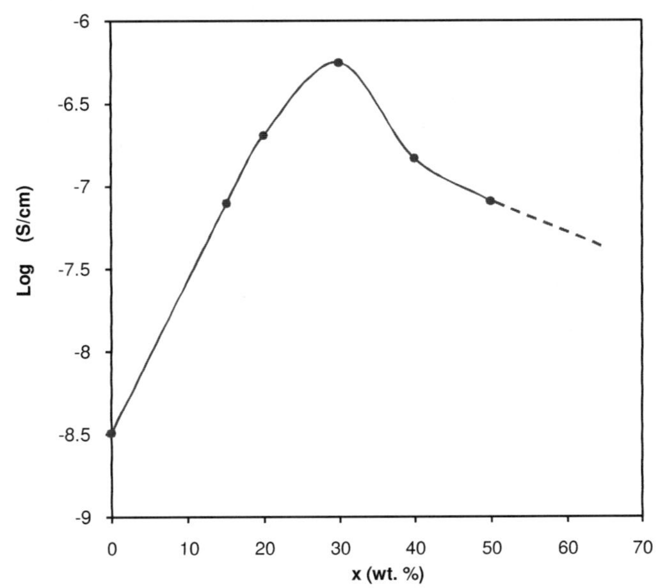

Figure 1. 'Log σ – x' plots for SPEs: (1-x)PEO: xKCl.

The temperature dependent ionic conductivity (σ) of hot-press synthesized SPE OCC: (70PEO:30KCl) is shown in figure 2. The conductivity increased linearly with temperature, except for an upward change in the slope at ~ 65-70 °C which corresponds to semi-crystalline to amorphous phase transition temperature of PEO. The increase in ionic conductivity with temperature can be interpreted as a hoping mechanism between coordinating sites and this can also be explained with the help of free

IOP Conf. Series: Materials Science and Engineering **798** (2020) 012007 doi:10.1088/1757-899X/798/1/012007

volume theory [4,12]. The linear portion below the transition temperature can be expressed by following Arrhenius equation:

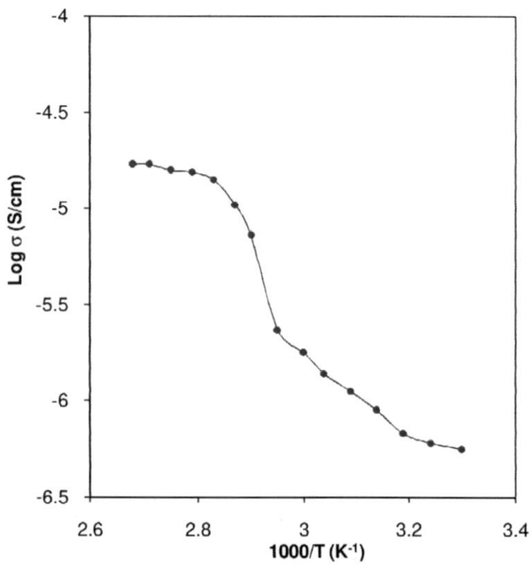

Figure 2. 'Log σ –1/T' plot for SPE OCC: (70PEO:30KCl).

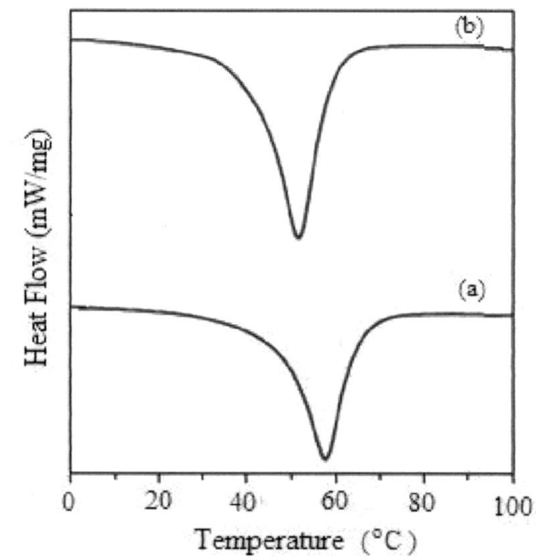

Figure 3. DSC thermograms: (a) pure PEO, (b) SPE OCC: (70PEO:30KCl).

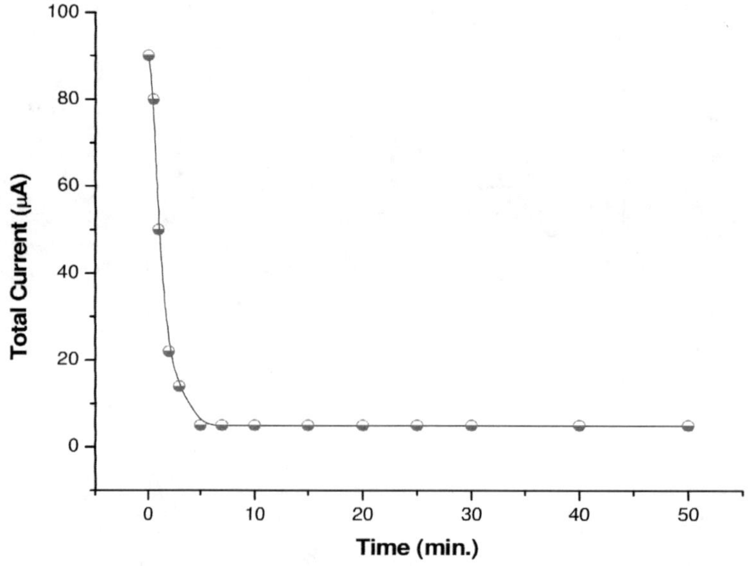

Figure 4. 'Total current vs time' plot for SPE OCC: (70PEO:30KCl) at room temperature.

$$\sigma (T) = 4.5 \times 10^{-3} \exp(-0.39/ kT) \quad [S.cm^{-1}] \qquad (3)$$

where 0.39 is the activation energy (E_a) in (eV). The low activation energy (E_a) in the present SPE system indicates that this can be used for all-solid-state device fabrications.
Figure 3 shows the DSC thermograms for the pure PEO and SPE OCC: (70PEO:30KCl). The broad endothermic peak was observed in SPE OCC at ~ 67 -70 °C and it is corresponding to the melting point temperature of pure PEO. It is clearly seen from the figure that the melting point temperature have been shifted towards lower side and this is due to the complexation of K^+ ions with the pure polymer PEO. This result confirms the polymer-salt complexation. Figure 4 shows the 'total current vs time' plots for hot-pressed SPE: (70PEO:30KCl) at room temperature. The rapid initial drop in the current is due to polarization of the mobile ions, as reported in our previous communications with other systems [13,14]. The ionic transference number (t_{ion}) of SPE OCC is ~ 0.95 (as shown in figure 4), which is indicated that the 95% of K^+ ions are mobile in this newly synthesized SPE and only a negligible contribution of Cl^- ions.

4. Conclusions

A new potassium ion conducting solid polymer electrolyte (SPE): (70PEO:30KCl) with ionic conductivity (σ) ~ 5.6×10^{-7} S/cm have been synthesized by a recently developed hot-press method. The lower activation energy in the present SPE indicated that this can be potentially used for all-solid-state device fabrications. t_{ion} measurement indicates that the present solid polymer electrolyte is a pure ionic system and this can be used for ion conducting device fabrications.

Acknowledgements

Authors gratefully acknowledges to CSVTU, Bhilai for providing financial assistance through the 'collaborative research project TEQIP-III' (CSVTU/CRP/TEQIP-III/81/2019). One of the co-author (Archana Chandra) thanks to DST, New Delhi, for providing financial assistance through a research project under 'women scientist scheme-A' (SR/WOS-A/CS-53/2018)'.

References

[1] Scrosati B 1993 *Applications of electroactive polymers* (London: Chapman and Hall)
[2] Gray F M 1997 *Polymer electrolytes* (Cambridge: Royal Society of Chemistry Monographs)
[3] Zhang X W, Wang C, Appleby AJ and Little FE 2002 *J. Power Sources* **112** 209
[4] Agrawal R C and Pandey G P 2008 *J. Phys. D: Appl. Phys.* **41** 223001
[5] Appetcchi G B, Croce F, Hasson J, Scrosati B, Salomon M and Cassel F 2003 *J. Power Sources* **114** 105
[6] Agrawal R C and Chandra A 2007 *J. Phys. D: Appl. Phys.* **40** 7024
[7] Chandra A and Chandra A 2010 *Hot-pressed polymer electrolytes: Synthesis and characterization* (Germany: Lambert Academic Pub.)
[8] Chandra A, Chandra A and Thakur A 2014 *Composites: Part B* **60** 292
[9] Chandra A 2016 *Indian J. Physics* **90** 759
[10] Chandra S, Tolpadi S K and Hashmi S A 1988 *Solid State Ionics* **28-30** 651
[11] Chandra A, Chandra A and Thakur A 2013 *Russian J. General Chem.* **83** 2375
[12] Chandra A 2019 *Polymer Bulletin* **76** 785
[13] Chandra A Chandra A and Thakur A 2013 *Chinese J. Polymer Science* **31** 302
[14] Chandra A 2017 *Polymer Bulletin* **74** 4815

IOP Conf. Series: Materials Science and Engineering **798** (2020) 012008 doi:10.1088/1757-899X/798/1/012008

Study of Effect of Y substitution on structural, dielectric, impedance and magnetic properties of Bismuth Ferrite

L Thansanga[1], Alok Shukla[1,*], Nitin Kumar[1], R N P Choudhary[2]

1 Department of Physics, National Institute of Technology Mizoram, Aizawl 796012, India
2 Multifunctional Materials Research Laboratory, Department of Physics, Siksha O Anusandhan (Deemed to be University), Bhubaneswar 751030, India
* Corresponding author.
e-mail address : aloks.nitmz@gmail.com

Abstract. Yttrium substituted Bismuth Ferrite of a composition, $Bi(Fe_{0.95}Y_{0.05})O_3$ (abbreviated as BFYO5) has been synthesized through a conventional mixed oxide route. The orthorhombic crystal structure and a pure-phase formation of the polycrystalline material was verified through X-ray diffraction pattern. The composition of its constituent elements and their concentration ratio were confirmed by using energy dispersive X-ray spectroscopy technique. Scherrer's equation was used to estimate the average crystallite size of the examined sample specimen and found to be 39 nm. Analysis of the BFYO5 compound's surface morphology was conducted using field emission scanning electron microscope, which clearly shows a densely packed specimen with a homogeneous distribution of various grain dimensions. The interrelationship between frequency and temperature in dielectric, impedance and conductivity parameters with the structure of a sample has also been reported. The impedance spectroscopy analysis shows the nature of negative temperature coefficient of resistance, grain and its boundary effect on the resistive and capacitive features and the presence of non-Debye type of electrical relaxation process in the studied material. The analysis of polarization versus electric field shows the improvement in the value of remnant polarization comparing with the parent compound bismuth ferrite. The substantial enhancement in the magnetization value of the prepared material was also confirmed by its magnetic hysteresis loop study. Based on an important enhancement in the electrical properties, remnant polarization and its magnetic properties, BFYO5 material could be considered as a promising candidate for some of the advance device applications.

1. Introduction

In recent years, multiferroic materials have greatly attracted a considerable attention of researchers and scientists due to their potential applications in data storage devices, spintronic devices, transducers, memory devices, ultrafast optoelectronic devices, sensors, etc. [1-3]. Multiferroic materials show coupled ferroelectric, ferromagnetic and ferroelastic ordering properties in a single phase [4]. Among all the multiferroic compounds, $BiFeO_3$ (termed as BFO) is the most studied compound as it has a great potential for device applications. It has a rhombohedral perovskite crystal symmetry with a ferroelectric and anti-ferromagnetic ordering at a transition temperature (T_c) of 810-830^{0}C and Neel temperature (T_N) of 350-370^{0}C [5]. The ferroelectric property of BFO is due to the existence of $6s^2$ lone pair electrons of Bi^{3+} ions while its ferromagnetic property is due to partially occupy d-orbital of Fe^{3+} ions (d^5) [6]. However, pure BFO has some inherent drawbacks for device application such as large current leakage, high tangent loss and structural distortion which affects its ferroelectric, dielectric and resistive properties. The high current leakage density is associated with the oxidation-reduction of iron ions $(Fe^{2+} \rightarrow Fe^{3+} - e^-)$ which results the creation of oxygen vacancies for the charge recovery [7,8]. A bulk bismuth ferrite has a magnetic spiral spin cycloid with a periodicity of around 620 Å [9]. Therefore, the rare earth ions (Y^{3+}) substitution is choosing to suppress the spiral structure in the present study.

In order to eliminate the above deficient properties of BFO, a lot of efforts have been paid by applying several additions at A/B sites or AB sites of other perovskite materials [10-17]. A substitution of BFO by rare earth ions such as Samarium (Sm), Dysprosium (Dy), Lanthanum (La), Gadolinium (Gd), etc., at Bi-site has also been carried out in the last decade [18-20]. C. Nayek et al. [21] reported that rare earth doping enhances the magnetization by decreasing dopant ionic radii. Mishra et al. [22] studied that an enhancement in dielectric and magnetic ordering and a reduction of dielectric loss was observed with increasing the doping concentration of Yttrium (Y). A. Perejon et al. [23] also reported that a structural phase transition from rhombohedral R3c to orthorhombic Pbnm was found in Y concentration of x = 0.07 and 0.10 at the Bi-site doping. Moreover, increasing Y concentration results a decrease in optical band gap and increase in an electrical conductivity is also added in their report. A. Mukherjee et al. [24] revealed that there is an increase in the band gap from 1.97 eV to 2.29 eV at the Bi-site doping of BFO. Doping of BFO by using rare earth elements has been recommended for the development of its ferroelectric and magnetic properties [25].

From the literature survey, the effect of ion substitution on magnetic and ferroelectric properties of BFO has not been clearly understood so far and therefore more research work is yet to be carried out to understand the multifunctional properties for Fe-site doping. So, it is worthwhile to study the Fe-site substitution of BFO by Y^{3+} to get a single phase formation and a significant enhancement in magnetic and ferroelectric properties of BFO. To the best of our knowledge, the Yttrium (x=0.05) substitution at Fe-site of BFO has not been studied in detail so far. Therefore, in this work, we are reporting our systematic studies of the effect of Yttrium (x=0.05) substitution at Fe-site (B-site) on structural, dielectric, electrical and magnetic properties of BFO.

2. Materials and experimental procedures

2.1. Chemicals and crucibles

A multi-step mixed oxide route was used to synthesize the polycrystalline sample with a chemical composition of $Bi(Fe_{0.95}Y_{0.05})O_3$. As a preliminary ingredients in a desired stoichiometric ratio, the high purity and analytical grade of oxides such as bismuth oxide (Bi_2O_3, Purity: 99%), iron oxide (Fe_2O_3, Purity: 98%) and yttrium oxide (Y_2O_3, Purity: 99%) were used. For the fabrication of BFYO5 compounds, the high purity alumina crucible of cylindrical, plate and tray design are used.

2.2. Synthesis and characterization procedure

The preliminary elements of different oxide materials were weighted precisely in a stoichiometric ratio and mixed thoroughly using agate mortar and pestle under the dry air for 2 hrs and then in a wet medium (high purity methanol) for 3 hrs. The homogeneous mixture of sample powder was calcined at an optimized temperature (750^{0}C) for 8 hrs. The phase formation of as-prepared compound was examined by using Powder X-ray diffractometer (Rigaku Ultima-IV) with CuKα radiation (λ=0.15405 nm) in a wide range of the Bragg angle ($20^0 \leq 2\theta \leq 80^0$) at a slow scan speed of 2^0 m^{-1}. After proper phase formation, the calcined sample powder was once more ground thoroughly using mortar and pestle. The subsequent powder was used for preparing a circular disc shape pellet of 12 mm diameter and 1.5 mm thick with the help of KBr hydraulic press by applying a pressure of 4 x 10^6 Nm^{-2}. The pellets were sintered at an optimized temperature (800^{0}C) for 6 hrs which increased the density of the pellets. A fine emery paper was used to polish the sintered pellets to make the surfaces of the pellets smooth and parallel. And then two sides of the pellet are painted with a silver conductive paint and dried for half an hour at 150^{0}C to remove moisture, if any. Lastly, electrical measurements such as dielectric permittivity, dielectric loss, impedance, electrical modulus and conductivity were performed using phase sensitive multimeter (PSM-1735) connected to a computer-interfaced furnace and sample holder in a wide range of frequency (1kHz-1MHz) and temperature (25-500^{0}C).

The surface morphology of the prepared sample and average particle size were calculated and analyzed by using field effect scanning electron microscope (FESEM). In addition, an energy dispersive X-ray spectroscopy (EDXS) was used to confirm the purity of a substance and its elemental analysis. The room temperature ferroelectric and magnetic properties of BFYO5 material were measured by using

P-E loop tracer (Marine India) and vibrating sample magnetometer (VSM) respectively. The schematic synthesis procedure of $Bi(Fe_{0.95}Y_{0.05})O_3$ compound is shown in figure 1.

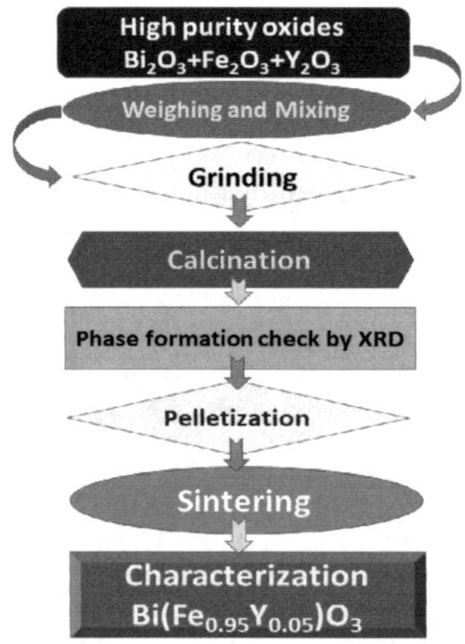

Figure 1. The schematic synthesis process of $Bi(Fe_{0.95}Y_{0.05})O_3$ compound by mixed oxide route.

3. Result and discussion

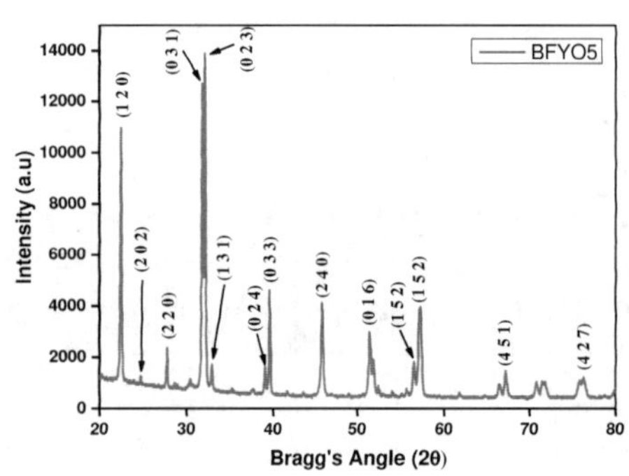

Figure 2. Calcined Powder-XRD pattern of $Bi(Fe_{0.95}Y_{0.05})O_3$ compound.

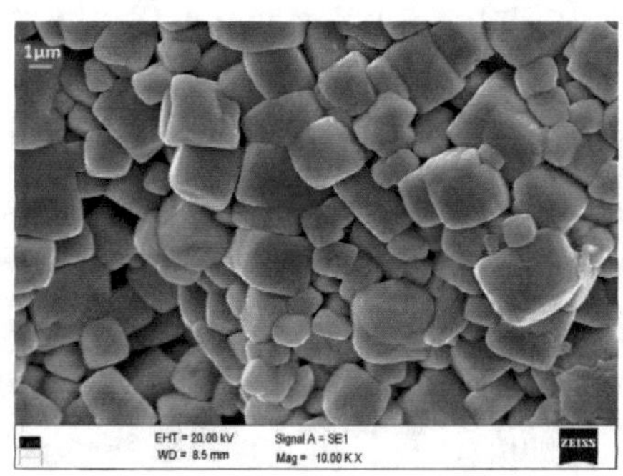

Figure 3. FE-SEM micrograph of sintered $Bi(Fe_{0.95}Y_{0.05})O_3$ compound.

3.1. Structural and microstructural studies

Figure 2 shows the powder X-ray diffraction peaks of $Bi(Fe_{0.95}Y_{0.05})O_3$ compound measured at room temperature (25^0C). The XRD data are analyzed by using "POWDMULT" indexing and refinement programmed for the calculation of the unit cell dimension of the studied material [26]. The crystalline nature of BFYO5 material was clearly confirmed by the sharp and well defined peaks of the compound. The orthorhombic phase symmetry of the prepared material was suggested by the value of observed interplanar distance (d_{obs}), calculated distance (d_{cal}) and the angle of diffraction for each reflection peak as shown in table 1. The values of different lattice cell parameters of a selected system are: a = 9.5490Å, b = 8.6769Å, c = 10.8778Å and Volume = 901.30Å^3. The average size of crystallite was determined using the equation of Scherrer; $D = k\lambda/\beta \text{ Cos } \theta$ from the XRD peaks and found to be 3 nm [27]. In this equation, D = crystallite size, k = constant that depends on the crystallite size, β = full width at half maximum of intensity versus 2θ plot, λ= wavelength of the CuKα radiation (=1.5418 Å) and θ = Bragg's diffraction angle.

Figure 3 displays the field emission scanning electron (FESEM) micrograph of the sintered sample pellet measured at a room temperature (25^{0}C). The micrograph clearly depicts the homogeneous distribution of grains of the sample which are densely packed with a few voids throughout the sample surface. The dimension of the grains are unequal in shape and size, which suggests the polycrystalline nature of BFYO5 material. Using ImageJ software, the average grain size of the present studied material is calculated and found to be 1.73μm. Since, the crystallite size is different from their grain size, the value of structural parameters are different. The X-ray diffraction is used for the study of bulk sample and it gives the crystallite size, whereas scanning electron microscopy (SEM) is used for the study of sample surface and it gives the grain size.

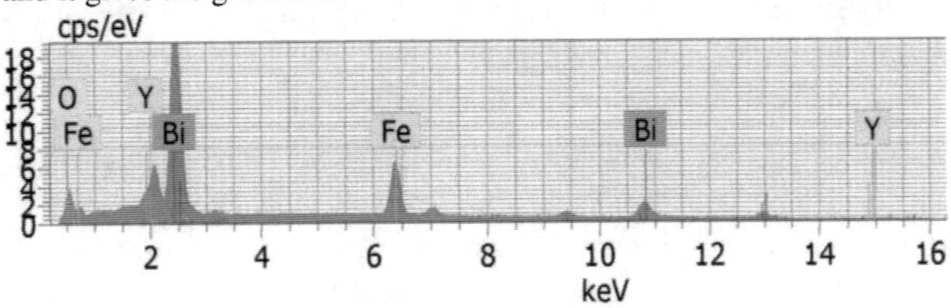

Figure 4. EDXS spectra of sintered $Bi(Fe_{0.95}Y_{0.05})O_3$ compound.

Figure 4 illustrates the energy dispersive X-ray spectroscopy (EDXS) of BFYO5 compound at 25^{0}C. From the EDXS study, it was found that the presence of all the constituent elements (bismuth, iron, yttrium and oxygen) close to their stoichiometric concentration with their respective atomic ratio. That clearly shows the purity of the specimen sample being prepared.

Table 1 : Comparison of measured and determined XRD peaks (in Å) with the observed relative strength (I/I_0) of BFYO5 material.

Peak	d-spacing (obs.)	d-spacing (cal.)	I/I_o	h	k	l	Two theta (obs±cal.) diff.
1	3.9482	3.9499	79	1	2	0	.010
2	3.5898	3.5881	9	2	0	2	-.012
3	3.2109	3.2109	17	2	2	0	.000
4	2.8064	2.7952	91	0	3	1	-.131
5	2.7809	2.7822	100	0	2	3	.015
6	2.7152	2.7195	12	1	3	1	.053
7	2.3041	2.3042	12	0	2	4	.002
8	2.2750	2.2800	33	0	3	3	.090
9	1.9762	1.9750	30	2	4	0	-.030
10	1.7755	1.7747	21	0	1	6	-.027
11	1.7628	1.7643	14	5	1	2	.049
12	1.6284	1.6290	13	1	5	2	.024
13	1.6117	1.6130	29	2	5	1	.050
14	1.6075	1.6080	29	4	0	5	.018
15	1.3919	1.3922	10	4	5	1	.015
16	1.2475	1.2474	9	4	2	7	-.008

Phase : Orthorhombic symmetry

3.2. Dielectric study

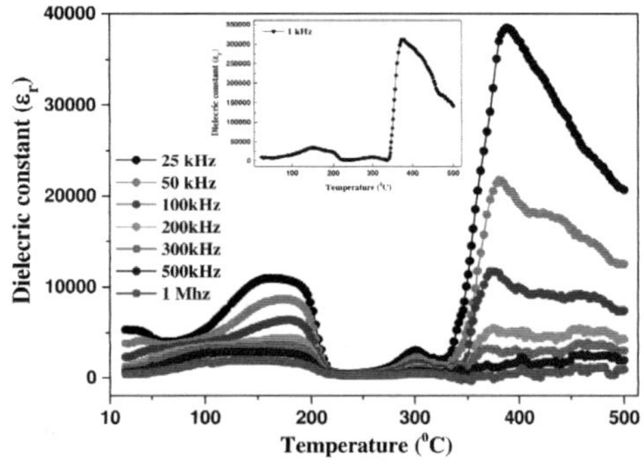

Figure 5. Variation of temperature dependent dielectric constant (ε_r) of BFYO5.

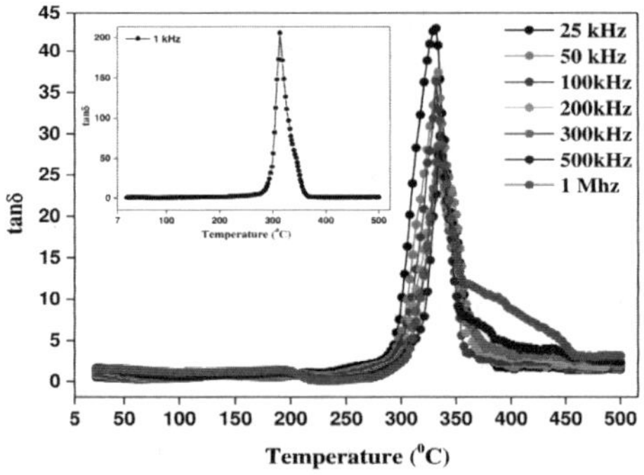

Figure 6. Variation of temperature dependent dielectric loss (tanδ) of BFYO5.

Figure 5 shows the temperature dependence characteristics of the BFYO5 dielectric constant calculated at a broad frequency range (1kHz-1MHz). The nature of dielectric constant spectrum shows a partial dependent in a frequency from the initial stage. The graph shows a gradual increase of dielectric constant on growing the temperature while there is a partial descent in a lower frequency spectrum. Two dielectrical variations are observed at lower frequency spectrum (<100 kHz) in the temperature range of 110-200^{0}C and 350-400^{0}C. The anomaly in a lower temperature region occurs probably due to the defects during the synthesis of the material, whereas that in a higher temperature region is associated with the existence of antiferromagnetic transition [28]. The dielectric spectrum gradually decreases at higher temperature regions (370-500^{0}C) for lower frequency spectra (1, 25, 50, 100 kHz) while the value of dielectric constant remains almost constant for higher operating frequencies (200, 300, 500, 1000 kHz).

Figure 6 shows the dielectric loss (tanδ) characteristic of BFYO5 at some selected frequencies in a wide range of temperature (25-500^{0}C). The spectrum shows the value of dielectric loss remains constant from the initial point to 270^{0}C. However, the anomaly of a spectrum occurs at an intermediate temperature (280-380^{0}C) for all the selected frequencies due to charge carriers and the presence of some defects in the synthesis process of the material and also possibly due to the antiferromagnetic transition. That consequently results a simultaneous increase of dielectric constant and loss value at this temperature region. And then the dielectric loss trend decrease at further higher temperature and remains constant. The dielectric constant and dielectric loss comparison at a certain frequency and temperature has been summarized in table 2.

Table 2 : Comparison of dielectric constant (ε_r) and dielectric loss (tanδ) of BFYO5 material at different temperature and frequency.

Sl. No.	Frequency	Temperature 25^{0}C		Temperature 500^{0}C	
		ε_r	tanδ	ε_r	tanδ
1	1 kHz	9748	0.719	141896	1.162
2	25 kHz	5329	0.547	20667	1.487
3	50 kHz	3795	0.780	12527	1.614
4	100 kHz	2288	1.101	7437	1.765
5	200 kHz	1258	1.427	4302	1.942
6	300 kHz	879	1.574	3028	2.104
7	500 kHz	597	1.650	1942	2.375
8	1 MHz	378	1.541	906	3.165

3.3. Impedance spectroscopy study

Complex impedance spectroscopy is used to perform a simultaneous study of the dielectric and electrical properties of advanced functional materials. In addition, this approach was extensively used to study the effect of grain, grain boundaries and interfaces in polycrystalline material and ionic conductors [29,30]. The present study reported a detailed analysis of impedance parameters (Frequency vs Z', Frequency vs Z'' and Z' vs Z'') for the study of different characteristics and relaxation mechanism of $Bi(Fe_{0.95}Y_{0.05})O_3$ compound.

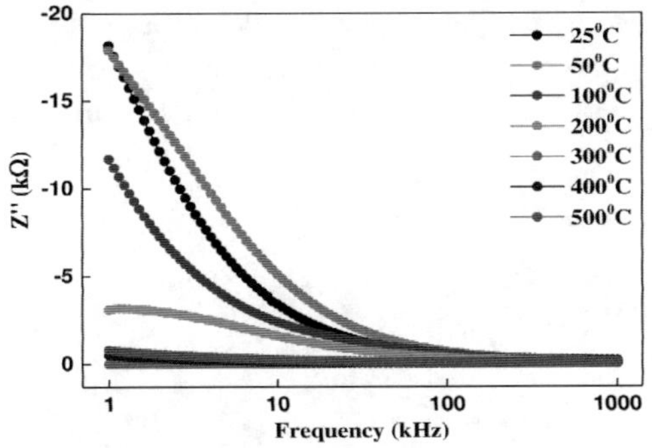

Figure 7. Variation of frequency dependent real impedance (Z') of BFYO5.

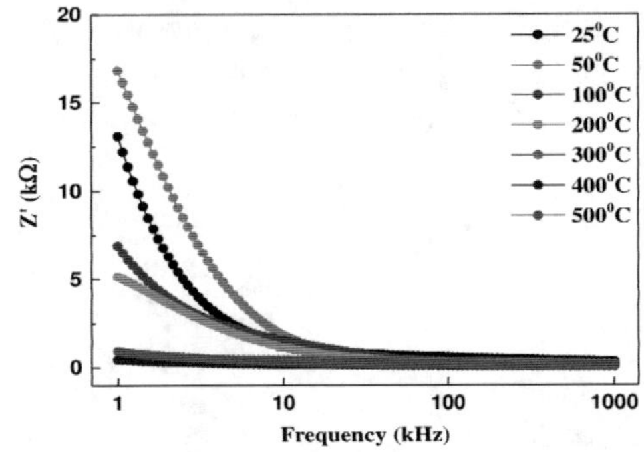

Figure 8. Variation of frequency dependent imaginary impedance (Z'') of BFYO5.

Figure 7 shows the variation of frequency curves of the real part of impedance (Z') at some selected temperatures for BFYO5. The figure clearly shows the steady downward trend in the value of the real impedance component with the temperature rise in the low frequency region. The pattern of a Z' spectra strongly suggested the existence of negative temperature coefficient of resistance in the as-prepared compound. Nevertheless, all the real impedance spectrum curves closed together to get nearly a constant value in the higher frequency region (60kHz-1MHz) regardless to temperature. That nature may be due to the release of a sample-to-electrode space charge [31]. This result is well in line with our previously reported impedance spectrum results [11].

Figure 8 describes the spectrum of hypothetical impedance component (Z") in a broad frequency range (1kHz-1MHz) at certain temperatures (25-500^{0}C). With the change in temperature as well as frequency, the magnitude of Z' decreases. The substantial increase of peak broadenings on increasing the temperature suggested the temperature dependent electrical relaxation mechanisms in the BFYO5 material. The asymmetric expansion of peaks with increasing frequency indicates the existence of several relaxation times in the material. The behavior of this electrical relaxation may be due to the occurrence of immobile charges at lower temperatures and some defects/oxygen vacancies at higher temperatures [32, 33]. The as-prepared material's relaxation frequency (F_r) and relaxation time (τ) can be calculated using the relationship; $\tau = 1/2\pi F_r$ with the help of peak points in the x-axis.

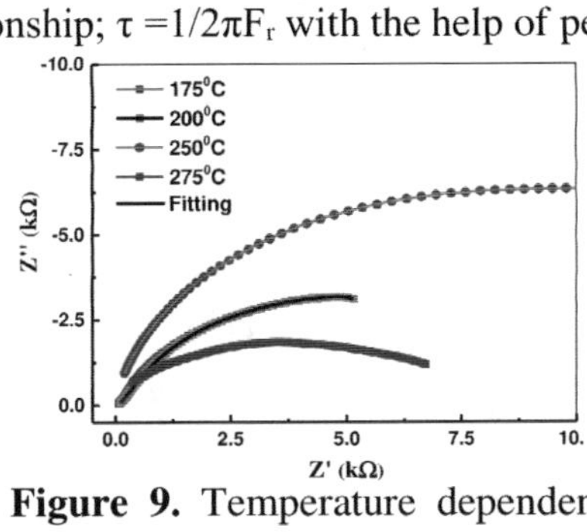

Figure 9. Temperature dependence Nyquist diagram of BFYO5 material.

Figure 9 displays the BFYO5 Nyquist plot drawn over a frequency range of 1-1000 kHz at different temperatures. A material's impedance properties are characterized by the creation of semi-circular arcs whose shape changes as temperature changes. The Nyquist features are not observed in lower temperature region (25-150^{0}C) while the semi-circular arcs are observed at medium and higher temperatures (175-300^{0}C) which exhibit the presence of a single conduction process in the present studied material. That indicates the influence of the grain boundary does not relate to the impedance and other electrical parameters [34]. Since, each of the semi-circular arcs shows some degrees of depression whose centre lie below the x-axis suggested the presence of non-Debye type of relaxation process in the as-prepared material. It has also been observed from the Nyquist plot that the radius of semicircle arcs decreases with the rise of temperature which confirms the NTCR behavior of the studied compound.

3.4. Electrical conductivity

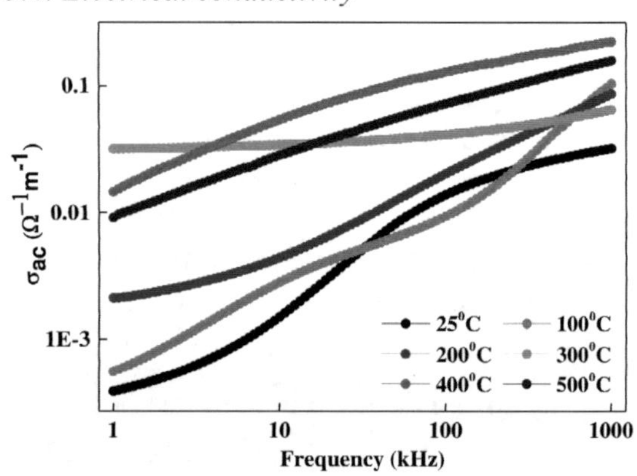

Figure 10. Difference of a frequency dependent ac-conductivity at some certain temperatures.

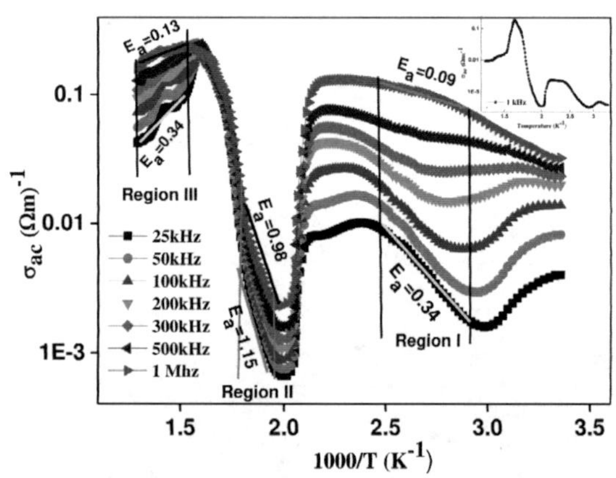

Figure 11. Arrhenius plot of ac-conductivity vs (1000/T) for $Bi(Fe_{0.95}Y_{0.05})O_3$ compound.

The electrical characteristics of a material have a significant importance in an electronic field. The study of electrical conductivity of a material is an essential technique as it provides reliable information which are responsible for its conduction process such as conduction mechanism and types of the charge carriers. Figure 10 shows the variation in ac-conductivity at selected temperatures in a wide range of frequencies. The nature of ac-conductivity spectra clearly displays the value of ac-conductivity gradually increases with the rise in frequency as well as temperatures. Whereas there is a partially independent of frequency at 300^{0}C spectrum. Hence, the plot of frequency dependent ac-conductivity of BFYO5 material also exhibits the negative temperature coefficient of resistance behavior. The characteristics of ac-conductivity plot could be investigated by using 'Jonscher's power law equation' as given below [35];

$$\sigma_{ac}(\omega) = \sigma_A + \sigma_B(\omega) = \sigma_C + P\omega^{\gamma} \tag{1}$$

σ_{ac}= total conductivity of the material, σ_A= frequency independent part, P = range of interaction between the mobile ions with the crystallographic lattice sites, γ = strength of polarizability.

Figure 11 displays the characteristics of ac-conductivity with inverse absolute temperature at a selected frequency (ranging from 1kHz to 1MHz). The plot was divided into three regions such as, Region-I (lower temperature), Region-II (intermediate temperature) and Region-III (higher temperature) for the study of the association between activation energy, inverse absolute temperature and frequency. In the lower temperature region, partial increases/decreases in the value of ac-conductivity was observed. The increasing trend of ac-conductivity becomes faster in the intermediate temperature region, whereas it decreases at further higher temperature region. Arrhenius general formula;

$$\sigma_{ac} = \sigma_0 * \exp(-E_a / K_B * T) \tag{2}$$

has been utilized for the calculation of activation energy at different temperature regions from the graph. Symbols have their usual meaning. The calculated activation energy of BFYO5 material at different temperature regions are summarized in table 3.

Table 3 : The value of temperature dependent ac-conductivity at different regions (region I, II and III) of $Bi(Fe_{0.95}Y_{0.05})O_3$.

Sl. No.	Frequency	Region - I	Region - II	Region - III
1	1 kHz	2.05	5.37	0.85
2	25 kHz	0.34	1.15	0.34
3	50 kHz	0.33	1.17	0.30
4	100 kHz	0.26	1.17	0.27
5	200 kHz	0.14	1.16	0.25
6	300 kHz	0.09	1.13	0.26
7	500 kHz	0.06	1.08	0.18
8	1000 kHz	0.09	0.98	0.13

3.5. Ferroelectric study

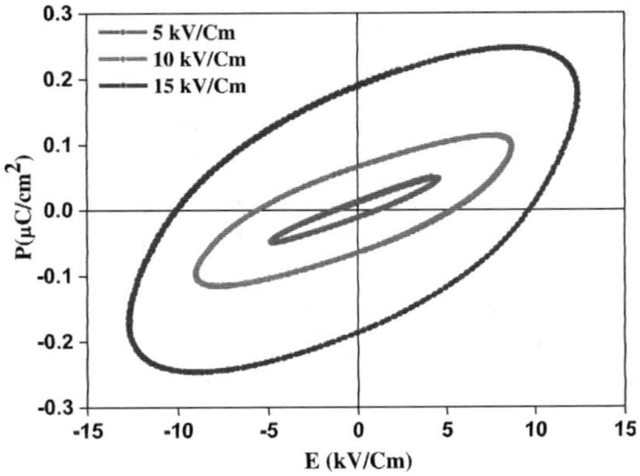

Figure 12. A room temperature polarization versus electric field hysteresis loop diagram of $Bi(Fe_{0.95}Y_{0.05})O_3$ material.

Figure 12 displays a room temperature (25^{0}C) polarization against electric field diagram of BFYO5 material. The plot of unsaturated hysteresis loop demonstrates the ferroelectric nature of the material. The estimated value of remnant polarization (P_r) and coercive field (E_c) of the studied material has been found to be ~0.19 μC/cm^2 and ~9.70 kV/cm respectively. Low value of remnant polarization, unsaturated loop and high coercive field of the material is possibly due to crystallites imperfection and non-homogeneous distribution of different grain dimensions [36]. The shape of the hysteresis loop, remnant polarization and coercive fields of the material are mainly affected by some factors, such as synthesis process, thermal treatment, charged imperfection, mechanical stresses, etc. [37,38]. In the present study, the material required higher value of the electric field for attaining a saturated hysteresis loop, however, it is unable to achieve due to experimental limitations. The value of remnant polarization and coercive field with their respective applied electric field are summarized in table 4.

Table 4: A comparison of P_r and E_c of BFYO5 compound measured at 25^{0}C in a sintered pellet sample.

Sl. No.	Applied electric field	Remnant Polarization (P_r)	Coercive field (E_c)
1	5 eV/cm	0.013 μC/cm^2	0.48 eV/cm
2	10 eV/cm	0.065 μC/cm^2	8.70 eV/cm
3	15 eV/cm	0.190 μC/cm^2	12.4 eV/cm

3.6. Magnetic study

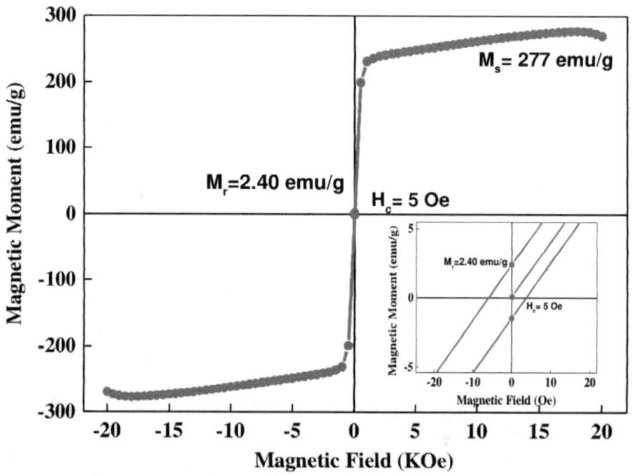

Figure 13. The magnetic hysteresis diagram of $Bi(Fe_{0.95}Y_{0.05})O_3$ material measured in a room temperature. Inset shows the enlarged view of magnetic moment versus magnetic field.

Figure 13 shows the magnetic features of BFYO5 material measured by using vibrating sample magnetometer at a room temperature (25^{0}C). The properties of magnetism of a material depends on various factors such as synthesis process, sintering temperature, size of the particles and distribution of cations. From the figure, the sintered sample have a coercive field (H_c) of 5 Oe. The remnant magnetization (M_r) value is found to be 2.40 emu/g. Based on the literature reports, the parent compound BFO has anti-ferromagnetic behavior without any remnant magnetization value [39,40]. Comparing with the literature survey, Yttrium substitution at Fe-site of BFO results a noteworthy enhancement in the value of remnant magnetization and coercive field, which subsequently effects the structural phase as well as the ionic radii between them. Thus, a partial substitution of bismuth ferrite by Yttrium (Y) at the Fe-site significantly increased and enhanced the magnetic properties as well as its electrical properties.

4. Conclusion

A single phase multifunctional material of the $Bi(Fe_{0.95}Y_{0.05})O_3$ composition was fruitfully synthesized via a mixed oxide route. The formation of phase and basic crystal structure of BFYO5 were obtained using the method of XRD. At the Fe-site of BFO, the Yttrium concentration of 5 percent substituted results in a crystal structure that changes from rhombohedral to orthorhombic symmetry. Through substituting 5 percent of Yttrium into BFO, microstructure, crystallite size and cell parameters are significantly affected. Using Scherrer's formula, the average size of the crystallite was measured and found to be 39 nm. The average grain size of the examined material measured with the help of ImageJ software is 1.73 μm. Study of the dielectric parameters (ε_r and tanδ) increases as the temperature rises and decreases as operating frequency increases. The impedance analysis shows the presence of grain and its boundary effects as well as the (NTCR) behavior and also the non-Debye type electrical relaxation mechanism in the BFYO5 compound. The temperature and frequency dependent ac-conductivity (σ_{ac}) analysis exhibits lower temperature region was dominated by hopping charge carriers while oxygen vacancies dominated the intermediate and higher temperature regions. Analysis of electrical polarization and magnetic hysteresis of the BFYO5 material reveals a noticeable improvement in the value of ferroelectricity, coercivity, remnant magnetization and saturation magnetization that are useful for some device applications.

Acknowledgement
Author Dr. Alok Shukla gratefully acknowledges the financial support received from SERB-DST, Government of India, New Delhi, in the form of Research Project no. EMR/2015/002420. Authors are also grateful to some people around us for their kind help and support.

Conflict of interest
The authors declared that they have no conflict of interest.

References
[1] Eerenstein W, Mathur N D and Scott J F 2006 *Nature* **442** 759
[2] Catalan G and Scott J F 2009 *Adv. Mater.* **21** 2463–85
[3] Cheong S W and Mostovoy M 2007 *Nat. Mater.* **6** 13–20
[4] Kumar N, Shukla A and Choudhary R N P 2018 Development of lead free multifunctional materials $Bi(Co_{0.45}Ti_{0.45}Fe_{0.10})O_3$ *Prog. Nat. Sci.: Mater. Int.* **28** 308–14
[5] Singh A, Pandey V, Kotnala R K and Pandey D 2008 *Phys. Rev. Lett.* **101** 247602
[6] Nan C W, Bichurin M I, Dongb S and Viehland D 2008 *J. Appl. Phys.* **103** 031101
[7] Wang Y P, Zhou L, Zhang M F, Chen X Y, Liu J M and Liu Z G 2004 *Appl. Phys. Lett.* **84** 1731–33
[8] Palkar V R, John J and Pinto R 2002 *Appl. Phys. Lett.* **80** 1628
[9] I. Sosnowska I, Peterlin-Neumaier T and Steichele E 1982 *J. Phys. C. Solid State Phys.* **15** 4835
[10] Wei J, Haumont R, Jarrier R, Berhtet P and Dkhi B 2010 Nonmagnetic Fe-site doping of $BiFeO_3$ multiferroic ceramics *Appl. Phys. Lett.* **96** 102509
[11] Thansanga L, Shukla A, Kumar N and Choudhary R N P 2019 Study of effect of Gd substitution at the Fe-site on structural, dielectric and electrical characteristics of $BiFeO_3$ *Appl. Phys. A* **125** 764
[12] Wen Z, Hu G, Fan S, Yang C, Wu W, Zhou Y, Chen X and Cui S 2009 *Thin Solid Films* **517** 4497–4501
[13] Kumar N, Shukla A, Choudhary R N P 2017 Structural, electrical and magnetic properties of (Cd, Ti) modified $BiFeO_3$ *Phys. Lett. A* **381** 2721–30
[14] Shukla A, Kumar N, Behera C and Choudhary R N P 2016 Structural and electrical characteristics of (Co, Ti) modified $BiFeO_3$ *J. Mater. Sci.: Mater. Electron.* **27** 7115–23
[15] Kumar N, Shukla A and Choudhary R N P 2017 Structural, electrical and magnetic characteristics of Ni/Ti modified $BiFeO_3$ lead free multiferroic material *J. Mater. Sci.: Mater. Electron.* **28** 6673–84
[16] Shukla A, Kumar N , Behera C and Choudhary R N P 2016 Structural, dielectric and magnetic characteristics of $Bi(Ni_{0.25}Ti_{0.25}Fe_{0.50})O_3$ ceramics *J. Mater. Sci.: Mater. Electron.* **27** 1209–16
[17] Kumar N, Shukla A, Kumar N, Sahoo S, Hajra S and Choudhary R N P 2018 Structural, electrical and ferroelectric characteristics of $Bi(Fe_{0.9}La_{0.1})O_3$ *Ceram. Int .* **44** 21330-37
[18] Pandit P, Satapathy S and Gupta P K 2011 *Physica B* **406** 2669
[19] Gabbasova Z V, Kuz'min M D, Zvezdin A K, Dubenko I S, Murashov V A, Rakov D N and Krynetsky I B 1991 *Phys. Lett. A* **158** 491
[20] Yi Du, Cheng Z X, Shahbazi M, Collings E W, Dou S X and Wang X L 2010 *J. Alloys Comp.* **490** 637
[21] C. Nayek C, Tamilselvan A, Thirmal Ch, Murugavel P and Balakumar S 2014 Origin of enhanced magnetization in rare earth doped multiferroic bismuth ferrite *J. Appl. Phys.* **115** 073902
[22] Mishra R K, Pradhan D K, Choudhary R N P and A. Banerjee A 2008 Effect of yttrium on improvement of dielectric properties and magnetic switching behavior in $BiFeO_3$ *J. Phys. Condens. Matter* **20** (4) 045218

[23] Perejon A, Gonzalez E G, Jimenez P E S et al 2015 Structural, optical, and electrical characterization of yttrium-substituted $BiFeO_3$ ceramic prepared by mechanical activation *Inorg. Chem.* 2015, **54** (20) 9876-84
[24] Mukherjee A, Hossain Sk M, Pal M and Basu S 2012 Effect of Y-doping on optical properties of multiferroics $BiFeO_3$ nanoparticles *Appl. Nanosci.* **2** 305–10
[25] Singh H and Yadav K L 2011 *J. Phys. Condens. Matter* **23** 385901
[26] Wul E, PowdMult, An interactive powder diffraction data interpretation and index program, Version 2.1, School of Physical Science Flinders University of South Australia, Australia
[27] Cullity B D 1978 Elements of X-ray diffraction 2nd edn. (Addison-Wesley, Bosten)
[28] Ma Y, Chen X M 2009 Enhanced multiferroic characteristics in $NaNbO_3$-modified $BiFeO_3$ ceramics *J. Appl. Phys.* **105** 054107–11
[29] Sen S and Choudhary R N P 2004 Impedance studies of Sr modified $BaZr_{0.05}Ti_{0.95}O_3$, ceramics. *Mater. Chem. Phys.* **87** (2) 256–63
[30] Brahma S, Choudhary R N P and Thakur A K 2005 AC impedance analysis of $LaLi\text{-}Mo_2O_8$ electroceramics *Physica B* **355**, 188–201
[31] Belboukhari A, Abkhar Z, Gagou Y, Belhadi J, Elmoznine R, Mezzane D, Ei Marssi M and Luk'yanchuk I 2012 *Eur. Phys. J. B* **85** (6) 1–9
[32] Jonscher A K 1977 The universal dielectric response *Nature* **267** 673–79
[33] Suman C K, Prasad K and Choudhary R N P 2006 Complex impedance studies on tungsten-bronze electroceramic $Pb_2Bi_3LaTi_5O_{18}$ *J. Mater. Sci.* **41** 369–75
[34] Kumar N, Shukla A, Behera C and Choudhary R N P 2016 Structural, electrical and magnetic properties of $Bi(Ni_{0.45}Ti_{0.45}Fe_{0.10})O_3$ *J. Alloys Compd.* **688** 858-69
[35] Jonscher A K 1996 Universal Relaxation Law (Chelsea Dielectrics Press, London)
[36] Kumar N, Shukla A and Choudhary R N P 2018 Structural, dielectric, electrical and magnetic characteristics of lead-free multiferroic $Bi(Cd_{0.5}Ti_{0.5})O_3$ $BiFeO_3$ solid solution *J. Alloy. Compd.* **747** 895-904
[37] Haertling G H 1999 Ferroelectric ceramics: history and technology *J. Am. Ceram. Soc.* **82** 797–818
[38] Damjanovic D 1998 Ferroelectric, dielectric and piezoelectric properties of ferroelectric thin films and ceramics, *Rep. Prog. Phys.* **61** 1267–324
[39] Zhang X Q, Sui Y, Wang X J,Wang Y and Wang Z 2010 Effect of Eu substitution on the crystal structure and multiferroic properties of $BiFeO_3$ *J. Alloy. Compd.* **507** 157–161
[40] Xue F, Fu Q F, Zhou D X, Tian Y H, Hu Y X, Zheng Z P and Zhou L 2015 Properties of $Bi_{0.8}Ln_{0.2}FeO_3$ (Ln=La, Gd, Ho) multiferroic ceramics *Ceram. Int.* **41** 14718–26

IOP Conf. Series: Materials Science and Engineering **798** (2020) 012009 doi:10.1088/1757-899X/798/1/012009

Luminescent characterization of $CaMgSiO_4:Dy^{3+}$phosphor for white light emitting diodes

B R Verma[1, 2*], **R N Baghel**[2], **D P Bisen**[2], **N Brahme**[2] **and A Khare**[3]
[1]Department of Physics, Govt. Nagarjuna PG College of Science, Raipur, 492010, India
[2]School of Studies in Physics and Astrophysics, Pt. Ravishankar Shukla University, Raipur, 492010, India
[3]Department of Physics, National Institute of Technology, Raipur, 492010, India

*Corresponding author's e-mail address: bhunesh.verma@gmail.com

Abstract. Different concentration of Dy^{3+} doped calcium magnesium silicate (CMS) phosphors are synthesized via solid state synthesis (SSS) method at high temperature. The structural properties are analyzed by X-ray diffraction (XRD). The surface microstructures are confirmed by scanning electron microscopy (SEM)and energy dispersive X-ray (EDX) spectra. The vibrational properties of the synthesized phosphor are carried out using fourier transform infrared (FT-IR) spectra. The luminescence properties are explained in terms of photoluminescence (PL) and thermoluminescence (TL) characteristics. The PL spectra showed maximum emission intensity peaking at 484 and 578 nm, which are attributed to the $^4F_{9/2} \rightarrow {}^6H_{15/2}$ and$^4F_{9/2} \rightarrow {}^6H_{13/2}$ transitions of Dy^{3+} ion respectively. The PL properties of the prepared phosphors are carried out at room temperature. The maximum PL intensity occurs at 0.5 mol% of doping concentrationof Dy^{3+} ions due to concentration quenching. The glow peak and TL parameters of the prepared phosphors are computed by means of peak shape method. The CIE color co-ordinates are originated (x= 0.3263, y= 0.3451) and correlated color temperature (CCT) are determined to be 5769 K. The results indicate that the prepared phosphor produces cool light in the white region.

1. Introduction

In recent years, research on different rare earths (REs) activated compounds have recognized widespread interest owing to their thermally stability, structural diversity, larger band gap, higher elemental stability, low expenditure and environmental individuality [1-3]. The investigation on silicate phosphor matrix is appropriate to substitute conventional light source in the present years. Silicate based phosphor materials with different REs activated have been enormous interest due to their various applications in field of solid state lighting. In the field of lighting technology, these phosphors can be employed as white light emitting diodes (WLEDs), display devices, optical memories, rewritable copy papers, smart windows and photo-switches [4-6]. It is well known that the $YAG:Ce^{3+}$is a phosphor emitted white light and $BaMgAl_{10}O_{17}:Eu^{2+}$ is phosphor emitted blue light which is available commercially [1]. However these phosphor matrix have the low color rendering index (CRI) and a less luminous efficiency [7-8]. Subsequently it is essential to develop the single rare earth ion doped phosphor matrix by simple preparation method having high efficiency, less electricity consumption, low preparation and production cost. To produce white light, dysprosium (Dy^{3+}) ion is used as adopant in various host matrix because it produces emission in the blue and yellow region of the visible spectrum. The blue region wavelength (470 nm-500 nm) and yellow region (570 nm- 600 nm)are mainly attributed to transition $^4F_{9/2} \rightarrow ^6H_{15/2}$ and $^4F_{9/2} \rightarrow ^6H_{13/2}$ respectively [9].

Calcium Magnesium Silicate ($CaMgSiO_4$) is a useful for solid state lighting as a luminescent materials with excellent structural and optical properties. Blasse et al. [10] and Bhatkar et al. [11]have synthesized and reported the luminescence properties of $CaMgSiO_4:Eu^{2+}$ phosphor. The two emission bands occurred at 446 and 524 nm were observed in their investigation and the emission spectrum dominated by the spectral band at 524 nm. The photoluminescence (PL) behavior of $CaMgSiO_4$ phosphors doped with Eu^{3+} were also studied and reported by Li et al. [12]. These phosphors showed an strong red emission from the $^5D_0 \rightarrow ^7F_J$ manifold transitions of Eu^{3+}, while the Eu^{2+}-doped phosphors showed an extremely weak emission peaked at 475 nm due to the $4f^65d^1 \rightarrow 4f^7$ transition of Eu^{2+}. The luminescent characteristics of $CaMgSiO_4$ phosphors doped with Dy^{3+} ions, have not been investigated systematically yet, which were fascinated to us for the preparation and characterization of phosphor.

In the present investigation, un-doped $CaMgSiO_4$ and $CaMgSiO_4$ doped with Dy^{3+} ion were synthesized by solid state synthesis (SSS) method. The phase identification, morphology, photoluminescence, thermoluminescence and colorimetric properties were investigated for optimum dopant concentration.

2. Experimental section

2.1 Synthesis

The phosphor $CaMgSiO_4:Dy^{3+}$ were prepared using solid-state synthesis (SSS) method at high temperature. The high purity of starting reagents $CaCO_3$ (99.9%), MgO (99%), SiO_2 (99.9%) and Dy_2O_3 (99.99%) were used for preparation of sample. First, the reagents are taken in stoichiometric amounts mixed by hand-grinding for 2 hour in an agate mortar. The mixture was placed in crucible and heated at 1050°C for 5 h in high temperature furnace to obtain the desired product. The heating and the cooling rate of furnace were fixed at 10 °C min^{-1}. Further, the obtained mixtures were reground in fine powder and used for further characterization. The following chemical reaction takes place during solid state synthesis process

$$CaCO_3 + MgO + SiO_2 \xrightarrow{1050\,^{o}C} CaMgSiO_4 + CO2\uparrow$$

$$CaCO_3 + MgO + SiO_2 + Dy_2O_3 \xrightarrow{1050\,^{o}C} CaMgSiO_4{:}\ Dy^{3+} + CO2\uparrow$$

2.2 Characterization

The phase identification and structure of the prepared phosphor $CaMgSiO_4$: $Dy^{3+}$was analyzed by Powder X-ray diffraction (XRD, 40 kV and 30 Ma, Cu Kα target, scanning step 0.001, PANalytical 3 kW X'pert Powder – Multifunctional X- ray Diffractrometer) and the data were collected in the diffraction angle 2 θ range from 10° to 60°. The surface morphology of powder was carried out by using scanning electron microscopy (ZEISS-EVO 60 m)and the elemental composition was quantified by energy dispersive X-ray spectrometer coupled with an Oxford Inca. The vibrational properties of the sample were performed by Fourier transform infrared (FTIR) spectroscopy. The photoluminescence (PL) spectra were obtained on a Shimadzu Spectrofluorophotometer (RF model 5301 PC) with a spectral resolution slit width of 3.0 nm. Thermoluminescence (TL) glow curve were also recorded using 1009I TL reader (Nucleonics, Hyderabad) after the sample exposed UV with 254 nm for 10 min.

3. Results and Discussion

3.1. X-ray Diffraction (XRD) Analysis

In order to determine the phase composition, purity and crystal structure of the as- prepared phosphor, XRD analysis was carried out. Figure 1 depicts the typical XRD patterns of the undoped and Dy^{3+} ion doped $CaMgSiO_4$. All the reflection planes observed in sample were matched with the JCPDS data file

(35-0590). From the XRD pattern of the sample, we can deduce that dopant ions Dy^{3+} were expected to occupy the Ca^{2+} sites in the $CaMgSiO_4$ host matrix due to the close radii and identical valence of the ions. $CaMgSiO_4$ has a orthorhombic structure with space group Pnma and its unit cell volume V= 338.42Å^3. The lattice parameters of refined values of $CaMgSiO_4$ are found as; a = 11.0510 Å , b =6.3640 Å, c = 4.8120 Å, $\alpha = \beta = \gamma = 90°$ and Z = 4, which signifies the proper preparation of the sample. The peaks at 24.60, 28.10, 30.65, 33.82, 34.83 and 50.44 are indexed the reflection planes of (111), (020), (301), (311), (121) and (222) respectively. Due to maximum diffraction intensity, the reflection plane (311) is used to calculate the crystallite size of the sample.The crystallite sizes (D) of samples can be calculated using Scherrer's Formula [13]:

$$D = \frac{K\lambda}{\beta \cos\theta} \quad ...(1)$$

where K is a Scherrer constant (taken as 0.94), λ =1.5418 Å is wavelength of X-rays, β is half maxima line width and θ is the diffraction angle. The crystallite sizes of host and Dy^{3+}doped sample $CaMgSiO_4$ were calculated to be as 52.34 nm and 63.62 nm respectively.

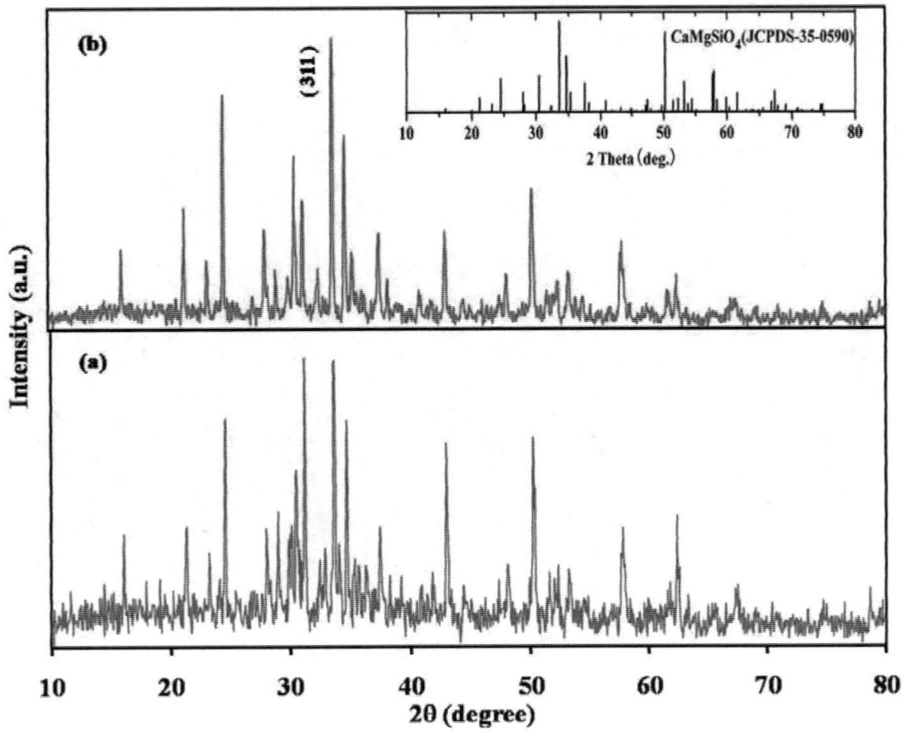

Figure 1. XRD patterns of (a) $CaMgSiO_4$and (b) $CaMgSiO_4$: Dy^{3+}(0.5 mol%) phosphors.

3.2 SurfaceMorphology

The morphologies of the prepared samples were investigated by SEM images. Figure 2 represents the SEM images of un-doped and Dy^{3+} doped sample, respectively. It is noticed from figure 2, the particles in undoped sample shows compact and irregular shape. On the other hand Dy^{3+} doped sample having particles more clear and uniform size, which may be due to that the surface energy distribution of sample is uneven, resulting in abnormal growth of crystals in certain crystal, leading to the

IOP Conf. Series: Materials Science and Engineering **798** (2020) 012009 doi:10.1088/1757-899X/798/1/012009

agglomerate of particles. From the images, it is clear that the approximate size of the particles is in micrometer range which is in favour of its applications in the field of solid state lighting.

Figure 2.SEM micrographs of (a) un-doped and (b) $CaMgSiO_4$: Dy^{3+} (0.5 mol%)phosphor.

3.3 Elemental Analysis

To verify the elemental composition and distribution of elements in the sample, EDX spectra were recorded and are shown in figure 3. It is found that the phosphors are composed of Ca, Mg, Si, O and Dy which are displayed in spectra in the form of peaks. No extra peaks are detected during the observation of EDX profile. The percentage of the elements in synthesized phosphor are tabulated in Table 1.

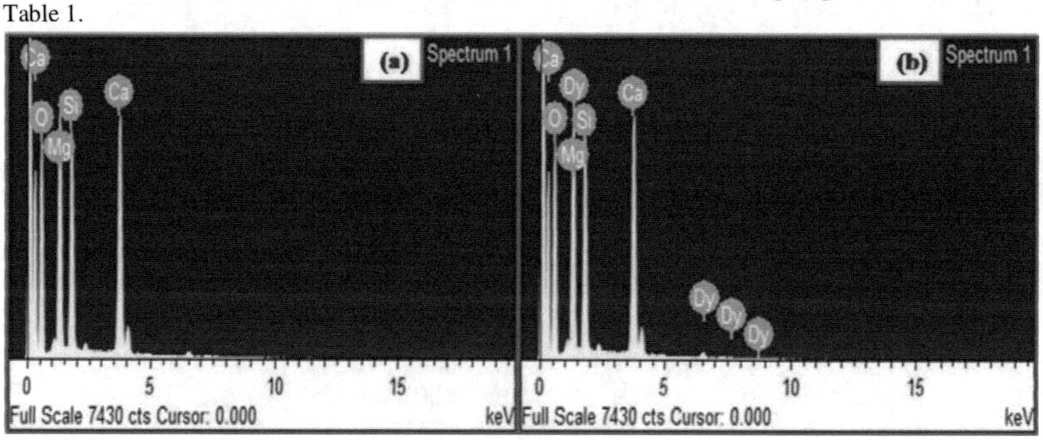

Figure 3. EDX profile of (a) Un-doped (b) $CaMgSiO_4$: Dy^{3+}(0.5 mol%) phosphor.

Table 1. Elemental percentages of $CaMgSiO_4$ and $CaMgSiO_4$: Dy^{3+} phosphor.

(a) Undoped $CaMgSiO_4$

S.No.	Standard	Elements	Weight %	Atomic%
1	SiO_2	O K	55.91	70.70
2	MgO	Mg K	13.80	11.48
3	SiO_2	Si K	11.70	8.43
4	CaWollastonite	Ca L	18.59	9.39
		Total	100.00	100.00

(b) $CaMgSiO_4$: Dy^{3+}(0.5 mol%)

S.No.	Standard	Elements	Weight %	Atomic%
1	SiO_2	O K	55.30	70.45
2	MgO	Mg K	14.05	11.13
3	SiO_2	Si K	11.68	8.63
4	CaWollastonite	Ca L	18.22	9.44
5	DyF_3	Dy L	0.75	0.35
		Total	100.00	100.00

3.4 FTIR analysis

The FTIR spectrum of $CaMgSiO_4$ for the 4000-400 cm^{-1} spectral region is given in figure 4. It is known form the literature that this spectrum is typical silicates. The most intense complex band at 964 cm^{-1} is assigned is the asymmetric stretching modes of the external Si-O-Si groups. The band at 588 cm^{-1} to the vibrational mode of bending. The band around 1265 cm^{-1} band may also be assigned to Ca-O bending vibrations. Mg-O bending is responsible for peak at 903 cm^{-1} due to asymmetric stretching. According to Jiang et al. Ca^{2+} is preferred for tetrahedral sites rather than Mg^{2+} because of the larger radius of Ca^{2+} than Mg^{2+}. When Ca^{2+} is located in tetrahedral sites, Ca–O bonds are exceedingly covalent in nature and dopant Dy^{3+} ion occupies the site created by Ca^{2+}owing to having similar ionic radii [14]. This might create deformation in the lattice resulting in 1514 and 1649 cm^{-1} vibration modes assigned to vibration in Ca^{2+} and Mg^{2+} ions respectively. The broad peak centered at 3414 cm^{-1} is visible because of stretching of O-H which supports the presence of moisture in the sample.

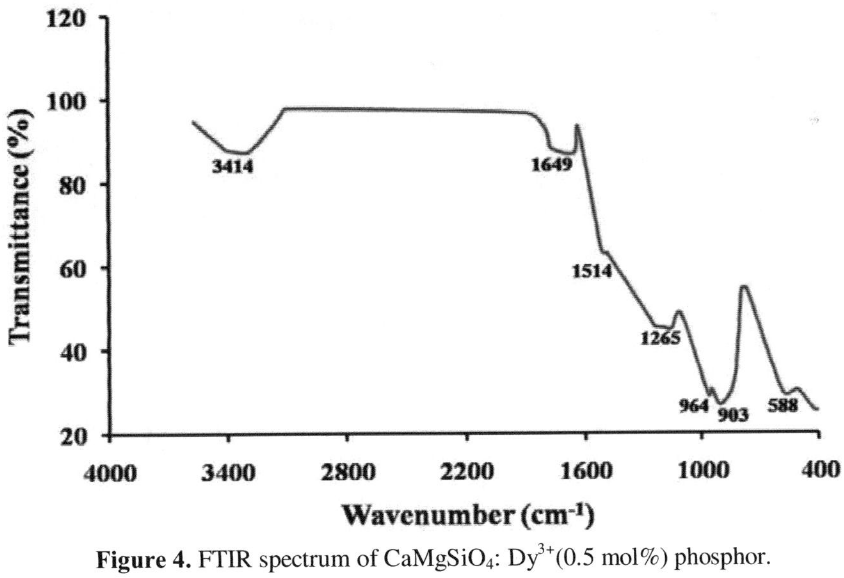

Figure 4. FTIR spectrum of $CaMgSiO_4$: Dy^{3+}(0.5 mol%) phosphor.

3.5 Photoluminescence (PL) studies

Figure 5 presents photoluminescence excitation (PLE) spectrum of the as-prepared $CaMgSiO_4$: Dy^{3+} (0.5 mol%) phosphor monitored at wavelength 577 nm ascribed to transition ($^4F_{9/2} \rightarrow {}^6H_{13/2}$). The sample shows broad excitation band from 340 to 360 nm and an intense band from 380 to 400 nm with a maximum at 384 nm transition ($^6H_{15/2} \rightarrow {}^4I_{13/2}$) due to the 4f-4f transition of the Dy^{3+} ions. Upon

IOP Conf. Series: Materials Science and Engineering **798** (2020) 012009 doi:10.1088/1757-899X/798/1/012009

excitation at 384 nm, the emission spectum of $CaMgSiO_4$: Dy^{3+} of the sample is shown in figure 6. It consists of the characteristic transition between Dy^{3+} levels. The emission spectrum exhibits main two groups of emission peaking at 484 and 578 nm, which are assigned to the $^4F_{9/2} \rightarrow {}^6H_{15/2}$ and $^4F_{9/2} \rightarrow {}^6H_{13/2}$ transitions of Dy^{3+}ion respectively. Obviously, the emission spectrum were dominated by the $^4F_{9/2} \rightarrow {}^6H_{13/2}$ (578 nm) transition of the Dy^{3+}, which is due to an electric-dipole-allowed transition and hypersensitive to the environment [15].

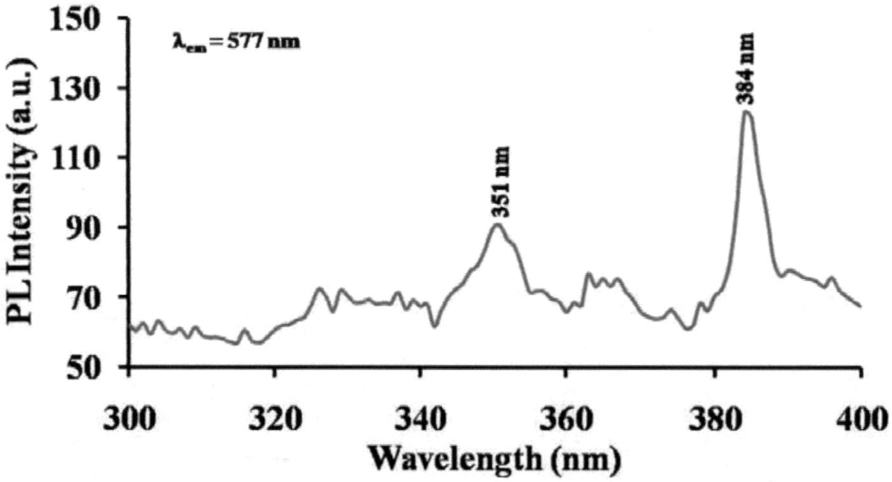

Figure 5. Excitation spectrum of $CaMgSiO_4$: Dy^{3+} (0.5 mol%) monitored at λ_{em}= 577 nm.

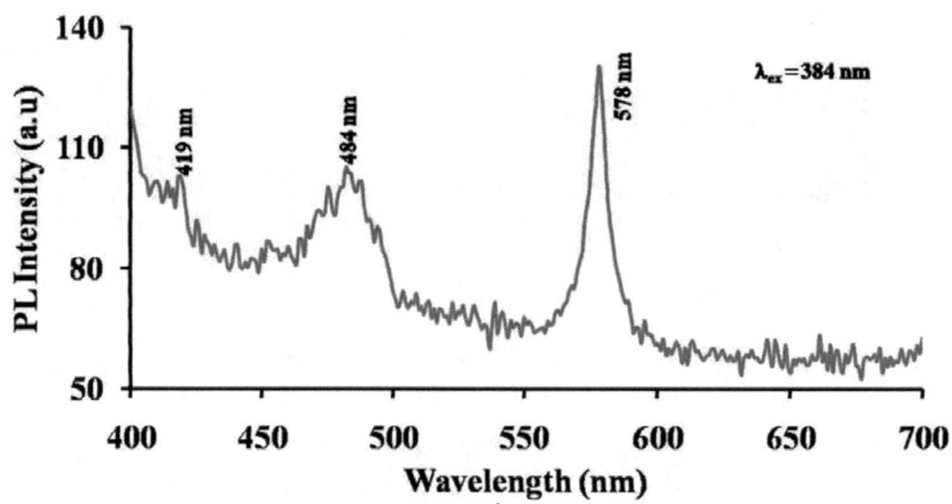

Figure 6. Emission spectrum of $CaMgSiO_4$: Dy^{3+} (0.5 mol%) excited at λ_{ex}= 384 nm.

In order to show the effect of doping concentration on PL luminescence properties, a series of $CaMgSiO_4$: Dy^{3+} (0.1, 0.2, 0.5, 1, 2 mol%) phosphors were prepared. Figure 7 shows the dependence of PL intensity on dopant concentration of Dy^{3+} ion.When the doping concentration of Dy^{3+} increases, the PL intensity increases gradually and reaches a maximum at 0.5 mol% of Dy^{3+}

ion. With further increasing of Dy^{3+}concentration, the emission intensity begins to decrease due to concentration quenching. According to the Dexter's energy transfer theory [16], concentration quenching is mainly caused by the non-radiative energy migration among the Dy^{3+} ions at the higher concentration.

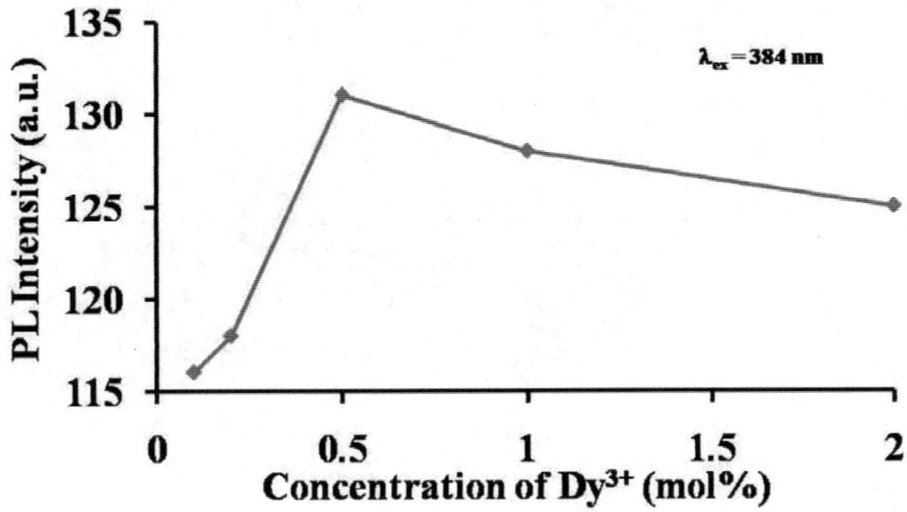

Figure 7. Variation of PL intensity with doping concentration of Dy^{3+} ion

3.6 CIE co-ordinates of $CaMgSiO_4$: Dy^{3+}phosphor

Figure 8 depicts the CIE diagram of synthesized $CaMgSiO_4$: Dy^{3+} phosphors. The chromaticity coordinate for the $CaMgSiO_4$: Dy^{3+} (0.5 mol%) phosphor was originated to be (x= 0.3263, y= 0.3451) which is close to the NTSC standard white light (0.33, 0.33). This parameter was estimated as per the emission spectra in $CaMgSiO_4$: Dy^{3+} (λ_{ex}= 384 nm). To further estimate the temperature of the light emission, the CCT was calculated using McCamy formula [17] and found to be 5769 K. Since the $CaMgSiO_4$: Dy^{3+} phosphor can display a bright white emission with good CIE chromaticity coordinate. The CCT of $CaMgSiO_4$: Dy^{3+} phosphor lies above 5000 K which indicates that the prepared phosphor produces cool light in the white region.

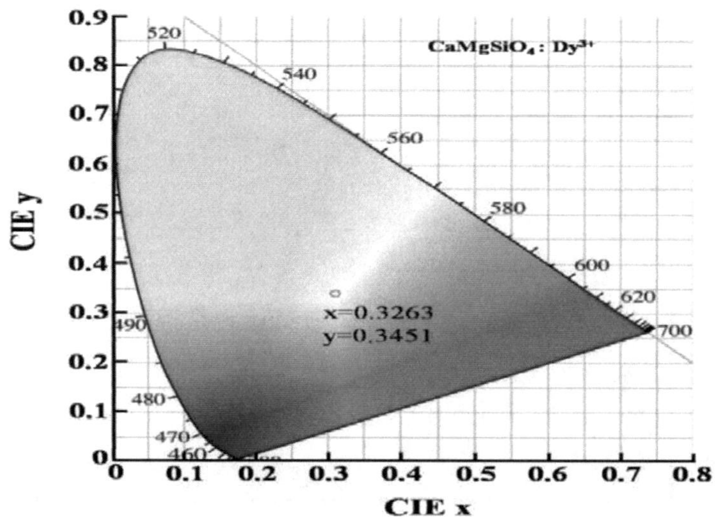

Figure 8. CIE diagram of $CaMgSiO_4$: Dy^{3+}(0.5 mol%) phosphor.

3.7 Thermoluminescence (TL) characteristics

Figure 9 shows the TL glow curves of $CaMgSiO_4$: Dy^{3+}phosphor materials exposed to UV excitation source at room temperature. The TL data were collected in the temperature range 50 to 300 °C with linear heating rate (LHR) of 5°C/s. The peak temperature remains almost same for all dopant concentration. In contrast to PL results, the TL intensity of $CaMgSiO_4$: Dy^{3+} was found to decrease with Dy contents. From the results presented here it is evident that TL of $CaMgSiO_4$: Dy^{3+} phosphor shows good enhancement in TL intensity up to the concentration of 0.5 mol% of Dy^{3+} subsequently it starts decreasing. The decrease in TL intensity may possibly be attributed to concentration quenching. The optimal TL intensity was observed at 0.5 mol% of doping concentration Dy^{3+} ion.

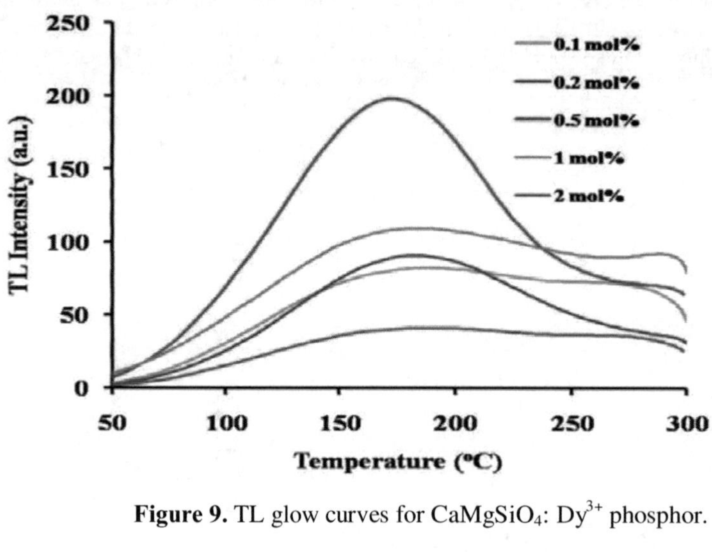

Figure 9. TL glow curves for $CaMgSiO_4$: Dy^{3+} phosphor.

To further evaluate the symmetry factor, trap depth and frequency factor, applying peak shape method proposed by Chen [18]. The value of trap depth energy (**E**) and frequency factor (**s**) for optimum concentration of dopant Dy (0.5 mol%) are presented in Table 2.

Table 2. Values of E (eV) and s(s^{-1}) for $CaMgSiO_4$: Dy^{3+} phosphor.

Dopant (mol%)	T_1(°C)	T_2(°C)	T_M(°C)	μ_g	E (eV)	s(s^{-1})
0.5	114.11	242.78	176.21	0.52	0.39	1.50×10^5

4. Conclusion

In summary, $CaMgSiO_4$: Dy^{3+} phosphors were synthesized via solid state synthesis method. The XRD spectra showed that the phosphors were matched with reported JCPDS data and the grain size of the phosphor in micron range and uniformity better. From the results of SEM and photoluminescence intensity, the property of the light-emitting substance was improved with the increasing of concentration. The as-obtained $CaMgSiO_4$: Dy^{3+} phosphors showed higher PL emission intensity at 484 and 578 nm. These peaks were attributed to electronic transition $^4F_{9/2} \rightarrow {}^6H_{15/2}$ and $^4F_{9/2} \rightarrow {}^6H_{13/2}$ of Dy^{3+} respectively. The optimal doping concentration of Dy^{3+} was obtained at 0.5 mol%. The chromaticity coordinate for the synthesized phosphor was found to be (x= 0.3263, y= 0.3451) and CCT was calculated and found to be 5769 K which produces cool light in the white region. The shape factor and trap depth energy were found 0.52 and 0.39 eV respectively. These results indicate that synthesized phosphor may be good promising candidate phosphors in the field of solid state lighting.

Acknowledgments

Authors acknowledge the Department of Metallurgical Engineering, National Institute of Technology, Raipur (CG) India for the XRD, EDX and SEM Characterization. Authors are also thankful to Mr. Suresh Dua and Mr. Vinod Kumar Ingole NIT, Raipur (CG) India for their cooperation at their center.

References

[1] Shukla D, Ghormare K B and Dhoble S J 2014 *Adv. Mat. Lett.* **5** 406-408.
[2] Poort S H M, Janssen W and Blasse G 1997 *J. Alloys Compd.* **260** 93.
[3] Sharma P K, Dutta R K and Pandey A C 2011 *Adv. Mater. Lett.* **2** 246.
[4] Huang C H *et al* 2012*J. Mater. Chem.* **22** 20210.
[5] Akiyama M 2010 *Appl. Phys. Lett.***97**181905.
[6] CohenA J and Smith H L 1962 *Science***137** 981.
[7] Wang Y, Wen Y and Zhang F 2010 *Mater. Res. Bull.***45** 1614.
[8] Chang C K and Chen T M 2007 *Appl. Phys. Lett.* **91** 081902.
[9] Jayasimhadri M, Ratnam B V and Jang K 2010 *J. Am. Ceram. Soc* **93** 494.
[10] Blasse G, Poort S H M, Reijnhoudt H M and Kuip H O T1996 *J. Alloys Compd.* **24**175.
[11] Bhatkar V B, Omanwar S K and Moharil S V 2007 *Opt. Mater.* **29** 1066.
[12] Li Y, Wang Y, Xu X, Yu G and Zhang F 2010 *J. Electrochem. Soc.* **157** J39-J43.
[13] Song L and Zhang S 2011 *Chem. Eng. J.* **166** 779–782.
[14] Jiang L, Chang C, Mao D and Feng C 2003 *J. Mater. Sci. Engg.* B **103** 271-275.
[15] Li Q, Liu Z P, Li X J and Dong L M 2016 *Dig. J. Nanomater. Biostructures* **11**313 – 319.
[16] Dexter D L 1954 *J. Chem. Phys* **22**1063.
[17] Kumam N, Singh N P, Singh L P and Srivastava S K 2015 *Nanoscale Res. Lett.* **10** 347.
[18] Pagonis V, Kitis G and Furetta C 2006 *Numerical and Practical Exercises in Thermoluminescence* (Berlin: Springer).

Synthesis and luminescence behavior of $Ba_2SiO_4:Dy^{3+}$ phosphors in presence of charge compensator ions

Ganesh Ram Banjare[1*], D.P. Bisen[1], N. Brahme[1], Chitrkant Belodhiya[1], A.K. Upadhyay[2]

[1]SoS in Physics & Astrophysics, Pt. Ravishankar Shukla University, Raipur (C.G.) India 492010
[2]Department of Physics, Government Engineering College, Raipur (Chhattisgarh) India 492015

*Corresponding author's e-mail address: banjare.ganesh@rediffmail.com, dpbsen@rediffmail.com

Abstract: A series of dysprosium (Dy^{3+}) doped barium ortho-silicate (Ba_2SiO_4) phosphors were synthesized by conventional Solid State reaction technique. In order to find out the phase purity and crystal structure, characterization of the prepared powder samples was done by X-ray diffraction (XRD) method. FTIR method confirms the presence of all elemental compositions. Photoluminescence (PL) properties were studied. The PL spectra was observed to study the effect of concentration of Dy^{3+} ion and it is found that the prepared phosphors show two strong emission peaks in the blue and yellow region at 478 nm and 573 nm, respectively, when excited by the near-ultraviolet (UV) of wavelength 348 nm. The critical distance between the Dy^{3+} ions was calculated using Blasse's equation. Effect of monovalent ion K^+ was examined as a charge compensator ion and observed that co-doping of charge compensator ion enhances the luminescence intensity. Chromaticity coordinates of the phosphor are found near to white light region, which confirms their applicability in solid state lighting applications; especially in white-LEDs.

Keywords: PL, charge compensator ion, LEDs.

1. Introduction

Phosphor materials play a key role in recent technologies; especially in lighting and display devices. They have outstanding chemical and physical properties as well as a stable crystal structure. Thus, a lot of attention has been drawn towards the study of phosphor materials because of their usefulness in the various fields such as Field Emission Displays (FEDs), sensors, Plasma Display Panels (PDP), X-ray imaging systems, optical fibres and white light emitting diodes (WLEDs). Among all these applications, WLEDs is considered to be the next generation lighting systems over the traditional lamps due to its reliability, long lifetime, high brightness, eco-friendly, and low power consumption properties [1-4]. It has been proved that Ultraviolet (UV) pumped phosphor converted WLEDs exhibit high Color Rendering Index (CRI) as well as high chromatic stability [5]. At the present time, phosphor based WLED was highly useful and a number of approaches have been developed for the emission of white light through LEDs.

In the different oxide based phosphor materials, silicates are of great interest as a host for their high chemical, physical stability and stable crystal structure [6]. In this context, rare earth doped alkaline earth ortho-silicates such as $SrCa_{1-x}SiO_4$: xEu^{3+} $Sr_{2-x}SiO_4$: xSm^{3+}, $Sr_xBa_{2-x}SiO_4$:Eu^{2+}, $Ba_{1.55}Ca_{0.45}SiO_4$:(Eu^{2+}, Mn^{2+})are popular commercial phosphors for WLEDs [7]. Recently, Dy^{3+} doped alkaline earth ortho-silicate phosphors were fascinated because of their high absorption efficiency in UV area, outstanding spectroscopic behavior, simple chemical compositions and single luminescence center [8]. Therefore, it's a huge demand for today's scenario to synthesis new novel efficient white light emitting phosphor. With this motivation, we choose barium ortho-silicate i.e. Ba_2SiO_4as a host due to its high quantum efficiency properties and lower harmfulness for the environment as well as health [9, 10].

In this paper, white light emitting phosphor materials Ba_2SiO_4: Dy^{3+} is synthesized by high-temperature solid-state reaction method and its photoluminescence properties are investigated. The main aim of this work is to improve the luminescence intensity of the phosphor materials by the incorporation of K^+ ion as charge compensator ion. The chromatic behavior of the prepared phosphors was studied in detail.

2. Synthesis and characterization

A series of trivalent dysprosium (Dy^{3+}) doped barium ortho-silicate namely $Ba_{2-x-y}SiO_4$: $_xDy^{3+}$,$_yK^+$ phosphors (x = 1.0, 2.0, 3.0, 4.0 and 5.0 %) were prepared by high temperature solid-state reaction method in presence of charge compensator ion K^+ (y = 1.0, 2.0, 3.0, 4.0 and 5.0 %). $BaCO_3$, SiO_2, Dy_2O_3, and K_2CO_3 were used as raw materials; all of the analytical grades (A.R.) with 99.9 % purity were utilized for the synthesis of phosphor under study. H_3BO_3 was used as a flux. Stoichiometric proportions of all raw materials were ground using an agate mortar and pestle for 3 hand transferred into alumina crucible, further kept into muffle furnace. Thereafter, mixed materials were fired at 1200°C for 4 h in air. The final products in powder forms were ground again, yielding the resulting powder form of the phosphors.

X-ray diffraction (XRD) measurements were carried out for the phase identification in powder samples using Bruker D8 Advance X-ray diffractometer with Cu Kα radiation (λ = 1.54 Å, 40 kV, 40 mA) at room temperature. For functional group identification Furrier transform infrared (FTIR) measurements have been performed using Shimadzu IRAffinity-1 Fourier Transform Infrared Spectroscopy. The photoluminescence (PL) spectra (excitation &emission) of the phosphors samples were carried out by using a Shimadzu, RF-5301 PC spectrofluorophotometer.

3. Results and discussions

3.1 XRD analysis

To identify the crystal structure and phase purity of the prepared samples, X-ray diffraction (XRD) technique was used.The XRD pattern of the $Ba_{1.94}SiO_4$: $_{0.03}Dy^{3+}$, $_{0.03}K^+$phosphoris being represented in figure 1. It was observed that all the peaks are well-matched with the JCPDS # 70-2113. No extra peaks are present in the pattern, which confirms the single phase of the $Ba_{1.94}SiO_4$: $_{0.03}Dy^{3+}$, $_{0.03}K^+$ phosphor. The single phase of the prepared samples indicates that dopant $Dy^{3+}$and co-dopant K^+ are fully dissolved into the host matrix. The powder samples shows an orthorhombic type crystal structure with the space group Pmcn,(62) and the cell parameters were determined to be a = 5.80 Å, b = 10.20 Å, c = 7.50 Å with α = β = γ = 90°. The position of atoms with the occupancy factors is listed in Table 2 [11]. Fig.2 (a) shows the tetrahedral structure of SiO_4 whereas fig. 2 (b) represents within a distance of 3.2 Å, both the sites namely Ba1 and Ba2 sites, are coordinated by ten and nine oxygen atoms, respectively.

IOP Conf. Series: Materials Science and Engineering **798** (2020) 012010 doi:10.1088/1757-899X/798/1/012010

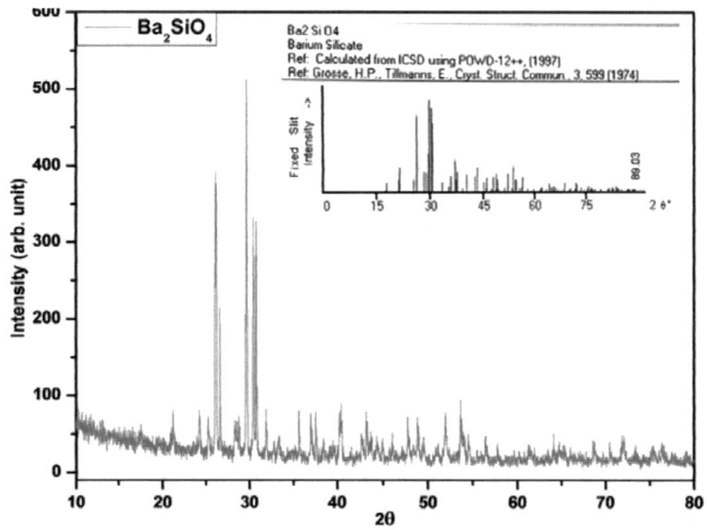

Figure 1. XRD pattern of Ba_2SiO_4 phosphor.

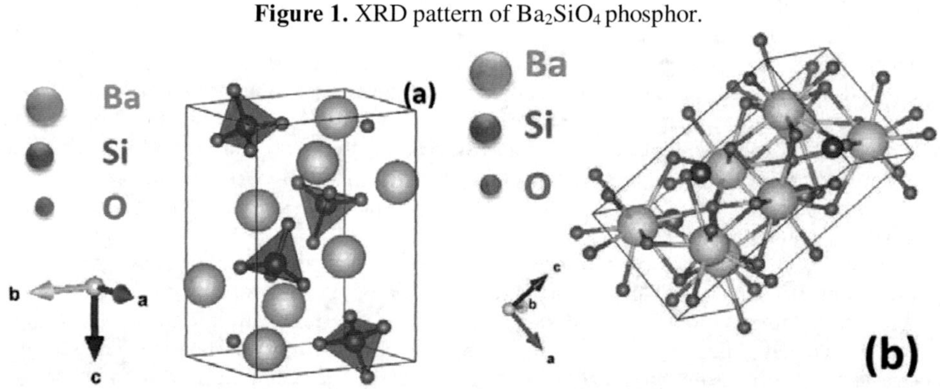

Figure 2. (a) tetrahedral structure of SiO_4 (b) coordination structure of Ba_2SiO_4 phosphor.

Table 1. Some important parameters of Ba_2SiO_4 phosphor.

Parameters	Ba_2SiO_4
Phase structure	Orthorhombic
Lattice type	Primitive
Space group	Pmcn(62)
Cell parameters	a = 5.80 Å b = 10.20 Å c = 7.50 Å $\alpha = \beta = \gamma = 90°$
Radiation	CuKα 1
Wavelength	1.54060 Å

Table 2. the position of atoms and occupancy factors.

Name of atoms	x	y	z	Sites	Occ.
Ba1	0.2499	0.0872	0.1608	4c	1
Ba2	0.2499	0.6955	-0.0085	4c	1
Si	0.2498	0.4198	0.2283	4c	1
O1	0.2498	0.5705	0.3097	4c	1
O2	0.0197	0.3480	0.3051	8d	1
O3	0.0251	0.4157	0.0130	4c	1

3.2 Fourier Transform Infrared Spectroscopy(FTIR) spectroscopy

The functional compositions of the synthesized phosphors were identified by FTIR spectroscopy. Fig. 3 shows the transmittance spectrum of the $Ba_{1.94}SiO_4$:$_{0.03}Dy^{3+}$, $_{0.03}K^+$ phosphor, which was recorded in the range of wave number 400 – 4000 cm^{-1}. The FTIR spectrum of the observed powder sample shows following six prominent bands which are located around (i) 3641.6 cm^{-1}(ii) 1408.08 cm^{-1} (iii) 972.12cm^{-1} (iv) 491.85 cm^{-1} (v) 432.05 cm^{-1} and (vi) 391.55 cm^{-1}. The absorption band around 3641.6 cm^{-1} ascribes the presence of hydroxyl group in the prepared phosphor. Thereafter, C=O in carbonyl functional group which have an anti-symmetric stretching vibrations shows a strong peak at 1408.08 cm^{-1}. The next broadband lying in the region 900–1200 cm^{-1}, and the single peak at 972.12cm^{-1} assigned due to the Ba^{2+} ions. The band centred at 491.85 cm^{-1} are combinations of a number of contributions of the asymmetric stretching modes due to the presence of silicates (SiO_4). The vibrations of the monovalent ion were found to produce a broad IR absorption band centred on 391.55 cm^{-1}.

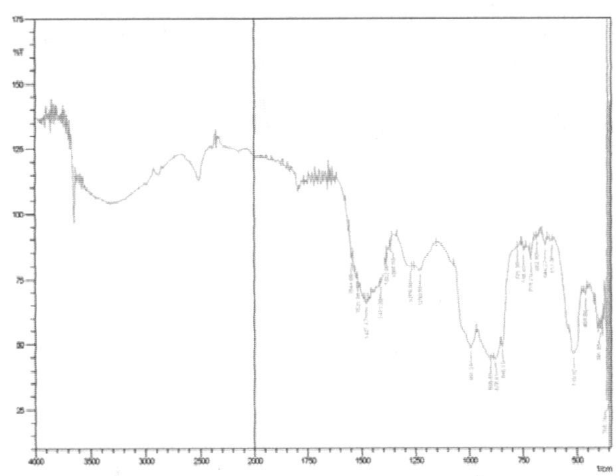

Figure 3. FTIR spectroscopy of the Ba_2SiO_4: Dy^{3+}, K^+ phosphor.

3.3 Photoluminescence (PL) studies

Photoluminescence behaviors of the prepared phosphors wereinspected. The excitation and emission spectra (see figure 4 a)were recorded in the wavelength range 200 – 700 nm at room temperature.The excitation spectra exhibit five excitation peaks centredat 295 nm, 325 nm, 352 nm, 364 nm and 390 nm.The sharp peaks are obtained due to 4f – 4f transition of Dy^{3+} ion. The sharp peaks originated from the ground energy state to higher energy state corresponding to following transitions $^6H_{15/2}\rightarrow{}^6P_{3/2}$, $^6P_{7/2}$, $^6P_{5/2}$, $^4I_{13/2}$ except the first peak. Among all these peaks, the peak at 352 nm is more dominant over the others.Z. Ci *et al* reported the suitable wavelength range from 350 nm to 410 nm for the strong absorption of ideal luminescent materials [12].In this study, the obtained absorption matches with the emission of commercially available NUV and blue LED chips light. Thus, it is observed that the prepared sample can be effectively excited by near-ultraviolet (n-UV) and blue light.

The emission spectra of $Ba_{1.97}SiO_4$: $_{0.03}Dy^{3+}$ phosphors under the n-UV excitation at the wavelength 352 nm corresponding to $^6H_{15/2}\rightarrow{}^6P_{7/2}$ transition are shown in fig. 4 (a). Two emission peaks were occurred at 476 nm (blue region) and 573 nm (yellow region), upon the n-UV excitation of 352 nm, which is a typical emission characteristic of Dy^{3+}ion.From figure 4 (a), it is clear that the intensity of emission peak at the blue region is stronger than emission obtained at the yellow region. The strong blue emission occurs due to $^4F_{9/2}\rightarrow{}^6H_{15/2}$ transition (magnetic dipole transition),which represents the presence of Dy^{3+} ion in a high symmetric site. However, the yellow emission is obtained as a result of the hypersensitive transition at 577 nm and which is electric dipole transition related to $^4F_{9/2}\rightarrow{}^6H_{13/2}$ transition. In this case, the PL intensity highly depends on the host matrix. Indeed, the hypersensitive transition is strongly influenced by the nearby atmosphere of activator i.e.Dy^{3+}ion [13]. From figure 4 (a), it is clear that the intensity of emission peak at the blue region is stronger than emission obtained at the yellow region. Thus, it is analyzed thatDy^{3+} ion present in the higher symmetry site with an inversion centre, whereas for yellow emission, dopants(Dy^{3+}) located at lower symmetry site with no inversion center [14].

3.3.1 Influence of concentration of Dy^{3+} ion

Varying concentrations of Dy^{3+} ions (1.0 – 5.0 %) were employed into the host. Initially, the luminescence intensity increases with the concentration and optimum occurred for 3.0 mol % of Dy^{3+}. Further increment of Dy^{3+} ion results inthereduction in PL intensity due to concentration quenching. Thus, x = 3.0 % is identified as critical concentration.According to Blasse, the critical distance (Rc) for Ba2-xSiO4: xDy^{3+}can be calculated using the following equation

$$Rc = 2\left[\frac{3V}{4\pi x_c Z}\right]^{\frac{1}{3}} \quad ...(1)$$

Where V is the volume oftheunit cell, x_c represents the critical concentration of the activator ion, and Z is the number of cations per unit cell. Here Z = 4, V = 387.233 Å and x_c = 0.03 for $Ba_{1.97}SiO_4$: $_{0.03}Dy^{3+}$ phosphor. Consequently, the critical distance (Rc) of activator ion (Dy^{3+}) was calculated by the formula (1) and obtained 24Å. The critical distance value determines interaction types between the activator ions on the basis of according to Dexter's theory, which is expressed as:

$$\frac{I}{X} = \mathrm{k}\left(1 + \alpha(\mathrm{X})\right)^{\frac{\theta}{3}})^{-1}$$

θ = 6, 8 and 10 then d –d, d – q and q –q are the corresponding interactions [15]. In this study, the value of θ is 8.37, which is determined by fitting the plot between log (I/x) and log (x) as shown in figure 6 (b). The value 8.37 is nearly equivalent to 8 determined the d – q interaction type between activator ions.

Thus, it's concluded that PL emission intensity is influenced by the concentrations of Dy^{3+} ions. In this study, it is investigated that introduction of Dy^{3+} ion into the host, replaces Ba^{2+} because the ionic radii of Dy^{3+}(91.2 pm) areless than Ba^{2+} (135 pm). Due to this replacement, charge imbalance (obtained due to the presence of Dy^{3+} ion in the symmetry site) creates defects, which is responsible for non – radiative transition. This transition reduces the PL intensity. Hence, the improvement of PL intensity is necessary.

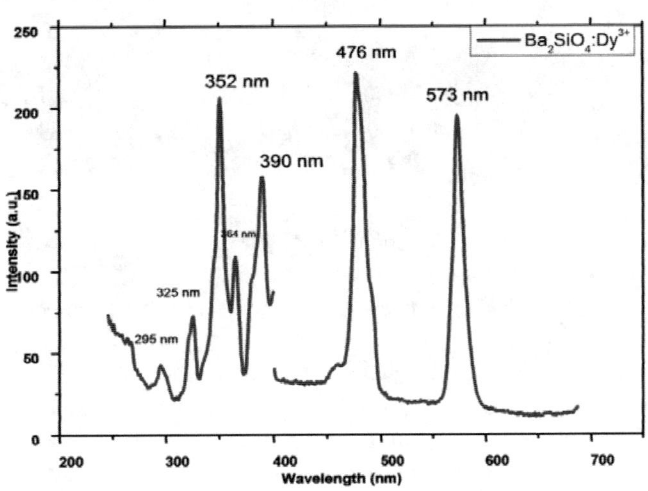

Figure 4. (a) PL excitation and emission spectra of $Ba_{2-x}SiO_4$: xDy^{3+} phosphors.

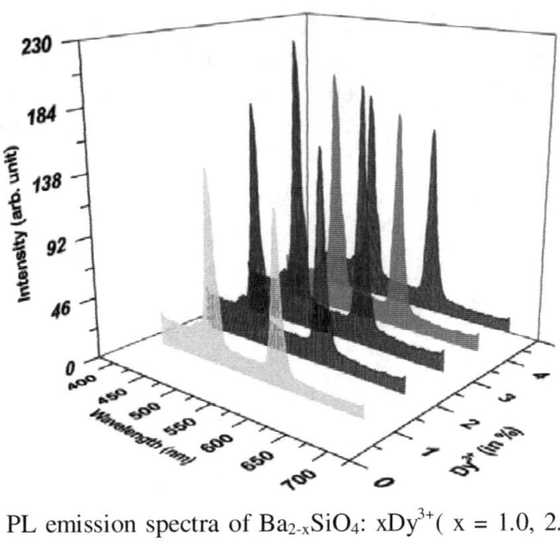

Figure 4. (b) PL emission spectra of $Ba_{2-x}SiO_4$: xDy^{3+}(x = 1.0, 2.0, 3.0, 4.0 and 5.0 %) phosphors.

3.3.2 Effect of charge compensator ion K^+

Doping of trivalent Dy^{3+} ions substitutes divalent Ba^{2+} sites will responsible in the formation of excess positive charge in the host. The excess charge resulting in the distortion of crystal structure and sample will become unstable;perturbed the total charge neutrality of the system [16]. The major effect of the above result is nothing but a reduction in PL intensity due to charge imbalance. In order to remove the above drawback, K^+ ions were incorporated into $Ba_{1.97}SiO_4$: $_{0.03}Dy^{3+}$ phosphor.

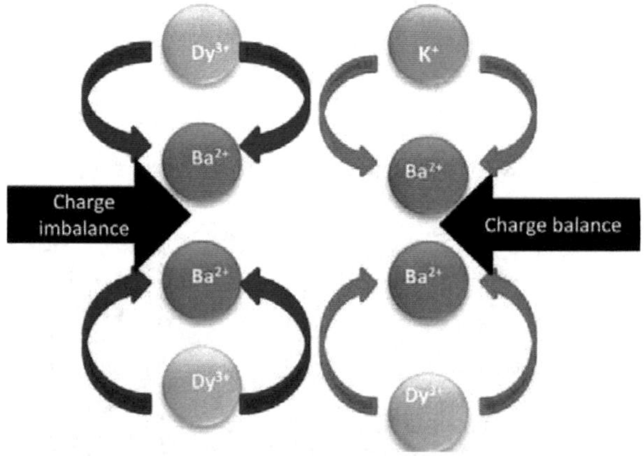

Figure 4. (c) Schematic diagram of the charge compensation mechanism.

It can be seen very clearly from the fig. 5 (a) that introduction of K^+ ion enhances the PL intensity because the K^+ ion reduces the lattice distortion and nullify the excess positive charge. In this case, two Ba^{2+} (ionic radius = 135 pm) site is substitutes by one Dy^{3+} ion (ionic radius = 91.2 pm) and one K^+ ion (ionic radius = 138 pm) [17]. Thus, it is investigated that due to the same size of the ionic radius, K^+ ion easily replaces the Ba^{2+} site. Here, K^+ ions are used as charge compensator ion. This is an important approach to enhance the luminescence intensity.Table 3 shows the relative PL intensity without K^+ ion and with K^+ ions. The PL intensity enhanced 1.75 times due to charge compensator ion (K^+ ion). Fig. 5 (b) represents the PL emission spectra for different concentration of K^+ (1.0 – 5.0 %) ions in the $Ba_{1.97}SiO_4$: $_{0.03}Dy^{3+}$ phosphor and maximum obtained for 3.0 mol % i.e. $Ba_{1.97}SiO_4$: $_{0.03}Dy^{3+}$, $_{0.03}K^+$ phosphor.

Table 3. Relative PL intensity without K^+ ion and with K^+ ion.

S. No.	Wavelength (nm)	PL intensity (without K+ ion) (arb. Unit)	PL intensity (with K+ ion) (arb. Unit)
1.	476	221	387
2.	573	194	340

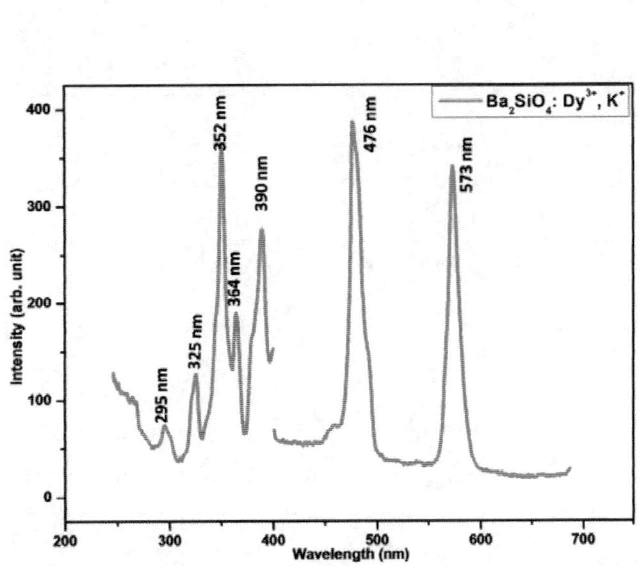

Figure 5. (a) PL excitation and emission spectra of $Ba_{2-x-y}SiO_4$: xDy^{3+} phosphors in presence of yK^+ ion.

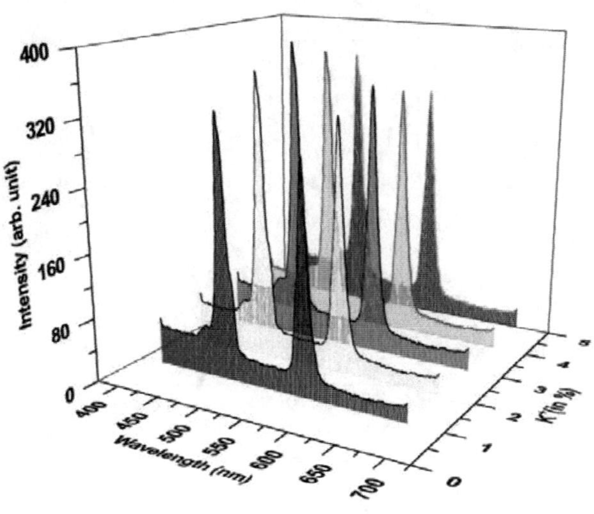

Figure 5. (b) Emission spectra of $Ba_{1.97}SiO_4$: $_{0.03}Dy^{3+}$ phosphors for varying concentration of K^+ ion (1.0, 2.0, 3.0, 4.0, and 5.0 %)

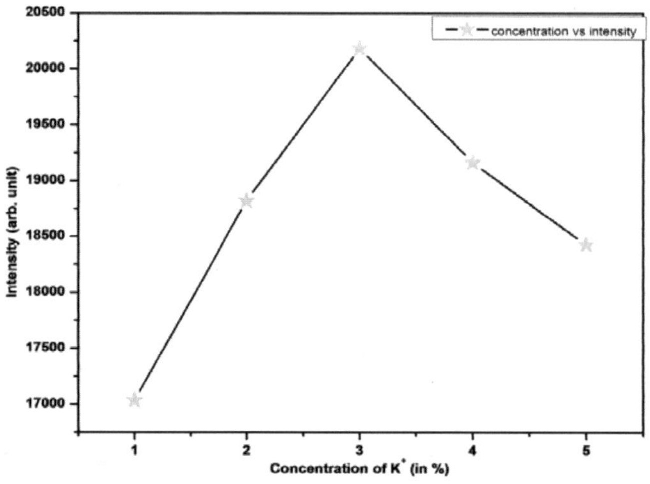

Figure 6. (a)Effect of concentrations of Dy^{3+} ions on PL intensity.

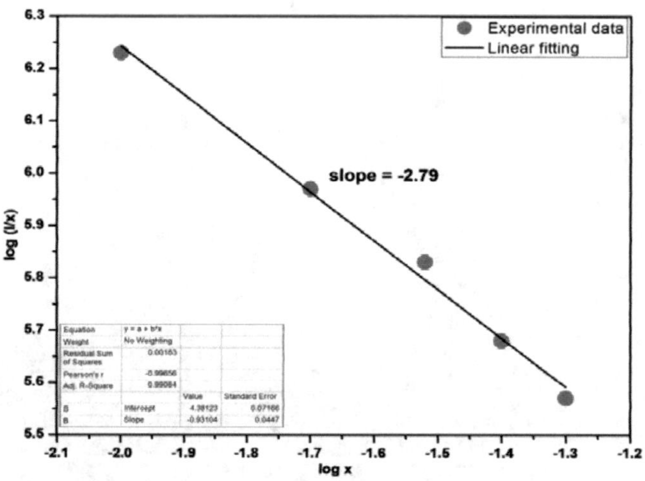

Figure 6. (b) Plot log (x) Vs log (I/x)

3.4 CIE coordinates

The CIE chromaticity coordinates of $Ba_{1.94}SiO_4$: $_{0.03}Dy^{3+}$, $_{0.03}K^+$ powders are calculated (see fig. 7). The values of chromaticity coordinates of the BOS powder were about (0.29, 0.31), which is very near to the typical value (0.33, 0.33) of white light. Therefore, this value directedthe high color stability and color saturation behavior of the phosphor. Hence, $Ba_{1.94}SiO_4$: $_{0.03}Dy^{3+}$ phosphor in presence of charge compensator ion (K^+) is a suitable candidates for solid state lighting and puts potentialaplicability in white light areas.

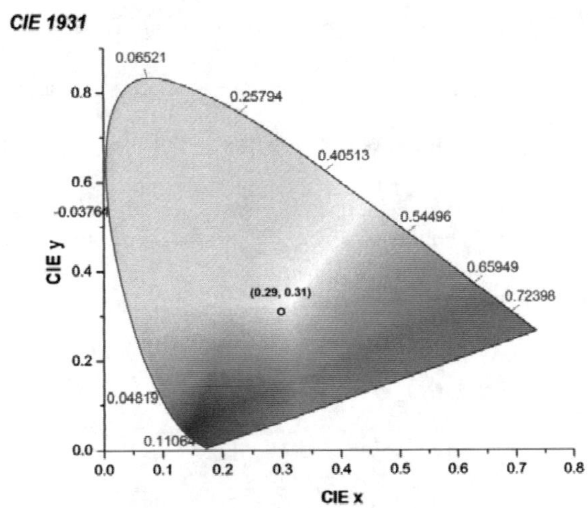

Figure 7. CIE coordinates of $Ba_{1.94}SiO_4$: $_{0.03}Dy^{3+}$, $_{0.03}K^{+}$phosphors.

4. Conclusion

Dysprosium (Dy^{3+}) doped alkaline (barium) ortho-silicate (Ba_2SiO_4) phosphors in the presence of K^+ ions were prepared via a high – temperature solid-state reaction route. The analysis of XRD patterns reveals the formation of single-phase crystal structure of the phosphor. The identified phase structure was orthorhombic, which is confirmed through JCPDS # 70-2113. All peaks are well-matched with this JCPDS file. Chemical compositions were identified by transmittance spectrum by FTIR analysis, which confirms the presence of Ba_2SiO_4: Dy^{3+}, K^+ phosphors. The maximum PL intensity was obtained for 3.0 mol% of Dy^{3+}. The two PL emission peaks occur at 476 nm and 573 nm related to $^4F_{9/2}\rightarrow{}^6H_{15/2}$ and $^4F_{9/2}\rightarrow{}^6H_{13/2}$ transitions when excited by 352 nm. The critical distance between dopants (Dy^{3+} ions) was calculated by using blasses equation and obtained 24Å. The Rc determines the dipole – quadrupole interaction between activator ions. K^+ ions are used as charge compensator ion; which enhances the luminescence intensity as well as crystallinity. Incorporation of K^+ ions enhances the PL intensity 1.75 times as compare to obtained with Dy^{3+} doping only.The CIE color coordinates were calculated (0.29, 0.31), which is very near to the standard value (0.33, 0.33). Thus, it is concluded that the synthesized phosphor is a fruitful candidate for white light applications (especially wLEDs).

Reference:

[1] Kumari Puja and Manam J. 2015, *RSC Adv.* **5** 107575.
[2] Kim C H et al. 2000 *J. Alloys Compounds* **311** 33
[3] Choe J Y et al. 2001 *Appl. Phys. Lett.* 78 3800.
[4] A D Deshmukh, S J Dhoble, S V Godbole, M K Bhide and D R Peshwe 2009 *Indian J. Phys.* **83** 423.
[5] Carreira J F C, Sedrine N B, Monteiro T and Rino L 2017 *J. Lumin.* **183** 251 – 258
[6] Yuan B, Song Y, Sheng Y, Zheng K, Hua Q, Xu X and Zou H 2014 *Powder Tech.* **253** 803
[7] Wei F and Q Jia 2015 *Superlattices and Microstructures* **82** 11-17
[8] Zhang L, Lu Z, Yang H, Han P, Xu N and Zhang Q 2012 *J. Alloys Compd.* **512** 5 –11
[9] Han J C, Hannah M E, Piquette A, Talbot J B, Mishra K C and McKittrick J 2015 *J. Lumin.* **161** 20–24
[10] Wang Z, Guo S, Li Q, Zhang X, Li T, Li P, Yang Z and Guo Q 2013 *Physica B* **411** 110–113.
[11] Lin L et al. 2017 *RSC Adv.* **7** 25685

.

[12] Ci Z, Sun Q, Qin S, Sun M, Jiang X, Zhang X and Wang Y 2014 *Phys. Chem. Chem. Phys.* **16** 11597

[13] Wang G Q, Gong X H, Chen Y J, Huang J H, Lin Y F, Luo Z D and Huang Y D 2014 *Optical Materials* **36** 1255–1259

[14] Tshabalala M A, Dejene F B, Pitale S S, Swart H C and Ntwaeaborwa O M 2014 *Physica B* **439** 126-129

[15] Dexter D L and Schulman J H 1954 *J. Chem. Phys.* **22** 1063–1071

[16] Tamboli Sumedha and Dhoble S J 2017 *Spectrochimica Acta Part A: Molecular and Biomolecular Spectroscopy* **184** 119–127

[17] Dabre KV and S J Dhoble 2015 *RSC Adv.* **5** 60409–60418

IOP Conf. Series: Materials Science and Engineering **798** (2020) 012011 doi:10.1088/1757-899X/798/1/012011

Photoluminescence behaviour of Gd doped barium hexaferrite

Jayashri Mahapatro and Sadhana Agrawal*

Department of Physics, National Institute of TechnologyRaipur, Raipur 492010, India

*Corresponding author's e-mail address: sagrawal.phy@nitrr.ac.in

Abstract. This paper reports synthesis of Barium Hexaferrite with different concentration of Gadolinium ion (0-2 mol %). Structural analysis of the prepared powders was done by X-ray diffraction pattern which reveals formation of hexagonal structure with space group $p6_3/mmc$. The FTIR and EDX spectra depict formation and composition of Barium Hexaferrite respectively. Agglomeration was observed from the SEM image of the prepared sample. Photoluminescence studies of the prepared samples show intensified PL emission peaks at 658 nm under ultraviolet (UV) excitation at 230 nm due to the optical transition $^6G_{7/2}$ to $^6P_{5/2}$ which arises because of the defects produced by doping of Gadolinium ion into the host lattice. Due to cross relaxation process PL intensity decreases after 1.5 mol% of doping concentration. CIE (commission International de) coordinates and CCT (correlated colour temperature) were also determined for the prepared samples which indicate that the powder samples show intense red cool emission.

1. Introduction

Hexaferrites having general formula $MFe_{12}O_{19}$ where (M=Pb, Sr, Baetc) were used in microwave devices, magnetic recording media, permanent magnetic materials because of their outstanding properties like high coercivity, high saturation magnetization, high curie temperature and good chemical stability and large uniaxial magnetic anisotropy [1]. Depending upon the crystal structure Hexaferrites were classified into 6 types such as X-type, W-type, M-type, Z-type and U-type [2].

Barium Hexaferrite (BaM) belongs to M-type hexaferrite with magnetoplumbite structure. Crystalline BaM particles have space group $p6_3/mmc$. The unit cell of the barium hexaferrite consists of 64 ions which are distributed in 11 different symmetric sites. Fe^{3+} ions are distributed in 1 trigonal bipyramidal (2b), 1 tetrahedral ($4f_1$), and 3 octahedral (12k, 2a, $4f_2$) sites. Barium Hexaferrites has been studied by a large number of researchers worldwide because of their superior magnetic properties and excellent electrical properties like low eddy current losses and high resistivity [3].

Rare earth (RE) activated materials now a day's grabbing attention because of their enhanced optical properties such as greater radiation stability, energy output, colour rendering index and enhanced luminescence intensity [4].

Keeping in view, the synthesis of Gadolinium doped Barium Hexaferrite by solution combustion technique and it's structural, morphological and photoluminescence behaviour has been reported in this paper.

IOP Conf. Series: Materials Science and Engineering **798** (2020) 012011 doi:10.1088/1757-899X/798/1/012011

2. Material and Synthesis

Barium Hexaferrite samples doped with different concentration of Gadolinium were synthesized by solution combustion technique. Barium nitrate, Ferric nitrate and Gadolinium nitrate all are supplied by sigma Aldrich with more than 99% purity were taken in appropriate stoichiometric amount and dissolved in deionized water. Further, the solution was placed on magnetic stirrer and citric acid as fuel added drop wise with continuous stirring. When temperature of the magnetic stirrer reaches 80^0c the self ignitation process starts and we get the nano porous powder. The porous powder was ground finely and placed in a high temperature furnace for calcination at 1200^0c for 3 hour.

3. Experimental Results and Discussion

3.1 Structural Analysis

The crystallographic characteristics of the synthesized powder samples were determined by the XRD pattern obtained from the PAN Analytical X-Pert Powder Diffractometer with Cu-K alpha radiation (λ=1.54A^0). Figure 1 represents XRD pattern of undoped and Gd^{3+} ions doped Barium Hexaferrite which confirms that the prepared samples have hexagonalstructure withspace group $p6_3$/mmc. The average particle size of the prepared samples calculated from Scherrer equation is 28.17nm.

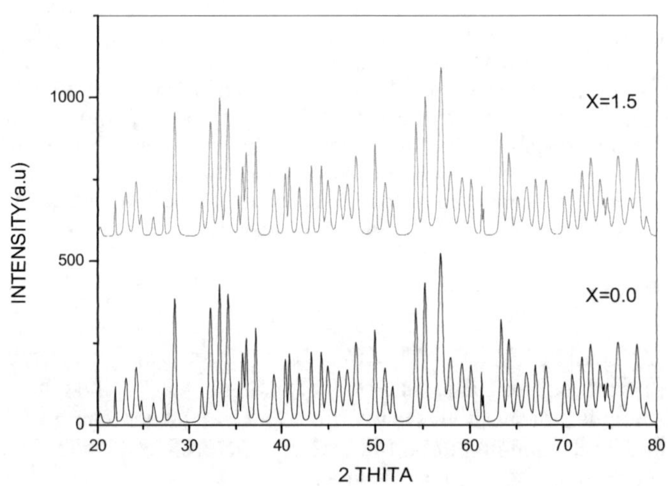

Figure 1. XRD pattern of undoped and Gd doped (1.5mol %) Barium Hexaferrite

3.2 Morphological Analysis

Figure 2 shows SEM image of prepared sample $BaFe_{12-x}Gd_xO_{19}$ (x=1.5mol %) which was taken by ZEISS Ultraplus SEM Analyzer. It is clear that the particles were agglomerated with irregular shape [5].

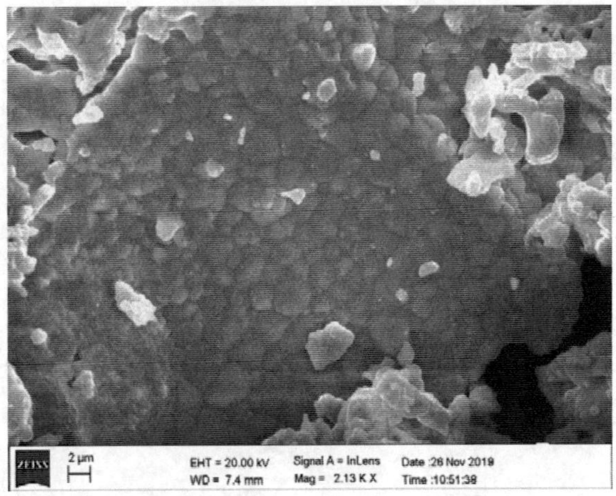

Figure 2. SEM image of Gd doped Barium Hexaferrite

3.3 Elemental Analysis

EDXS of the prepared sample is as shown in figure 3. The intense peak corresponding to Oxygen (O), Iron (Fe), Barium (Ba) and Gadolinium (Gd) confirms that all the starting materials are present in the prepared sample. The atomic and weight percentage of the starting materials are given in table 1.

Table 1. EDX Spectrum analysis data of the prepared sample.

Element	Weight%	Atomic %
O	31.38	65.48
Fe	50.30	30.07
Ba	18.03	4.38
Gd	0.29	0.06

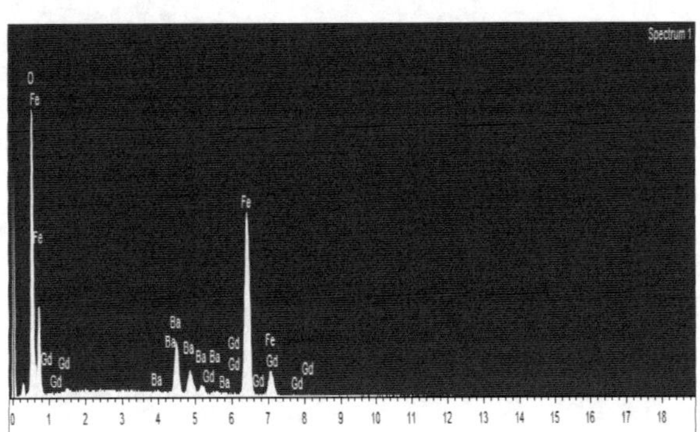

Figure 3. EDX Spectrum of the prepared sample.

IOP Conf. Series: Materials Science and Engineering **798** (2020) 012011 doi:10.1088/1757-899X/798/1/012011

3.4 FTIR Analysis

FTIR analysis has been performed to confirm the formation of phase of the prepared samples. Figure 4 represents FTIR spectra of the prepared samples which were recorded by Bruker FTIR spectrometer. The band corresponds to 500 cm^{-1} to 400cm^{-1} confirms the formation of Barium Hexaferrite [6].

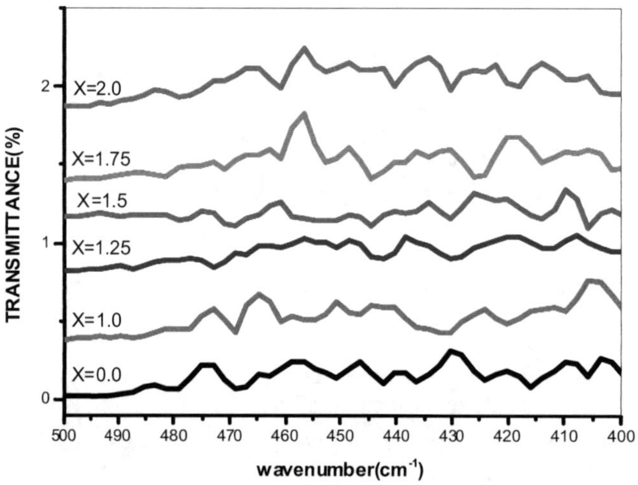

Figure 4. FTIR Spectra of the prepared samples.

3.5 Photoluminescence (PL) Studies

Photoluminescence emission spectra of Gd^{3+}ions doped Barium Hexaferrite samples were recorded by Shimadzu RF-5301PC spectrofluorophotometer. Figure 5 represents PL emission spectra of Gadolinium doped Barium Hexaferrites under UV excitation of 230nm. The most intense peaks were observed at 658nm due to the optical transition $^6G_{7/2}$ to $^6P_{5/2}$ which were originated from the defects level of Gd^{3+} ion. From the figure it is clear that with increasing the concentration of Gd^{3+} impurity upto 1.5 mol %, the intensity of the PL emission increases after that PL emission intensity gradually diminishes because of the cross relaxation process which arises from the non radiative transition between the dopant ions[7].

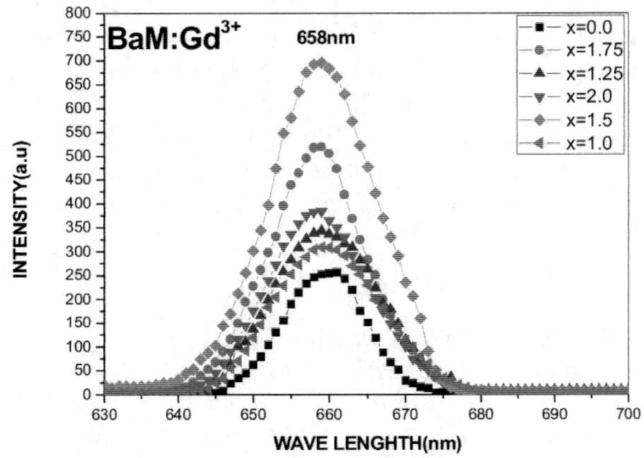

Figure 5. PL emission spectra of the prepared samples

3.6 Chromaticity Characterization

The CIE (Commission International de) coordinates and CCT (correlated colour temperature) of 1.5 mol% Gd^{3+} ions doped barium hexaferrite for which the PL intensity is maximum is shown in the table 2 to determine the accurate colour and nature of the prepared sample [8].

Table 2. Chromaticity coordinates of 1.5mol% Gd^{3+}doped Barium Hexaferrite.

Sample	Concentration of Gd^{3+}	X coordinate	Y coordinate	CCT(K)
$BaFe_{12}O_{19}$:Gd^{3+}	1.5 mol %	0.730	0.270	4605

It is clear from the figure that the sample shows deep red emission and the evaluated CCT value indicates that the prepared sample belongs to cool emission.

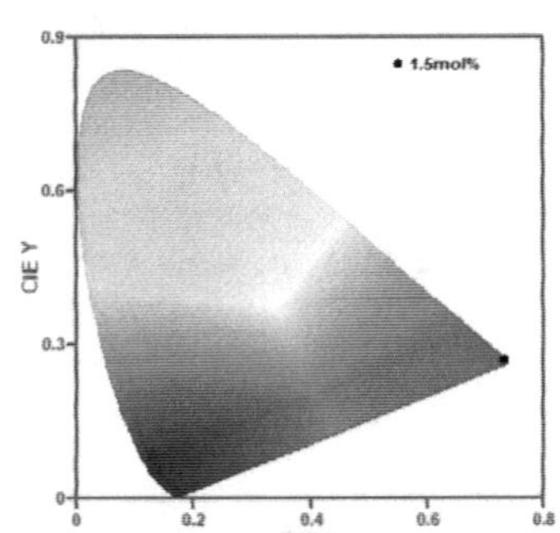

Figure 6. CIE Diagram of 1.5 mol% Gd doped Barium Hexaferrite.

4. Conclusion

Barium Hexaferrites doped with Gd^{3+} ions were prepared by solution combustion technique. Particles of the prepared samples have hexagonal structure with space group p63/mmc as confirmed by XRD pattern analysis. Gd^{3+} ions were successfully incorporated into the host lattice as confirmed by EDX analysis. Photoluminescence spectroscopic studies reveal that $BaFe_{12-x}Gd_xO_{19}$(x=0 to 2 mol %) show PL emission spectra at 658nm corresponding to $^6G_{7/2}$ to $^6P_{5/2}$ optical transition of gadolinium ion. The chromaticity investigations confirm that the prepared samples can be applied to various display devices because of their cool red emission.

References

[1] Sharma M and Kashyap S C 2019 *J. Ceramint.* **45** 11226.
[2] C Pullar R C 2012 P*rogress in Material Science* **57** 1191.
[3] Malana M A and Qureshi R B 2016 *J. Ceramint.* **42** 2686.
[4] Katyayan Shambhavi and Agrawal Sadhana 2017 *J Mater Sci: Matter Electron* **28** 18442.
[5] Kale Sanjay and Thube Dilip 2017 *Int. Res. J. Science & Engineering* A1 : 87.
[6] Packiaraj G, Nital Panchal, Jotania R B 2010 *Journal of Biomedical and Bioengineering* **1** 01.
[7] Wegh R T and Meijerink A 1996 *Acta Physics Polonica A* **90** 333.
[8] Verma Tarkeswari and Agrawal Sadhana 2018 *J Mater Sci: Materials in Electronics* **29** 7832.

Structural, morphological and photoluminescence behaviour of $Sr_{10-x}Ba_x(PO_4)_6(OH)_2$ hydroxyapatite

Ananya Rout, Sadhana Agrawal*

Department of Physics, National Institute of Technology Raipur, 492010, India

*Corresponding author's e-mail address: sagrawal.phy@nitrr.ac.in

Abstract. This paper reports synthesis of $Sr_{10-x}(PO_4)_6(OH)_2$ with different concentration of Ba^{2+} ion by co-precipitation method. To enhance the intensity of photoluminescence properties of strontium hydroxyapatite, barium (Ba^{2+}) ions are incorporated into hydroxyapatite structure. XRD pattern analysis shows single phase hexagonal structure along with space group P63/m. SEM image depicts the deposition of apatite layer on the surface of agglomerated spheroidal grains. FTIR analysis reveals the presence of functional group (PO_4) in the synthesized hydroxyapatite. Photoluminescence spectra shows an intense green light emission at 523 nm under 230 nm UV light irradiation.

1. Introduction

Hydroxyapatite containing phosphate group have general formula $A_{10}(PO_4)_6X_2$, where A site belongs to the family of larger cations of alkali, alkaline earth or lanthanide metal (Ca^{2+},Pb^{2+},Ba^{2+},Sr^{2+},Mn^{2+}etc.) and X site consists of anions of alkali halides (F^-,Cl^-,O^{2-} etc.) respectively.The crystal structure of hydroxyapatite was given by Mehmal and Naragszabo [1] in 1930. Its structure is hexagonal close packed bunch of spheres with space group $P6_3/m$, in which each sphere specify tetrahedral ion with octahedral interstial site that form channel between cationic and anionic sites respectively. It is formed by the occupancy of two cationic sites A(I) and A(II). Four A_4(I) are fixed at the centre of narrow channels (4f sites), six A_6(II) are fixed around large channels (6h sites) [2].

In the midst of the family of apatite, hydroxyapatite with strontium as a host have captivated the interest of scientist and researchers in the field of environmental science, optoelectronics, and medical science [3]. Ba^{2+} ion doped hydroxyapatite have been used as WLEDs because of its supremacy in high brightness, tensile design, strength and eco-friendliness [4]. In addition it shows several beneficial properties such as long fluorescence lifetime, high quantum yield, good luminescence efficiency and sharp emission peaks [5].

Apart from various wet and dry synthesis methods, co-precipitation is a wet synthesis technique which is used for the synthesis of nano dimension powder due to its reliability in controlling the stoichiometric ratio of the material [6]. Co-precipitation method have several advantages like it requires less time, temperature, solvents etc. [7].

2. Experimental Method

$Sr(NO_3)_2.4H_2O$, $Ba(NO_3)_2.4H_2O$, $(NH_4)_2HPO_4$ were taken as starting material. All the materials were taken in stoichiometric amount, mixed in 50ml of distilled water to form desired solution. The solution was placed on magnetic stirrer at 400 rpm for 1 hour. NH_4OH was added to the solution during stirring to maintain the P_H value of solution at 10 [8]. After continuously stirring the precipitate was formed, which was filtered and washed several times by using ethanol. The obtained precipitate was placed in oven at 80^0C for 3 hours, the evaporated precipitate was ground for 30 min and kept in high temperature furnace for calcination at 900^0C for 4 hours (See figure1).

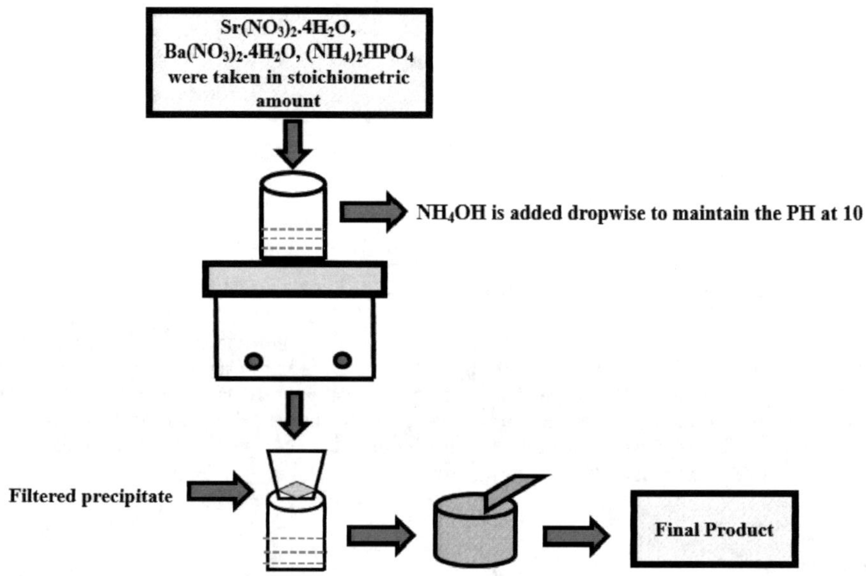

Figure 1. Co-precipitation method

3.Results and discussions

3.1Crystallography Analysis

The crystal structural analysis of $Sr_{10-x}Ba_x(PO_4)_6(OH)_2$ samples as shown in figure 2 were taped by using 3kW PANalytical-XPERT-Powder X-ray diffractometer. The analysis of PXRD patterns confirm the formation of single phase hexagonal closed packed structure with space group $P6_3/m$ [9]. The average crystallite size of the particle calculated by Debye Scherer formula is 31nm.

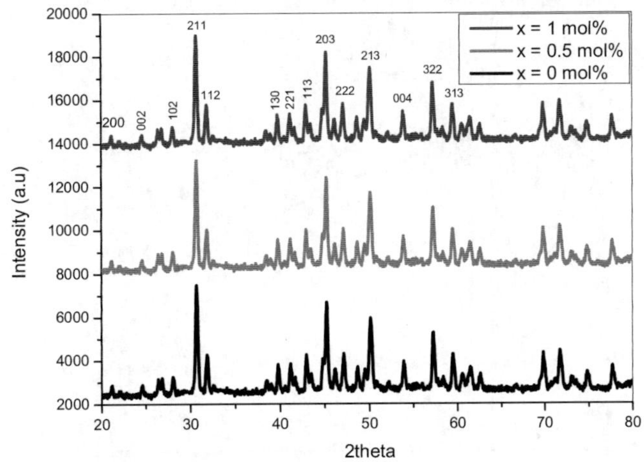

Figure 2. XRD patterns of $Sr_{10-x}Ba_x(PO_4)_6(OH)_2$ (x = 0,0.5,1 mol%) hydroxyapatite

3.2 Scanning Electron Microscopy (SEM) Study

SEM micrographs for the sample $Sr_{10-x}Ba_x(PO_4)_6(OH)_2$ (x = 1 mol%) were taken by using SEM analyser - ZEISS Ultraplus is shown in figure 3. It depicts the deposition of apatite layer on the surface of spheroidal grains [10].

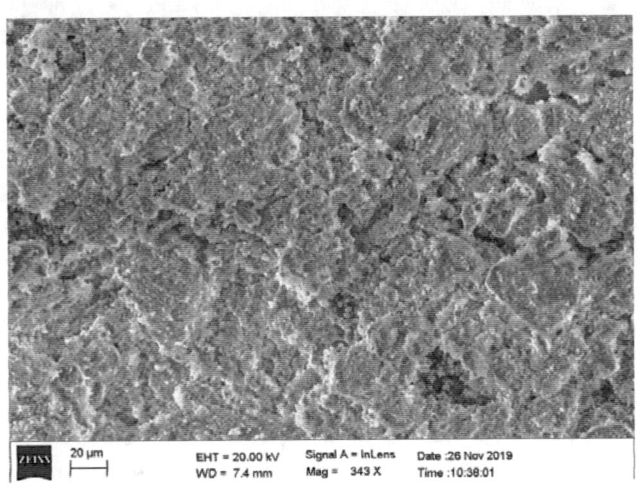

Figure 3. SEM image of $Sr_{10-x}Ba_x(PO_4)_6(OH)_2$ (x = 1 mol%) hydroxyapatite

3.3 Fourier Transform Infrared (FTIR) Spectroscopy

The FTIR spectrum of $Sr_{10-x}Ba_x(PO_4)_6(OH)_2$ (x = 0,0.5 and 1 mol%) were recorded by using Bruker FT-IR Spectrometer in the range 500-4000 $cm^{-1}$is shown in figure 4. The doublet bands obtained at 560 cm^{-1} and 593 cm^{-1} are symmetrical bending modes (v_2) of PO_4^{3-} functional group [11]. Further because of symmetrical stretching (v_1) mode O – P, the bands were obtained at 854cm^{-1} and 941 cm^{-1}. The observed bands gathered at 1021 cm^{-1} and 1075 cm^{-1} corresponds to the unsymmetrical stretching

IOP Conf. Series: Materials Science and Engineering **798** (2020) 012012 doi:10.1088/1757-899X/798/1/012012

(ν_3) mode of PO_4 group. Uncertain bands obtained at 2850 cm^{-1} and 2924cm^{-1} are because of O-H stretching [12]. Presence of OH group is observed by the band obtained at 3594 cm^{-1}.

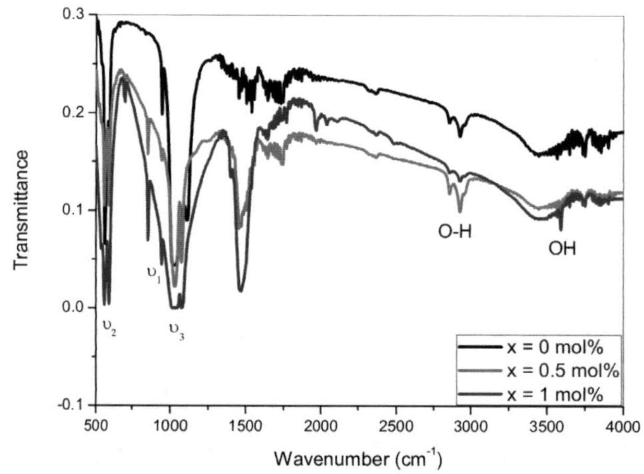

Figure 4. FTIR spectra of $Sr_{10-x}Ba_x(PO_4)_6(OH)_2$ (x = 0,0.5 and 1 mol%) hydroxyapatite

3.4 Photoluminescence (PL) Spectroscopy

Figure 5 display the photoluminescence spectra for $Sr_{10-x}Ba_x(PO_4)_6(OH)_2$ (x = 0,0.5 and 1 mol%) phosphors, obtained by using RF-5301 PC spectrofluorophotometer. The broad emission peak obtained at 523 nm in the green region corresponds to $^6S_{1/2}$-$^6P_{1/2}$ transition [13].

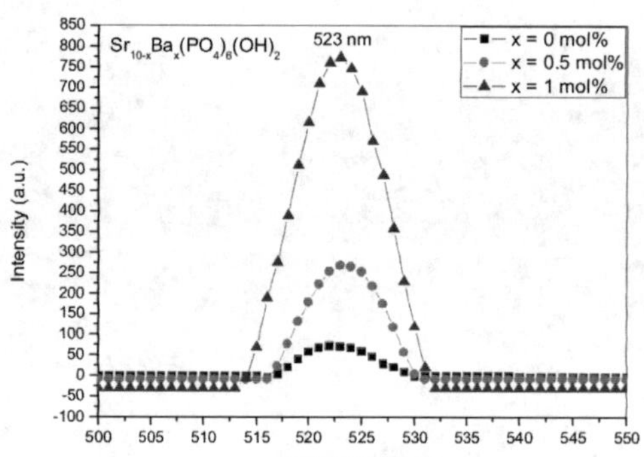

Figure 5. PL spectroscopic spectra of $Sr_{10-x}Ba_x(PO_4)_6(OH)_2$ (x = 0,0.5 and 1 mol%)

The CIE chromaticity coordinates (1931) of $Sr_{10-x}Ba_x(PO_4)_6(OH)_2$ (x = 1 mol%) was done by GoCIE software shows a good understanding with NTSC (National television standard committee) values. The broad emission peak obtained at 523 nm in the green region as shown in the figure 6 with coordinates value CIE X = 0.3 and CIE Y = 0.7.

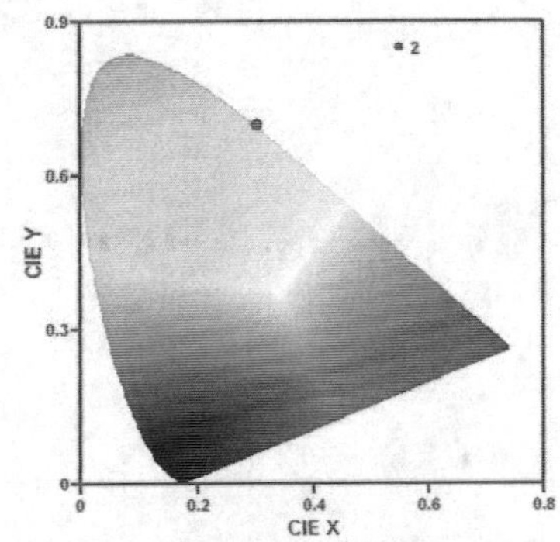

Figure 6. CIE Chromaticity diagram for $Sr_{10-x}Ba_x(PO_4)_6(OH)_2$ (x = 1mol%)

4.Conclusion

The Ba^{2+} ions doped $Sr_{10-x}(PO_4)_6(OH)_2$ hydroxyapatite were prepared by co-precipitation technique. PXRD analysis explains the design hexagonal close packed structure with space group $P6_3/m$. SEM image depicts the formation of apatite layer over the spheroidal grains. FTIR spectroscopy shows the asymmetric and symmetric stretching and bending modes of phosphate (PO_4) group. PL spectroscopy reveals that the emission spectrum analogous to $^6S_{1/2}$-$^6P_{1/2}$ transition consist of an intense line at 523 nm in the green region. CIE coordinate values show a keen agreement with NTSC value showing a cool emission in green region.

References

[1] Ravinder K and Kanchan Lata S 2016 *Advances in Applied Science Research* **7** 66
[2] ZhihongY, Tao G, Yuanlei Z, Shenggui M, Mingli Y and Piheng C 2017*The Royal Society of Chemistry* **73** 0310
[3] Guo C, Luan L, Ding X, Zhang F, Shi G, Gao F and Liang L 2009 *Appl. Phys. B: Lasers Opt.* **95** 779
[4] Kottaisamy M, Jagannathan R, Jeyagopal P, Rao R and Narayanan R 2001 *J. Phys. D: Appl. Phys.* **27** 2210
[5] Zhang W, Shen Y, Pan H, Lin K, Liu X, Darvell B, Chang J, Deng L and Wang D 2011 *Acta Biomater.***7** 800
[6] Maji B, Jena H, Krishnaiah M and Asuvathraman R 2016 *J. Therm. Anal. Calorim.***124** 857
[7] Kai Li, Kunhuan He, Quanwen Li, Xia B, Wang Q and Zhang Y 2018 *Chem. Res. Chin. Univ.* **34** 700
[8] Rabiee S, Moztarzadeh F and Solati-Hashjin M 2010 *J Molstruct* **969** 172
[9] Homg S, Hsu M and Liu T C 2004 *Journal of Magnetism & Magnetic Material* **279** 401
[10] Benito G., Morales M P, Requena J, Raposo V, Vazquez M and Moya J S 2001 *Journal of Magnetism & Magnetic Material* **234** 65
[11] Albert J, Bergman C and Fischer H 2014 *Ceram. Int.* **409** 195
[12] Ruixue S, Kezheng C, Xiangfeng W, Dandan Z and Zhenzong S 2013 *Cryst. Eng. Comm.***15**3442
[13] Kivrak N and Taş A C 1998 *Journal of the American Ceramic Society* **81** 2245

IOP Conf. Series: Materials Science and Engineering **798** (2020) 012013 doi:10.1088/1757-899X/798/1/012013

Development of ZnS:Mn mechanoluminescent film for impact sensor

Piyush Jha* and Ayush Khare

Department of Physics, National Institute of Technology, GE Road, Raipur – 492 010, India

*Corresponding author's e-mail address: piyushjha22@rediffmail.com

Abstract: This paper reports the elastico-Mechanoluminescence (EML) induced by the impact of a steel ball onto the composite film of ZnS:Mn phosphor. The film is prepared by pasting the mixture of ZnS:Mn phosphor and optical resin on polycarbonate substrate. The EML induced by impulsive excitation using a steel ball has been measured. The ML intensity linearly increases with square of impact velocity of the steel ball. The t_m (temperature corresponding to maximum ML intensity) does not change with increasing impact velocity. The ML intensity is observed to recover after every impact of steel ball (height of fall = 20 cm) and does not require any irradiation. The outcome of present investigation is expected to be useful in extracting information about ML when a projectile with small contact area, such as steel ball makes an impact on the ML sensitive composite film.

1.Introduction

Pure zinc sulfide (ZnS) is a non-centric [1] and mechanoluminescent [2] material emitting the lightning lines characteristics of fracto-mechanoluminescence (FML). When doped with small percentage of strongly luminescent transition metal ions, such as copper, manganese [3-7] and even silver [4], zinc sulfide becomes strongly mechanoluminescent under deformation exhibiting a spectrum most similar to the electroluminescence (EL) spectrum caused by the dopant. ZnS doped with Mn is found to drastically enhance the mechanoluminescence (ML) properties [8].

ML or triboluminescence (TL) has been known for almost 400 years. One of the earliest references of this phenomenon was given by Francis Bacon's in his writing "Advancement of Learning" in 1605. ML is categorized in three ways: elastico-mechanoluminescence (EML), plastico-mechanoluminescence (PML) and FML [9, 10]. PML and FML are destructive classes of ML, while EML is a non-destructive technique and is recoverable. In reality, before 1950 there were reports only on FML [11] in which the ML material was used for one time. So, it was realized to develop ML materials, which could be used for non-destructive applications. Recently, ZnS:Mn has been rigorously investigated for applications in the field of ML.

Xu et al. [12] demonstrated the EML from impact of a steel ball on ZnS:Mn film fabricated by ion plating or sputtering method. They reported that the prepared film was very sensitive to mechanical stimuli, and could act as a self-diagnosis material to detect mechanical stress remotely by visible light emission.

Recently, Jeong et al. [13] reported ML flexible composite films consisting of ZnS:Cu and polydimethylsiloxane (PDMS). This film was quite durable for repeated load and could bear repeated mechanical stress of over ≈100000 cycle. In this film, the decrease in intensity after ~ 100 000 cycles was only ~35%; thus, maintaining 65 % intensity of the initial ML.

Wang et al. [14] gave an outstanding example of EML to the scientific community in 2015. They made a flexible pressure sensor matrix (PSM) using ZnS:Mn, polyethylene terephthalate and ethylene–vinyl acetate, which were very useful in hand writing mapping. This flexible PSM recorded pressure/force applied at each pixel during signing.

Literature suggests that limited studies have been carried out on the EML of ZnS:Mn film induced by the impact of the moving steel ball till date [12, 13]. Therefore, there it is imperative to study EML produced by impact of the steel ball in more detail. In last three decades, many EML materials, such as $Sr_3Al_2O_6$:Eu,Dy, $SrAl_2O_4$:Eu,Dy, $CaAl_2SiO_7$:Ce, $SrAMgSi_2O_7$ (A = Ca, Sr, Ba), $CaZr(PO_4)_2$:Eu, CaZnOS:Mn, etc. showing potential for light sources, sensors, colored displays, etc. [15-28] have been studied and developed [29-35]. Among other ML materials, ZnS:Mn is a very cheap in preparation, easily available, highly intense [8] and reproducible [12, 29].

The present paper reports the EML induced by impact of a steel ball on the ML of a film prepared by mixing ZnS:Mn with optical resin on polycarbonate substrate. The outcome of this study is useful to develop impact sensor where projectile with small contact area, such as steel makes an impact on the ML sensitive composite film.

2.Experimental

The starting materials ZnS, $MnCl_2$, and NaCl were used for the preparation of microcrystalline ZnS:Mn phosphor. The particular amount of activator in the form of$MnCl_2$wasmixed with the appropriate amount of ZnS. 4 % NaCl was mixed into the mixture using a pestle and a mortar. The so prepared charge was transferred to a clean silica boat and fired at 1150°C for 1 h in nitrogen atmosphere. After combustion, the phosphor was taken out of the furnace and cooled to the room temperature (RT). The synthesized ZnS:Mn sample was cleaned for two to three times with distilled water and dried again.

Thereafter, the powder sample is mixed with optical resin in the ratio 1:1 and deposited on Lexan polycarbonate substrate using doctor blade method [36]. For the measurement of EML, a small ball having mass 5.02 g and diameter 1.04 cm was dropped from different heights (10 cm - 30 cm).

The photoluminescence (PL) spectrum at an excitation wavelength of 365 nm was recorded on Carry eclipse fluorescence spectrophotometer.

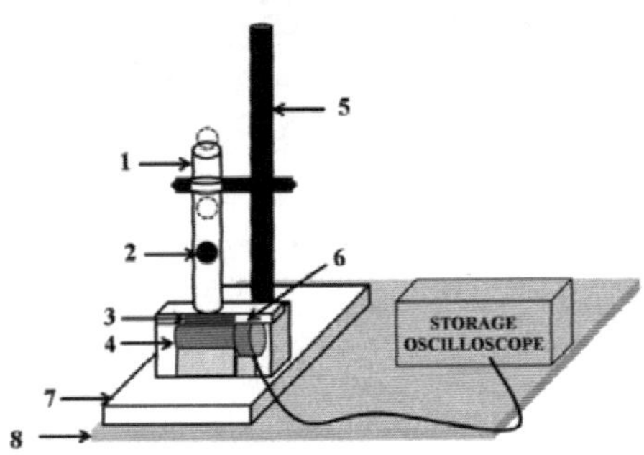

Figure 1. Schematic diagram of the experimental arrangement used for measuring the time dependence of ML in crystals (1-PVC pipe, 2- steel ball, 3-mechanoluminescent film, 4-photomultiplier tube, 5- stand, 6- sample holder, 7-wooden block, 8-iron base mounted on a table).

The EML was produced by dropping steel ball on ZnS:Mn film placed on the sample holder. The schematic diagram of experimental set-up employed for EML measurement is shown in Figure 1. The light produced due to ML and transmitted through the transparent polycarbonate substrate was fed to an RCA-931A photomultiplier tube (PMT) placed just below the sample holder. The response time of PMT system was nearly 5 μs. The output of the PMT was then fed to a storage oscilloscope [27]. All the experiments were carried out at RT.

3. Results and Discussions

Figure 2 illustrates the time dependence of EML intensity for ZnS:Mn film at different impact velocities of steel ball. After the impact of a free-falling ball, initially the compression takes place at a faster rate, slows down in middle phase, and lastly goes into the elastic region. Correspondingly, as seen in Figure 2, initially the EML intensity increases with time, attains a peak value at a particular time t_m, and finally follows decay for longer time. In the time vs EML plot, only one peak was observed and the fast and slow decay times are recorded to be 0.511ms and 1.028ms, respectively[37].

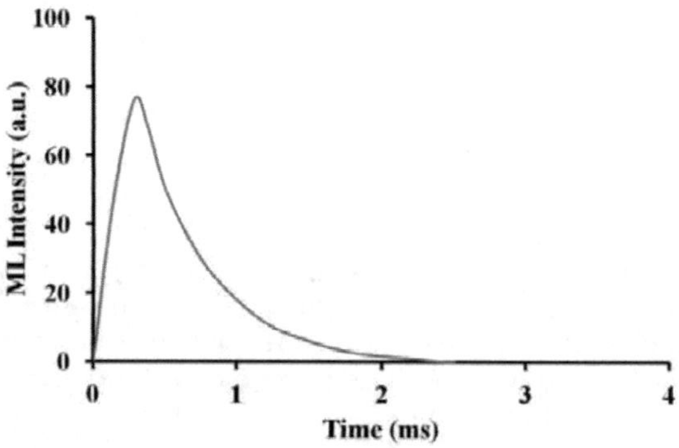

Figure 2. Time dependence of the ML intensity for ZnS:Mn film during impulsive excitation of the sample.

The steps involved in the EML emission from ZnS:Mn crystal are as follows [31]: (i) when the steel ball hits the ZnS:Mn film, the ML emission results from the contact area only (ii) piezoelectric field is produced because the crystal – structure of ZnS is non-centrosymmetric [1] (iii) due to the piezoelectric field, band bending occurs as a result of which de-trapping of electrons from filled-electron traps to the conduction band increases (iv) non-radiative emission of energy during the electron-hole recombination (v) release of energy during electron-hole recombination and excitation of Mn^{2+} ions, and (vi) generation of light during the de-excitation of excited Mn^{2+} ions.

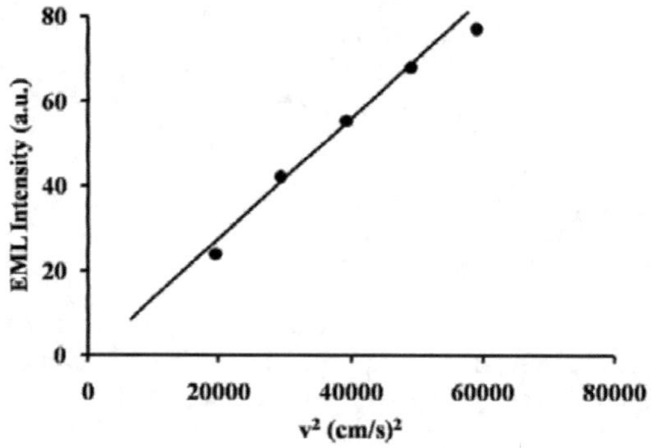

Figure 3. Variation of ML intensity with the square of impact velocity for ZnS:Mn film.

Figure 3 presents the variation of EML intensity with the square of impact velocity of the steel ball (v^2). The linear relation is found between EML intensity and square of impact velocity of the steel ball. The EML intensity was measured by dropping the steel ball 10 times from each height. This finding establishes the suitability of present film to be developed as an impact sensor [27, 38].

The variation in EML intensity with successive number of dropping of steel ball from 20 cm height for ZnS:Mn film as shown in Figure 4. In case of FML of ZnS:Mn phosphor [39], the FML intensity is observed to diminish due to the decrease in newly created surface area with successive number of drops. But, in case of EML, the ML intensity is recovered after every impact of steel ball and does not require any irradiation. The recovery of EML intensity can be explained by the fact that charge carriers that lay in the vacant traps and are produced due to the detrapping caused by the piezoelectric field takes place by trapping of the charge carriers that drift in the field produced piezoelectric effect.

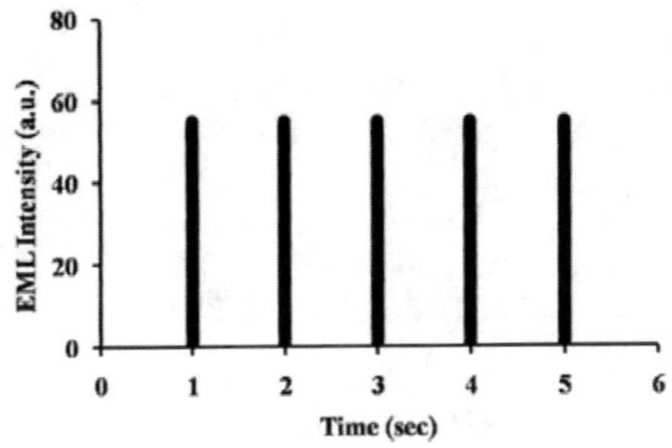

Figure 4. Dependence of ML intensity on repetitive number of droppings of a steel ball for ZnS:Mn film.

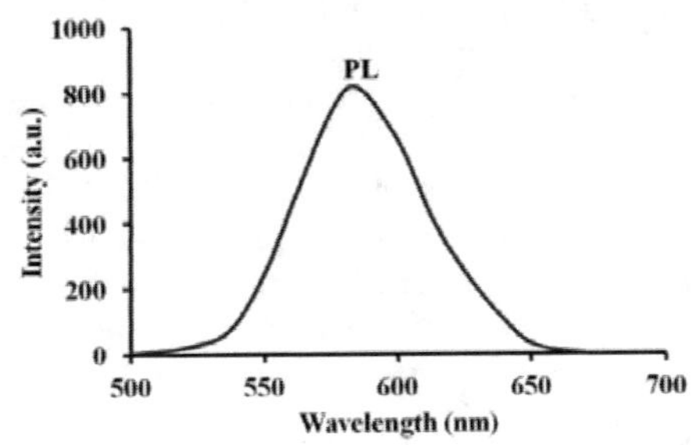

Figure 5. PL spectra of ZnS:Mn film (for PL λ_{ex} =365 nm).

Figure 5 presents the PL spectrum of ZnS:Mn film excited by the light of wavelength= 365 nm in the same plot. The observed peak is corresponding to wavelength = 584 nm, which is attributed to $^4T_1$-$^6A_1$ transition of Mn^{2+} ions [40].

4. Conclusions

The EML induced by steel ball on ZnS:Mn film has been investigated. When the ball impacts the ZnS:Mn film, initially the EML intensity increases with time, attains a peak value and then decreases. The linear relation is found between EML intensity and square of the impact velocity of the steel ball. This result reveals that in ML process, some part of the mechanical energy or piezoelectric energy is converted into the light energy, which is beneficial in the development of impact sensors. The EML intensity does not change significantly with number of droppings (up to 5 drops) of the steel ball. The PL spectrum is found to peak at 584 nm, which is attributed to $^4T_1$-$^6A_1$ transition of Mn^{2+} ions.

Acknowledgement

The authors acknowledge the financial support received from Science and Engineering Research Board (SERB), a statutory board under Department of Science and Technology, Government of India (Grant No. PDF/2018/000335) under National Post-Doctoral Fellowship (N-PDF) scheme.

References

[1] Scarmozzino R 1970 *Lett. Nuovo Cim.* **4** 825
[2] Bredhikhin S I and Shmurak S Z 1974 *JETP Lett.* **19** 367
[3] Chandra B P, Dubey P K and Datt S.C. 1986 *Phys. Stat. Sol. (a)* **97** K59
[4] Meyer K, Orbikat D and Rossberg M 1970 *Krist. Tech.* **5** 5
[5] Scarmozzino R, 1971 *Sol. St. Commun.* **9** 1159
[6] Alzetta G, Chudacek I and Scarmozzino R 1970 *Phys. Stat. Sol. (a)* **1** 775
[7] Vardanyan R A, Veselko S G and Kirakosyan G G 1989 *Sov. Phys. Sol. State* **31** 24
[8] Xu C N, Watanabe T, Akiyama M and Zheng X G 1999 *Mater. Res. Bull.* **34** 1491
[9] Chandra B P 1998 *Luminescence of Solids*, ed D. R. Vij (Plenum Press, New York) p. 36
[10] Chandra B P 2011 *Mechanoluminescent Smart Materials and their Applications*, in *Electronic and Catalytic Properties of Advanced Materials*, ed A. Stashans, S. Gonzalez and H. P. Pinto, (Transworld Research Network, Trivandrum, Kerala, India), p.1
[11] Jha P and Chandra B P, 2014 *Lumin.* **29** 977
[12] Xu C N, Watanabe T, Akiyama M and Zheng X G 1999 *Appl. Phys. Lett.* **75** 1236
[13] Jeong S M, Song S, Lee S K and Choi B 2013 *Appl. Phys. Lett.* **102** 051110-1
[14] Wang X, Zhang H, Yu R, Dong L, Peng D, Zhang A, Zhang Y, Liu H, Pan C and Wang Z L 2015 Advanced Materials **27** 2324
[15] Akiyama M, Xu C N, Nonaka K and Watanabe T 1998 *Appl. Phys. Lett.* **73** 3046
[16] Xu C N, Watanabe T and Akiyama M 1999 *Appl. Phys. Lett.* **74** 2414
[17] Xu C N, Liu Y, Akiyama M, Nonaka K and Zheng X G 2000 *SPIE Proc.* **4220** 344
[18] Terasaki N, Xu C N, Imai Y and Yamada H 2007 *Jpn. J. Appl. Phys.* **46** 2385
[19] Jeong S M, Song S, Lee S K and Choi B 2013 *Appl. Phys. Lett.* **102** 051110-1
[20] Akiyama M, Xu C N, Matsui H and Nonaka K 1999 *Appl. Phys. Lett.* **75** 2548
[21] Zhang H, Terasaki N, Yamada H and Xu C N 2009 *Jpn. J. Appl. Phys.* **48** 04C109-1
[22] Zhang J C, Xu C N and Long Y Z 2013 *Opt. Exp.* **21** 13699
[23] Zhang J C, Xu C N, Kamimura S, Terasawa Y, Yamada Hand Wang X 2013 Opt. Exp. **21** 1276
[24] Jeong S M, Song S, Lee S K and Ha N Y, 2013 *Adv. Mater.* **25** 6194
[25] Jeong S M, Song S, Joo K, Kim J, Hwang S H, Jeong J and Kim H 2014 *Energy Environ. Sci.* 73338
[26] Xu C N, Yamada H, Wang X and Zheng X G 2004 *Appl. Phys. Lett.* **84** 3040
[27] Jha P and Chandra B P 2013 *J. Lumin.* **143** 280
[28] Jha Piyush 2016 *Lumin.* **31** 1302
[29] Chandra B P, Xu C N, Yamada H and Zheng X G 2010 *J. Lumin.* **130** 442
[30] Chandra B P 2011 *J. Lumin.* **131** 1203
[31] Chandra V K, Chandra B P and Jha P 2013 *Appl. Phys. Lett.* **102** 241105-1
[32] Chandra V K, Chandra B P and Jha P 2013 *Appl. Phys. Lett.* **103** 161113-1
[33] Chandra B P, Chandra V K and Jha P 2014 *Appl. Phys. Lett.* **104** 031102
[34] Chandra B P, Chandra V K and Jha P 2015 *Physica B* **461** 38
[35] Chandra B P, Chandra V K and Jha P 2015 *Physica B* **463** 62
[36] Berni A, Mennig M, Schmidt H, 2004 *Sol–Gel Technologies for Glass Producers and Users*, (Dordrecht: Kluwer Publications) p 89–92
[37] Chandra B P, Chandra V K, Jha P and Sonwane V D 2016 *Physica B* **491** 12
[38] Chandra B P, Chandra V K and Jha Piyush 2015 *Sens. Actuators A* **230** 83
[39] Chandra B P, Chandra V K, Jha Piyush, Pateria Deepti and Baghel R N 2016 *Lumin.* **32** 67
[40] Osipyan Y A, Petrenko V F, Zaretskii A V and Whitworth R W 1986 *Adv Phys.* **35** 115

IOP Conf. Series: Materials Science and Engineering **798** (2020) 012014 doi:10.1088/1757-899X/798/1/012014

Synthesis of blue fluorescent carbon dotsfor sensitive and selective detection of glucose in biological samples

Vinayak Sahu[1], Fahmida Khan[1*], Pradeep Kumar Dewangan[1], Komal Kashyap[1], Khilawan patel[1], Rama devi[1], Vipin Kumar Soni[1]and Sonalika Agrawal[1]
[1]National institute of technology, Raipur (C.G.), 492010, India

Email: vinayaksahu2208@gmail.com
*Corresponding author, Email:fkhan.chy@nitrr.ac.in

Abstract. This Article describes a technique for the preparation of blue fluorescent carbon dots (CDs) showing fluorescence emission at 450nm. CDs were prepared using ultra-sonication technique using citric acid and glycerol as precursors. Since glucose quenches the fluorescence of CDs, depending upon this fact, a fluorescent sensor for glucose determination has been established. Under optimized conditions, linear response in therange of 0.1-100μM was obtained for glucose concentration. Limit of detection is found to be 0.08μM. This method was promisingly used for the glucose determination in serum albumin samples and urine samples from different human beings.

Keywords
Fluorescence, Glucose, Carbon dots, UV-Visible, FTIR, HR-TEM.

1. Introduction

Glucose is an essential biomolecule which plays a vital role as a source of energy in cell function and regulation of metabolism. Glucose is an important metabolic intermediate of other complex biomolecules. Glucose level in urine and blood is a measure of human fitness states [1-3]. Investigation of urine and blood glucose level is very crucial because abnormal level of glucose can related to hyperglycaemia, hypertension,and visionproblem; increase the risk of heart disease and stroke, kidney disease, and diabetes (type-II) [4]. Diabetes mellitus, now a day a frequent disease occurs due to malfunction of pancreas and is spreaded worldwide. Reports of W.H.O. states that 347 million diabetic people were found in 2013 year and 3.8 million people leads to death every year, this number will be doubled in 2030 [5]. Blood sugar (glucose) levels both in diabetic human being and healthy people lies in the millimolar range, biological contains potential interfering ions and analogues , thus very sensitive method is required for detection of glucose in the low concentration range [6]. Several techniques for determination glucose have been existed like mass spectrometry [7], fluorescence [8], electrochemistry [9], Surface Enhanced Raman Scattering (SERS) [10], chemiluminiscence [11], and colorimetry [12].Out of these methods fluorescence based detection have attracted great attention due their low cost, good stability, easeness of handling, and excellent sensitivity. Alots of fluorescence sensors have been applied for determination glucose like organic dyes [13], nanoclusters [14], and Quantum dots [15].However, most of the existing methods require complex modification, more cost of experiment, unappreciable limit of detection, less sensitivity and not applicable for biological samples. So, to develop the fluorescent sensor with high sensitivity, lower limit of detection, having high selectivity, cost-effective method is of great significance. Quantum dots have been also used as fluorescent sensor but toxicity, high cost, and complex method of preparation, limits their applicability.Carbon quantum dots (CDs) having various advantages over QDs, like biocompatibility, water solubility,chemical inertness, extremely tunable photoluminescence and electrochemiluminiscence [16], That's why CDs are used now a days in the place of toxic QDs and organic dye mediated sensors.

In this study, fluorescent probes (CDs) have been synthesized without surface modification for glucose determination in human serum and urine samples. The Carbon dots were prepared using citric acid and glycerol as precursors using ultra-sonication. Synthesized fluorescent probe was successfully used for detection of glucose in the real samples.

2. Experimental section

2.1. Chemicals and reagents

Citric acid and glycerol were procured from Sigma Aldrich Pvt. Ltd. Mumbai. Glucose, ribose, maltose, fructose, xylose, galactose, lactose, arabinose were purchased from Himedia Chemical Pvt. Ltd. Standard solution of different metal ion were purchased from Merck Chemical Pvt. Ltd. During all the experimental procedures ultrapure water was used from Millipore Merck water purification system. All chemical employed are of AR grade, without any refinement.

2.2. Apparatus

HR-TEM images (High Resolution Transmission electron microscopy) were recordedfrom JEOL JEM-2100 transmission electron microscope (JEOL, Japan). UV-Visible spectra were measured using Thermo-Fischer spectrophotometer in the 200-600 nm of wavelength range.Fourier transform infrared spectra were recorded on Thermo-Fischer DRS-FTIR instrument, in the range of 1000-4000cm^{-1}.Fluorescence spectra were recorded using Carry-Eclipse (Agilent technologies) from 350-550 nm wavelength range. pH measurement was made by using Hanna digital pH meter. Eszetultrasonicator was applied for utrasonication process.

2.3. Synthesis of Blue fluorescent carbon dots (CDs)

The CDs were synthesized from glycerol and citric acid using ultra-sonication technique. In the experiment 1g citric acid and 10 ml glycerol are mixed with 10 ml distilled water resulting solution was stirred for 10 minutes, then transferred in a ultra-sonication chamber for 6 hours, then the reaction mixture was cooled at room temperature.The obtained raw CDs after reaction was dialyzed in a tubing membrane of 1000 Da, mol. Wt. cut-off for 48 hours at room temperature for removing impurities. Obtained purified CDs were kept at 4^0C for experiment.

2.4. Characterization of CDs

A UV-Visible band between 200nm to 360nm is the characteristic of carbon nanomaterial. A UV-Visible band at around 240nm is because ofthe π-π* transition of sp^2 aromaticcarbon atom. CDs shows fluorescence band at 450 nm as shown in fig. 1(a). Dilute solution of CDs is light yellow in colour and shows blue fluorescence when exposed to long wavelength ultraviolet light (365nm). CDs show excitation wavelength dependent fluorescent behaviour.

Fluorescence excitation was started from 305nm and ends at 360nm of excitation wavelength, fluorescence intensity increases on increasing wavelength then reaches to maximum and decreases further, fluorescence intensity found maximum at 350nm excitation wavelength as shown in fig. 1(b).These shifts in fluorescence band and intensity aredue the different surfaces of CDs. The steadiness of nanoparticle was also analysed at wide range of pH (2-10.0), CDs show tremendous stability at all pH.Functional groups can be identified using FTIR spectrophotometer. Broad bands at around 3450 cm^{-1} and 3100 cm^{-1} are due to O-H and C-H stretching. Bands at 1450 and 1680 cm^{-1} are due to C-H bending and C=O functional groups. These observations shows that surface of CDs consists of –COOH, -OH and amide groups[17] as shown in fig. 1(c). Morphology of synthesized CDs was analysed by HR-TEM (High Resolution Transmission electron microscope). Particles of CDs are of spherical shape having sizes between 1nm to 5nm with average diameter of 3.2nm respectively as shown in fig. 1(d).

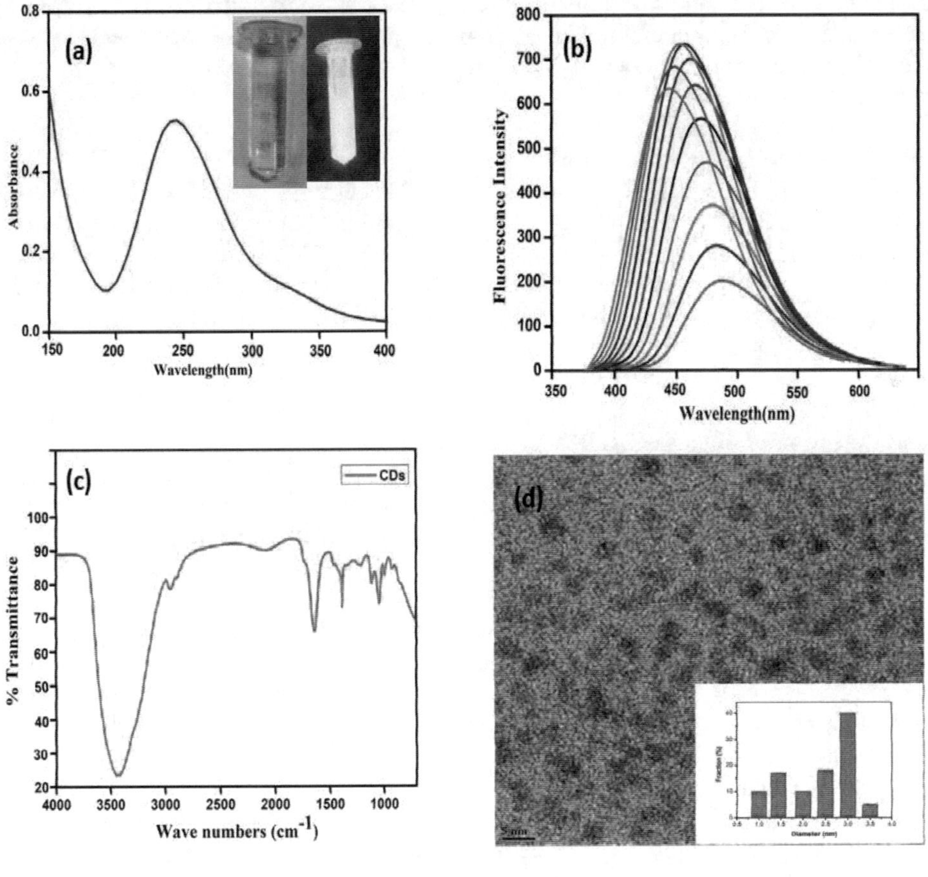

Figure 1 (a) UV-Visible absorption spectra of synthesized CDs, (inset: photographs of CDs under ordinary light and ultra-violet light respectively); (b) Excitation spectra of Fluorescent CDsunder various wavelengths; (c) FTIR spectra of CDs; (d) image of the CDs under HR-TEM (inset: histogram showing size distribution of CDs).

2.5. Calculation of Quantum yield ofCDs

The quantum yield calculation of CDs is carried outusing equation given below-[18]

$$Q = Q_R \frac{I}{I_R} \frac{OD_R}{OD} \frac{n^2}{{n_R}^2}$$

Where Q is quantum yield of CD,$Q_R$represent the quantum yield of reference, I stand for optical density of CDs and $I_R$is optical density of reference which is obtained by UV-Vis spectrophotometer. The quantum yield of CDs was found to be 4.8 upon 350 nm excitationwavelength inreference quinine sulphate which has quantum yield 54% in0.1M H_2SO_4(n=1.33).

2.6. Detection of glucose using CDs fluorescent probe

In the experiment different concentrations of glucose solutions were added to the solution of CDs and incubated at room temperature for 30 minutes, and then fluorescence spectrum was measured employing excitation wavelength350nm.Concentrations of glucose vary in each sample from 0 μM to 1000 μM.

2.7. Glucose detection in biological samples

The urine samples of diabetic human beings were collected from Pt. Jawaharlal Nehru Medical College Raipur(C.G.) and urines of healthy human beings were obtained from healthy individuals after and before oral glucose tolerance test.Before the analysis obtained samples were stored at -20°C in deep freezer. Samples were centrifuged at 15000 rpm for 15 minutes, and then supernatant was diluted 10 times. 100 μL of following samples was employed for detection of glucose as described above.

3. Results and discussions

3.1. Limit of detection of glucose (LOD)

When various solutions of glucose were added to CDs solution then fluorescence of CDs was quenched linearly. It was found that with increasing concentrations of glucose fluorescence of CDs decreased. Concentration of glucose ranges from 0-1000 μM, indicating that quenching of fluorescence is related to concentration of glucose added. The calibration graph was plotted by using fluorescence quenching efficiency **(F_0-F)/F_0**versus glucose concentration (fig. 2 a, b, c) where F_0 is the fluorescence intensity in absence of glucose F is the fluorescence intensity in the presence of glucose. A good linear correlation obtained from 0.1 to 100 μM concentrations of glucose. By using regression equation LOD was found to be 0.08 μM of glucose, with a correlation coefficient R^2 of 0.996. These results prove that the fluorescent probe CDs is a suitable sensor for highly sensitive detection of glucose.

For the improved detection of glucose by proposed method, reactions conditions are optimized such as temperature, pH, incubation time, and concentration of CDs. Quenching efficiency was checked at different pH values it is observed that between 7-8 pH quenching efficiency was maximum i.e. more interaction between fluorescent sensor and analyte molecule(glucose).

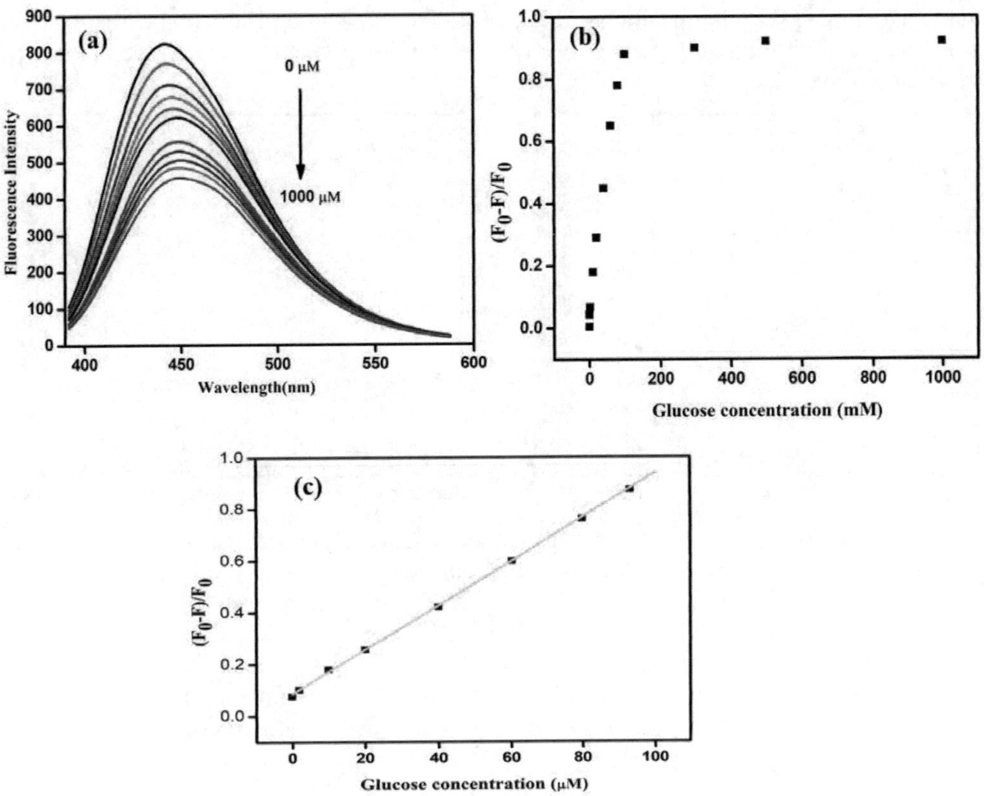

Figure 2 (a) Quenching of fluorescencein the presence of various concentrations of glucose (0-1000 μM); (b) Graph between normalized fluorescence intensity versus concentration of glucose; (c) Linear calibration graph between normalized fluorescence intensity versus concentration of glucose (0.1-100 μM)

3.2. Mechanism of quenching for fluorescent probe

Possible mechanism of fluorescence quenching may be dynamic or static quenching, in dynamic quenching there is FRET (Froster Resonance Energy Transfer) between excited state of donor and acceptor molecules. In static quenching there is complex formation between fluorofore and analyte molecule. Here the possible mechanism of quenching is static quenching because –COOH groups which are present on the surface of CDs gives esterification reaction with –OH groups of glucose molecule.

3.3. Selectivity investigation of the fluorescent probe (CDs) for detection of glucose

For assessment of selectivity of fluorescence probe(CDs) for glucose, the effects of certain biomolecules(ribose, arabinose, maltose, lactose, xylose, fructose, mannose, galactose) and some other interfering materials which exists in human urine (urea,Mg^{2+},uric acid,Na^{+}ascorbic acid , K^{+},Zn^{2+},Cl^{-},Ca^{2+}) were tested. On addition of these biomolecules and interfering ions to CDs solution, fluorescence quenching is negligible. Fluorescence quenching efficiency **$(F_0-F)/F_0$**was compared with glucose, it is found that 100-times more concentration of such ions does not cause appreciable quenching the fluorescence of CDs.The high selectivity imputed to the specificity of CDs for glucose. These results reveal that CDs is very selective for differentiating glucose from other ions and molecules which are present in the biological samples.

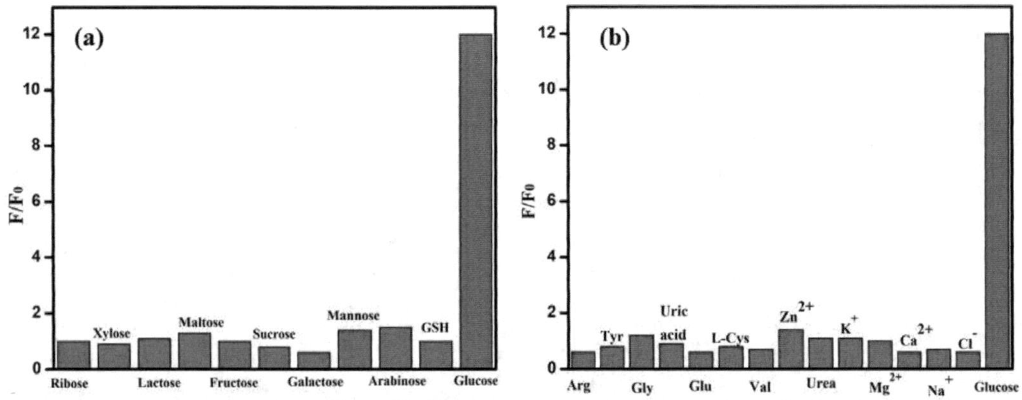

Figure 3 Selectivity of glucose (100 μM); (a) in the presence of various analogues of glucose (1000 μM); (b) in the presence of various potential interfering analogues.

3.4. Real sample analysis

To evaluate the applicability of fluorescent probe (CDs) for the detection of glucose in biological samples, glucose concentration in human urine and serum samples from healthy and diabetic patients were analysed. Concentrations of glucose in human urine were detected by glucometer and compared with present method; results show that there is good agreement between results obtained by CDs and glucometer. To further investigate the accuracy of presentmethod following experiments were performed. Recovery experiment suggests that recoveries of glucose concentrations 98.9 - 102.6% were obtained and relative standard deviation (RSD) was found to be below 5%, these results are acceptable for quantitative analysis of glucose in biological samples.

Table 1.Glucosedetermination in human serum samples and recovery experiment

Spiked(mM)	Detected (mM) (mean[a]±SD[b])	Glucometer(mM) (mean[a]±SD[b])	Average recovery(%)
0	5.45± 0.07	5.51± 0.06	
2.0	7.34± 0.06	7.49± 0.11	99.6
4.0	10.38± 0.05	10.46± 0.05	99.3
8.0	13.42± 0.09	13.52± 0.08	98.9
10.0	15.43± 0.12	15.30± 0.12	102.6

[a]Mean of three values of standard deviation[b]Standard deviation

Determination of glucose in also carried out using fluorescent probe (CDs) and compared there results with convenient glucometer which is used for glucose estimation. Samples were collected three diabetic human beings then samples were also collected from healthy human beings before and after 2hr for taking oral doses of 50g tolerance test. Results obtained shows that fluorescent probe (CDs) is very sensitive and here incubation time was also minimized to 30 min.

Table 2. Glucose determination in human urine from diabetic and healthy human beings

Samples$^{(x, y)}$	Glucometer (mM)	Present Method (mM)	RSD (n=3, %)
DH 1	6.82	6.78	1.66
DH 2	8.12	8.02	2.21
DH 3	18.36	18.40	3.86
HH 1	<LOD undetected	0.48	2.64
HH 2	<LODundetected	0.70	1.89
HH 3	3.46	3.48	3.62
HH 4	4.32	4.28	2.04

^{x}Samples from DH1 to DH3 are urine from diabetic human beings
^{y}Samples from HH1 to HH4 are urine from healty human beings, before (HH1, HH2) taking oral doses of 50g and after 2hr (HH3, HH4) for taking oral doses of 50g og glucose.

We compared our results with previously existing nanoparticle based sensing method for glucose detection including C-dots/AgNPs[19],SiNPs[20], dsDNA-CuNPs-SG1[21], AuNPs-GOQD[22], MnO2-nanosheet[23], CeO2 NPs[24], ZnO nanoparticles [25], and CDs[1], all these method includes complex preparation and low sensitivity than present method expect MnO2-CD nanocomposite[26], whose LOD is lower than our method.

4. Conclusions

In the present study fluorescent nano-biosensor was synthesized using green ultrasonication method. Synthesized CDs is dispersible in water, show strong blue fluorescence, it is photostable, CDs is very sensitive and selective for glucose than other analogues of glucose and potential interfering ions which are present in biological sample (serum and urine), this method is accurate enough for glucose detection,and this method shows various advantages than most of existed methods. CDs were used without any further modifications, so it reduces cost and time both. This method provides an alternate way to determine glucose in biological samples.

Acknowledgement

One of theauthor Vinayak Sahu is thankful to CSIR(Council for Scientific and Industrial Research) Pusa, New Delhi for financial support in the form of SRF with file no..09/1116(0003)/2017-EMR-I. Authors also gratefully acknowledge the Director of National Institute of Technology, Raipur (C.G.) for providing required research facilities.

References

[1] Mutuyimana FP, Liu J, Na M, Nsanzamahoro S, Rao Z, Chen H and Chen X 2018 *Microchim.Acta.* **11**518.

[2] Wang G, He X, Wang L, Gu A, Huang Y, Fang B, Geng B, and Zhang X 2013*Microchim.Acta.* **180**161-186.

[3] SuL, Feng J, Zhou X, Ren C, Li H, and Chen X *Anal. chem.*2012 **84**5753-5758.

[4] Ferrannini E 1998*Endocrine Reviews***19**477-490.

[5] Shaw JE, Sicree RA, andZimmet PZ.2010*Diabetes research and clinical practice***87**4-14.
[6] Zachariou M, and Hearn MT. 2000*Journal of Chromatography A***890**95-116.
[7] Tsao CW, and Yang ZJ. 2015*ACS applied materials & interfaces***7**22630-22637.
[8] Ding L, Gong Z, Yan M, Yu J, and Song X. 2017 *Microchim. Acta*. **184**4531-4536.
[9] Dhara K, and Mahapatra DR.2018*Microchim. Acta*.**185**49.
[10] Bi XSZ, Du X, Jiang JJ, and Huang X.2015 *Anal. chem*.**87**2016-2021.
[11] Hallaj T, Amjadi M, Song Z, and Bagheri R.2018 *Microchim. Acta*.**185**67.
[12] Ortiz-Gómez I, Salinas-Castillo A, García AG, Álvarez-Bermejo JA, de Orbe-Payá I, Rodríguez-Diéguez A, andCapitán-Vallvey LF.2018 *Microchim. Acta***185**47.
[13] Luo PG, Sahu S, Yang ST, Sonkar SK, Wang J, Wang H, LeCroy GE, Cao L, Sun YP. 2013*J. Mat. Chem. B*. **1**2116-2127.
[14] Wang HB, Chen Y, Li N, and Liu YM.2017 *Microchim. Acta*.**184**515-523.
[15] Wang L, Zheng J, Li Y, Yang S, Liu C, Xiao Y, Li J, Cao Z, Yang R. 2014*Anal.chem*.**86**12348-12354.
[16] Zuo P, Lu X, Sun Z, Guo Y, and He H. 2016 *Microchim. Acta*.**183**519-542.
[17] Bourlinos AB, Trivizas G, Karakassides MA, Baikousi M, Kouloumpis A, Gournis D, Bakandritsos A, Hola K, Kozak O, Zboril R, Papagiannouli I. 2015 *Carbon* **83**173-179.
[18] Shi L, Yang JH, Zeng HB, Chen YM, Yang SC, Wu C, Zeng H, Yoshihito O, Zhang Q 2016 Nanoscale. **8**14374-14378.
[19] Ma JL, Yin BC, Wu X, and Ye BC. 2016 *Anal. chem.* **89**1323-138.
[20] Ding L, Gong Z, Yan M, Yu J, Song X.2017 *Microchim. Acta*.**184**4531-4536.
[21] Chen J, Ji X, and He Z. 2017 Anal. Chem. **89**3988-3995.
[22] Na W, Liu H, Wang M, Su X. A. 2017*Microchim. Acta*.**184**1463-1470.
[23] Wang HB, Chen Y, Li N, and Liu YM.2017 *Microchim. Acta*.**184**515-23.
[24] Liu B, Sun Z, Huang PJ, and Liu J 2015 *J. Am. Chem. Soc*.**137**1290-1295.
[25] Sodzel D, Khranovskyy V, Beni V, Turner AP, Viter R, Eriksson MO, Holtz PO, Janot JM, Bechelany M, Balme S, Smyntyna V.2015 *Microchim. Acta*.**182**1819-1826.
[26] Ma H, Liu X, Wang X, Li X, Yang C, Iqbal A, Liu W, Li J, Qin W. 2017 *Microchim. Acta*.**184**177-185.

IOP Conf. Series: Materials Science and Engineering **798** (2020) 012015 doi:10.1088/1757-899X/798/1/012015

A study on the structure-property relationship of microwave irradiated Sunn Hemp fiber reinforced polymer composite

Chinmayee Dash[1] and Dillip Kumar Bisoyi[*]

[1, *]Department of Physics and Astronomy, National Institute of Technology Rourkela, Rourkela, Sundergarh, Odisha-769008, India

Corresponding author's e-mail address- dkbisoyi@nitrkl.ac.in

Abstract. The current investigation has been performed on the microwave irradiated plant-derived cellulosic Sunn Hemp fiber and its composites. The fibers are pretreated by microwave irradiation at a power of 160 watts with different exposure times (2, 4, 6 and 8 minutes). The significant change in fiber and hence composite due to microwave treatment are characterized both in structural and mechanical property. A reasonable increment in cellulose crystallinity with crystallite size, rough surface with fine fibrillation in fiber may lead to improve the mechanical strength in the composite for 4min treatment. The macromolecular structure of the fiber is unaltered by the microwave radiation as indicated by Fourier Transmission Infrared Spectra.

1. Introduction

Production of polymeric composite material from Sunn Hemp natural fiber have gained enough attention by reserachers and industry when environemt and health become a major concern. This help to reduce CO_2 level from atmosphere so ecofriendly and moreover it is biodegradable, abundant and recycle. Sunn Hemp fibers are widely being popular due to their possession of high aspect ratio (500-600), cost effective and as reinforcing material in order to fabricate composite[1].Sunn Hemp as reinforcing material also provide good strength to weight ratio, high toughness, resistance to fatigue and corrosion. But the major drawback include in NFRC is their inherent hydrophilic nature. Due to the intra- and inter- molecular hydrogen bonding with hydroxyl group (-OH) of plant fiber, they are very much likely to absorb air molecules and moisture from the atmosphere. In contrast, the polymer matrix in hydrophobically crosslinked, so there is a huge mismatch between reinforcementing fiber and polymer matrix than can adhere to the properties of composite material. To avoid this problem, various kind of modifications(physical/chemical) are adopted in order to modify the structure of fiber[2,3,4]. In the regard of physical modification, microwave irradiation pretreatment has gained a substaancial attention to provide a better interface for fiber-matrix composite. The physical method for fiber treatment is choosen in order to get ride over the toxicity and water when compared to various chemical process [5]. Microwave frequencies (300 MHz-300 GHz) lies in between radio and infrared frequencies in electromagnetic spectrum and is polularly used in communication as well as radar purposes. The present research work is focused on the effect of microwave pretreatment on sunn hemp fiber and hence it's composite. Microwave heating of material is very much effective over conventional heating as the material under microwave get energy and convert it into heat where the conventional heating to the material are concentrated on the surface[6]. There are some literatures says the microwave techniques towards fibers is one of the effective physical pretreatment to modify the structure of the fibers[7,8,9].

2. Materials and Methodology

Sunn Hemp fibers are collected from Central Research Institute for Jute and Allied Fibers (CRIJAF), Barrackpore, Kolkata-700120, India in bundled form..The general purpose matrix used in this study is thermosetting "EPOXY L-12" polymer having density- 1.17g/cm^3 and the catalyst to accelerate the curing process is "HARDENER K-6" having density- 0.98 g/cm^3 which were procured from Himedia Laboratories pvt. Ltd, Mumbai, India. For the physical modification, cooking microwave oven is used.

2.1. Fiber pretreatment

Sunn Hemp fibers were washed with normal detergent to remove the dirts from the surface. For purification, the fibers are dewaxed with the help of a couple of chemical ethanol:benzene (1:2) subjected to water bath under temperature variator electric heater The purpose of dewaxing the fiber is to remove extra impurities (wax/oil/fatty substances) from the fiber surface and stack the cellulose layers maintain the space between them. The fiber are then collected,washed thoroughly with running tap water and dried under vacuum at 80°C. The fiber is named as raw sunn hemp (RSH) fiber. In later process, the dried fibers are placed under microwave oven at 160 watt power setting for different exposure times (2, 4, 6 and 8 minutes). The fibers were then collected, cooled and stored in desiccator. The treated specimen are named as 160W 2m, 160W 4m, 160W 6m and 160W 8m where 160 denote the power setting and the suffix denote as the different exposure timings.

2.2. Composite preparation

The composites of treated and untreated reinforced fiber were developed by using Hand lay out technique. Two different kinds of rectangualr specimens of dimension (90×12×5)mm^3 and (165×19×5)mm^3 were fabricated for the 3- point bending and tensile test according to the ASTM standard of D7254 and D3039 respectively. A measured (weight of 100 gram and 48 gram) amount of epoxy was taken and was presonicated for 1 hr followed by degassing under vacuum for 8 hr at 80°C and after a mild cooling the catalyst was mixed (epoxy:hardner=10:1) in order to crosslink the curing process. Finally the chopped fiber of 3mm length was mixed in the solution and poured into the mould which was designed using glass slabs and sparyed using an heavy duty silicon sparay prior to pouring for easy removal of the composite. The mould was covered using an plastic sheet and a roller was used to remove the extra resin and air bubbles from the mixture. After 24 hours of curing, the specimen were collcted from the mould and ready for the mechanical tests.

2.3. Instrumentaions

For fine structural characterization of fiber, X-Ray Diffraction (XRD), Fourier Transmission Infrared (FTIR) spectroscopy and Field Emission Scanning Electron Microscopy (FESEM) were carried out to investigate the crystallographic information, functional group and surface morphology respectively. X-Ray spectral patterns of both untreated and microwave treated Sunn Hmep were recorded by by Ultima IV- Rigaku, Japan diffractometer using CuKα radiation. Data were taken at Brags angle range from 10°- 50° and scanning rate of 3°/min with step size 0.05. Functional groups of untreated and treated fiber were investigated by FTIR PerkinElmer Spectrum with averaged over 8 scans using a nominal resolution of 4 cm^{-1}. The surface morphology were carried out by using (Nova Nanosem 450, Japan) FESEM at 10KV accelerating voltage. The mechanical test was executed by a Universal testing Machine (Instron-5967, UK) with environment chamber by applying a cross head speed of 3mm/min.

3. Results and Discussions

3.1. XRD analysis

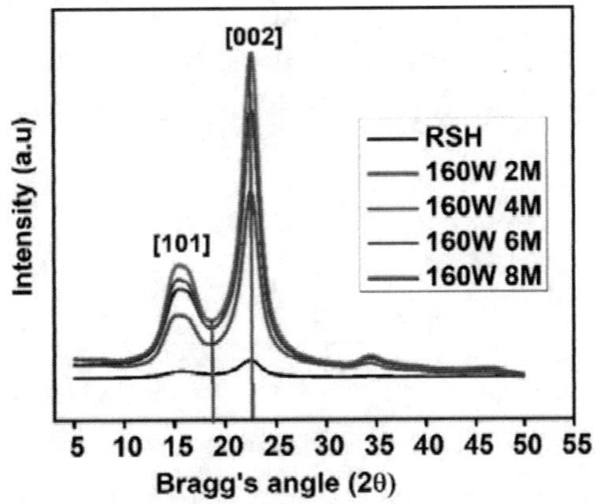

Figure 1. XRD pattern of untreated and treated fiber

The XRD pattern of untreated and microwave treated Sunn Hemp fiber is shown in Figure 1. There are two peaks are observed at the Braggs angle of 15.6° and 22.6° of miller indices [101] and [002] which are ascribed to crystalline cellulose I region and pure cellulosic component in nature. Cellulose known to be a biphase material which consist of both cryatalline and amorphous phases. To know the relative amount of crystalline material present in the cellulose, Segal and his co-workers found a technique called 'Peak height method' where they approached the height of 100% intensed peak into the consideration and substarcted it from the amorphous phase. In this pattern, the region between [101] and [002] peak is considered as cellulosic amorphous components. The microwave treatement does not seem to chage the crystalline structure of the cellulose chain rather than change the arrangement. The crystallite size is calculated using Schere's formula. The relative amount of crystalline material which is also named as Crystalline index (CI) and Crystallite size are found to be high in case of 160W 6m followed by 160W 4m, 160W 8m, 160W 2m and RSH. All the values are listed in tabular form shown in the Table 1.

Table 1. Crystallographic parameters of untreated and microwave treated Sunn Hemp fibers

Fiber name	Crystalline Index (CI) in %	Crystallite size (L) in
RSH	72.5	31.2
160W 2m	78.1	37.6
160W 4m	78.1	40.3
160W 6m	79.9	40.5
160W 8m	77.2	39.0

The reason of appreciable increment in treated fiber may be due to the arrangement of amorphous cellulosic chain arrangement in the crsyatalline domain. Treatment with irradiation time upto 6m leads to formation of reasonable modification in bond chain than leads to formation of new crystallites. Such behaviour reduce the possibility of distrortion and enhance the crystallite dimenisons. It also remove the extra amount of impurities and non cellulosic compounds from fiber. Treatment with higher time lead to degradation of cellulose polymer chain and increased the amorphous cellulose content amout along with crystalline cellulose.

3.2. FTIR analysis

FTIR spectra of untreated and microwave treated Sunn Hemp fiber is shown in Figure 2. It was observed that all the spectra follow an equivalent pattern as the treatment looks like unaltered the macromolecular structure of the cellulose. But there is a disappearance of 2852 cm^{-1} peak upon higher order timing exposure in 160W 8m. This peak is specially meant to wax presence in the fiber surface which is mainly

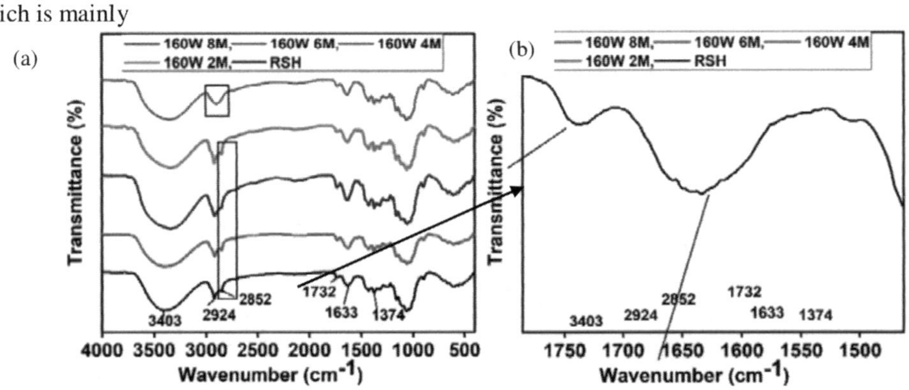

Figure 2. FTIRcspectra of (a) untreated and treated fiber (b) zooming of 1732 cm^{-1} peak

amorphous impurity. It is also noticed that the transmittance peak at 3401 cm^{-1} of RSH peak that is attributed to –OH group show the presence of water molecule is shifted to 3341 cm^{-1}, 3343 cm^{-1} and 3351 cm^{-1} attribute to lowering of water molecules due to absorbance by radiation[10]. There are no other major change observed in the transmittance peak which suggest that the microwave irradiacation of fiber did not change the molecular arrenagment of polymer chain.

Table 2. Comparison of wavenumbers of untreated and treated Sunn Hemp fibers

RSH (cm^{-1})	160W 2M (cm^{-1})	160W 4M (cm^{-1})	160W 6M (cm^{-1})	160W 8M (cm^{-1})
3403	3401	3341	3343	3351
2924	2920	2921	2922	2901
2852	2851	2851	2849	-
1732	1739	1735	1744	1734
1633	1633	1633	1633	1638
1374	1383	1372	1372	1371

3.3. FESEM analysis

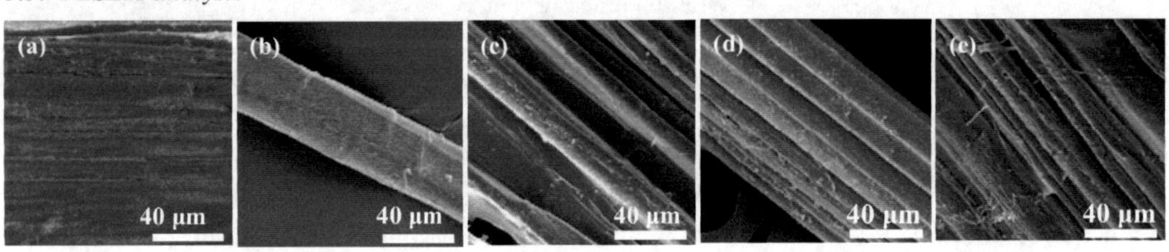

Figure 3. FESEM images of (a) untreated(b) 160 2m (c)160 4m (d)10 6m (e)160 8m fibers

The visibility in surface change of fiber in comparison to untreated was shown in figure 3. It was obsereved that increase in high irradiation time increase the surface roughness of fiber. The single fiber was splitted into many microfibrills in case of 160W 4m, 160W 6m and 160W 8m with making space in between them. The surface area is effectively increased at the higher irradiation time but got degradation on the suraface in case of 160 8m. Such damaging effect appear may ascribe to degradation of internal macromolecular chain structure. The surface roughness and increased surface area of fiber make an easy flow of the resin and strengthen the fiber-matrix interface.

IOP Conf. Series: Materials Science and Engineering **798** (2020) 012015 doi:10.1088/1757-899X/798/1/012015

3.4 Mechanical analysis

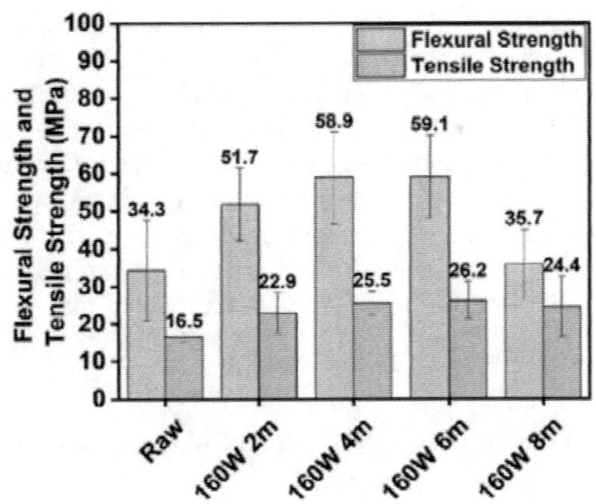

Figure 4. Mechanical strength of untreated and treated composite

3-point bending test or flexural test and tensile strength of untreated and treated composite are shown in figure 4. Increasing relative amount of crystallinity and crystallite size strengthened the fiber. Sunn Hemp fiber possess a higher amount of cellulose (70-78%) after cotton (85-95%). A higher quantification of cellulose always make the fiber strong and stiff. Moreover upon treatment, the increment in surface roughness and surface area facilitated the mechanical binding between fiber-matrix. Failure in mechanical strength at higher irradiatioh time of 8m degraded the cellulose structure which also lead to form a poor adhesion between fiber-matrix.

3.5 Microstructure

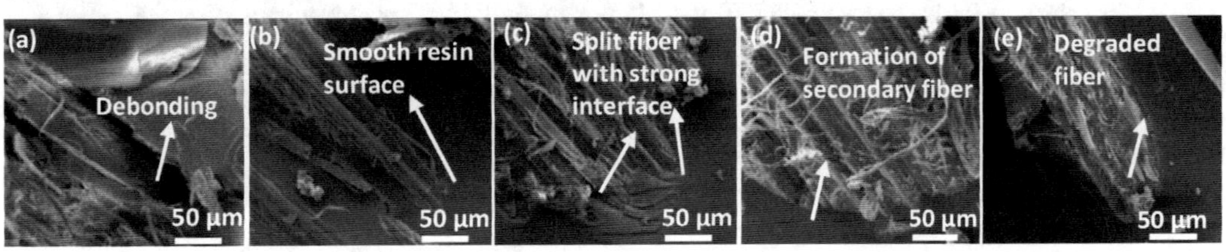

Figure 5. FESEM images of (a) untreated(b) 160 2m (c)160 4m (d)10 6m (e)160 8m composite

The adhesion between fiber-matrix of both untreated and microwave treated composite was obsereved from the fracture micrograph in figure 5. Due to weak bonding between fiber-matrix , RSH fiber experience debonding around the fiber periphery and also the matrix region is layered over by over. This suggest a sign of poor adhesion at interface. But upon treatment, the adhesion got stronger due to the increment in cellulose crystallinity in fibe which lead to a stronger interface. The splitting of fiber was observed in all treated case but there is a formation of seconadary fibers was noticed in case of 160W 6m. This is obviously a scenario of initiation of the fiber weakening but due to strong adhesion, no debonding was marked rather than breakage at the fiber edge. Treatment at 8m exposure time lead to degradation of fiber due to absorbing a lot of electromagnetic radiation which provide a poor wetting of fiber into matrix.

4. Conclusions

Sunn Hep fiber was physically modified by microwave irradiation technique for making it applicable for various potential applications such as composites widely. The treatment unaltered the macromolecular strcture of cellulose which is confirmed from FTIR spectra. Higher crystallinity and crystallite size were found in case of 160W 6m fiber which suggests the rearrangement of cryalline cellulosic polymer chain and increment in crystallite dimensions. Suraface roughness are observed in all treated fibers but a better splitting with adequate surface area was the main focus for 160 6m. Mechanical strength of Sunn Hemp composite was found to be higher in case of 160 6m due to better wetting and adhesion between fiber-matrix. So the microwave irradiation of fiber at power 160 watt with different exposure times was could have the best mechanical strength at 6 minute. This technique of microwave irradiation is quite doable in the favrication of obtaining safer and effective treatment for heating rather than conventional heating.

Acknowledgment

The author is acknowledged Director, National Institute of Technology Rourkela, Rourkela-769008, Odisha, India for providing the fellowship and to perform this research.

References

[1] Liu M, Thygesen A, Summerscales J and Meyer A.S 2017 *Ind. Crops Prod***108**, 660
[2] Fiore V, Bella G.D, Valenza A 2015 *Comp. Part B***68** 14
[3] Zhou, Feng Cheng, Guangxu Jiang, Bo 2014 *Appl. Surf. Sci.***292** 806
[4] Bozaci, Ebru Sever, Kutlay Sarikanat, Mehmet Seki, Yoldas Demir, Asli Ozdogan, Esen 2013 *Comp. part B***45** 565
[5] Oliveira M. D, Cioffi M.O.H, Benini K.C.C de C.B and Voorwald H.J.C, 2013 *Procedia Engineering***200** 357
[6] Basics of Microwave, 2019, *Micro Denshi Co. Ltd.* url: https://www.microdenshi.co.jp/en/
[7] Schwarzova I, Nadezda S, Terpakova E and Junak J 2015 *Proceedings of REHVA Annual Conference "Advanced HVAC and Natural Gas Technologies"*, 206
[8] Krishnaiah Prakash, Ratnam C. T and Manickam S 2017 *Ultras– Sonochemistry***34** 729
[9] Gassan J, Gutowski, V.S 2000 *Compos. Sci. and Tech.* 60 2857
[10] Patra A, Bisoyi D.K,·Manda P.K,·Singh A.K 2012 *Appl. Phys. A***112** 1063

IOP Conf. Series: Materials Science and Engineering **798** (2020) 012016 doi:10.1088/1757-899X/798/1/012016

Microbial-induced phase transition of GeO_2 and its study by XRD and Raman spectrum

D K Golhani[1], B Gopal Krishna[2], M Jagannadha Rao[3] and Ayush Khare[1]

[1]National Institute of Technology, Raipur (CG)
[2]SOS in Electronics and Photonics, Pt. Ravishankar Shukla University, Raipur (CG)
[3]Department of Geology, Andhra University, Andhra Pradesh, India
*Corresponding author's e-mail address: dkgolhani@gmail.com, krishna_burra85@yahoo.com

Abstract. In this paper, we report the formation of germanium dioxide nanoparticles obtained from the action of certain microbes on Germanium crystal under low magnetic field at ambient temperature. The atomic oxygen is chemisorbed in the presence of catalyst by forming GeO_2. The evolution of the Germanium dioxide from Germanium and microbes-induced phase transitions in GeO_2 is studied by the means of X-ray diffraction method and Raman spectroscopy. The formation of GeO_2 from Germanium is also prone to surface oxidation which may compromise surface sensitive measurements. The X-ray diffraction method is not sensitive to anharmonicity-induced dynamical effects due to which there would be a lack of clarity about the phases and transitions intervening in the phase transition of GeO_2. The detection of phase transition of GeO_2 was being studied by broadening and shifts of diffraction angle in X-ray diffraction. Raman scattering establish the microbial induced phase transitions and identify phases of germanium oxide. The detection of phase transition was being studied by substantial broadening and softening of Raman modes as to identify the transition regions and possible intermediate phases.

1. Introduction

Germanium dioxide (GeO_2) is a versatile, wide band gap material having unique thermal, optical and electrical properties, such as high thermal stability, high dielectric constant, and refractive index. Due to these properties, GeO_2 is a promising candidate for different applications, such as vacuum technology, solar cells, catalysis, Li-ion batteries, especially in optoelectronic devices; core of optical fibers [1, 2]. It is found in crystalline, liquid and amorphous phases.

At ambient temperatures, crystalline GeO_2 occurs in three stable polymorphs, i.e., (i) α-quartz like trigonal or hexagonal structure with tetrahedral coordination, (ii) Tetragonal or Rutile structure with octahedral coordination, (iii) Amorphous or Vitreous (Glassy). α-quartz and rutile forms exhibit similarity in hardness, transparency, and colourlessness but show considerable difference in chemical and optical properties, such as solubility, density and indices of refraction [3-6].

Tetragonal structure of GeO_2 has denser and stable phase at ambient conditions but transforms to hexagonal phase at 1306 K with less density [7]. The tetragonal structure is insoluble in water and hexagonal phase is slightly soluble in water (0.4 g/100 g H_2O at 300 K). These differing structural properties induce physical and chemical properties of the GeO_2 and in turn also affect possible applications of both modifications. Interdisciplinary studies on GeO_2 are necessary to bridge the gap between its research applications and commercialization. The structural analogy of GeO_2 is close to SiO_2 and the study of the transformation of different quartz phases occurring during reactions is important. Polymorphism of GeO_2 influences the material characteristics and hence utilized for different applications.

The primary aim of this research work is to study the phase transitions related to GeO_2 nanoparticles (NPs) which is biologically synthesized by germanium crystal. GeO_2 NPs exhibit unique physical and chemical properties as compared with the bulk material. X-ray diffraction (XRD) and Raman spectroscopy was employed to study the phase transition of the GeO_2. In the present research, germanium dioxide NPs was fabricated biologically from Ge crystal by using different bacteria in presence of magnetic field at ambient temperature. The phase of GeO_2 NPs was studied by analyzing X-ray diffraction patterns. The spectral properties could also be used to identify the α-quartz, rutile and amorphous forms of GeO_2, which were recorded by Raman spectroscopy.

2. Methodology

Germanium dioxide nanocrystals were fabricated in the presence of culture of bacteria Pseudomonas putida and low magnetic field of about 0.1 Gauss at ambient temperature. The exposure time for the reaction was kept at 6 hrs. The formation of germanium dioxide nanocrystals occurs due to the interaction of bacteria with the Ge crystal. Bacteria are well known natural living organism for synthesis of NPs as well as they can control and modify various reaction parameters, such as pH, temperature and chemical potential of the reaction etc., to catalyze the oxidation of germanium for the formation of germanium dioxide nanocrystals [8]. Magnetic field induces directional movement of microbes and show unique characteristics [9].

The newly reported magnetically induced direct method of biosynthesis may be more effective than currently employed physical and chemical methods for synthesis of GeO_2. In this mechanism, pre- and post- physical/chemical/biological treatments like preparation of metal salt solution, filtration, sonication, etc. for the synthesis and retrieval of nanoparticles are not required. It is a faster and more economical than other methods. More research is required to explore this new route of biogenesis.

3. Experimental results and Discussions

The characterization of biologically prepared sample of GeO_2 is performed by X-ray diffraction [PANalytical, operated at 40 kV, 30 mA, Cu-K_α radiation (λ= 1.5405Å)] and Raman spectroscopy.

3.1. X-ray diffraction of pure Germanium crystal (Ge) and biologically synthesized GeO_2

XRD of pure Ge shows an intense peak at 45.5° which matches with standard data as shown in figure 1. X-ray diffraction data gives all the structural information of GeO_2 due to difference in chemical sensitivity of Ge-O and Ge-Ge pairs. The structural information of GeO_2 can be resolved better with X-ray diffraction [10,11]. The germanium dioxide nanoparticles were studied using XRD with series of intensive peaks at 20.5°, 26.1°, 27.6°, 36.1°, 37.9°, 38.0°, 39.0°, 40.5°, 43.3°,45.5°, 56.5°, 57.3°, 65.6°, 76.4° corresponding to germanium dioxide lattice reflections as shown in figure 2. The Ge atoms in GeO_2 are arranged in basic tetrahedral units with quartz like polymorph. X-ray diffraction shows that the O − Ge − O intratetrahedral angle is more distorted in vitreous GeO_2.

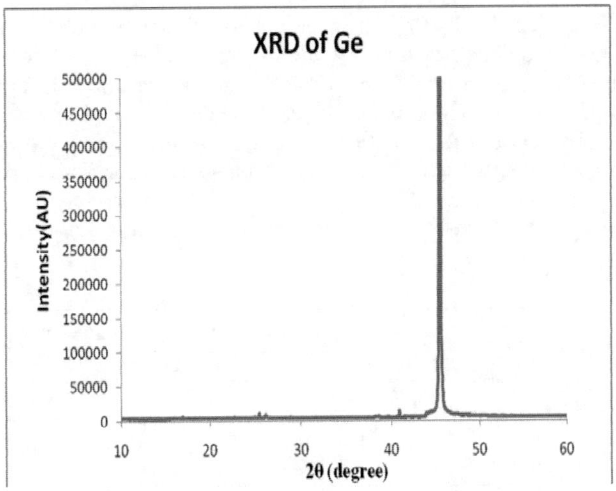

Figure 1. XRD of Pure Ge

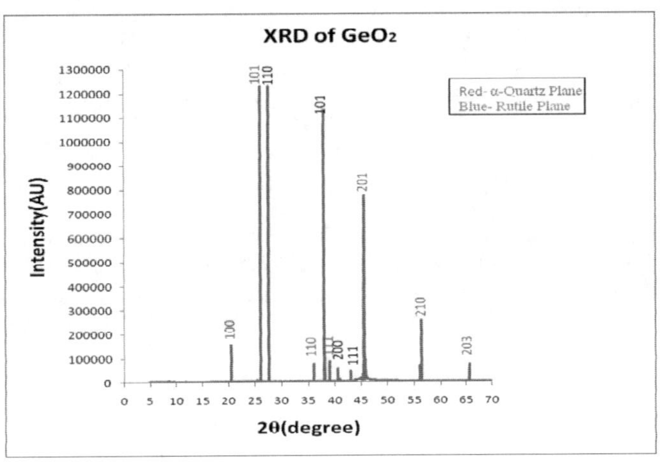

Figure 2. XRD of GeO_2

From the XRD data, the lattice planes (100, 101, 110, 102, 111, 201, 210 and 203) of the quartz GeO_2 are found at 20.5°, 26.1°, 36.1°, 38.0°, 39.0°, 45.5°, 56.5°and 65.6° respectively. The lattice planes (110,101,200,111 and 002) of the rutile GeO_2 are found at 27.6°, 37.9°, 40.5°, 43.03°, and 57.3° respectively. The lattice planes (100 and 102) of the vitreous GeO_2 is found at 36.1° and 56.5° respectively.

IOP Conf. Series: Materials Science and Engineering **798** (2020) 012016 doi:10.1088/1757-899X/798/1/012016

3.2. Raman Spectra

The structure of GeO_2 as depicted by Raman spectra at room temperature occurs in α-quartz, rutile and amorphous forms [12]. In quartz structure, Ge atoms are arranged at the center of the corner linked GeO_4 tetrahedra. The structure is trigonal with three Ge atoms per unit-cell. For quartz structure, there are low, mid and high frequency vibrational modes for GeO_2. The high frequency region ranges from 857 to 970 cm^{-1} which are assigned to Ge-O stretching motions within the GeO_4 tetrahedra groups as shown in figure 3. In this range, there are four weak bands ranging from 886 to 958 cm^{-1}. The mid-frequency region ranges from 442 to 595 cm^{-1} which are assigned for Ge-Ge stretching motions (figure 4). The low frequency region ranges from 122 to 327cm^{-1} which are assigned to Ge-O bending modes as shown in figure 5[13].

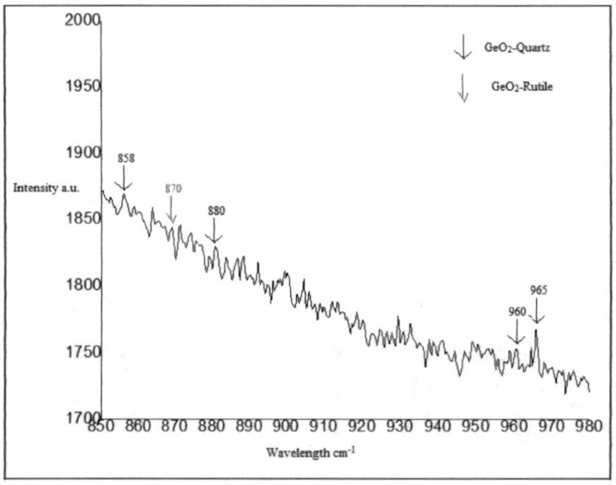

Fig. 3 High frequency range Raman spectra Spectraspectra

In rutile structure, Ge atoms are arranged at the center of the edge linked GeO_4 tetrahedra. The structure is tetragonal with two Ge atoms per unit cell. For rutile structure, the highest frequency corresponds to 700 cm^{-1} which is assigned to Ge-O stretching modes within GeO_6 octahedral group [14].

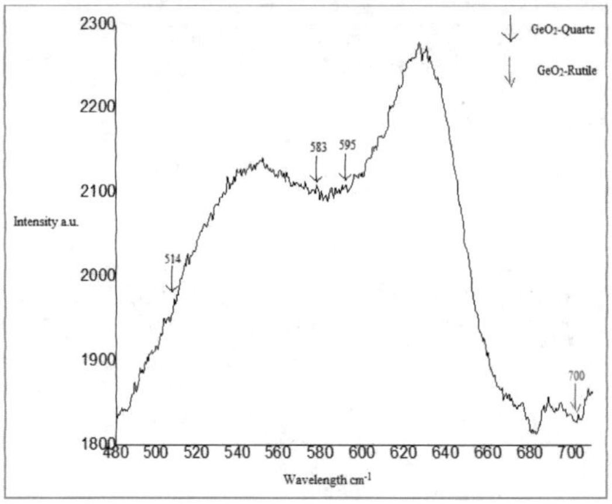

Figure 4. Mid-frequency range Raman spectra.

The Raman spectra of vitreous GeO_2 exhibit similar spectrum as that of quartz form. This similarity between vitreous and quartz form implies that germanium is four coordinated in the vitreous GeO_2.

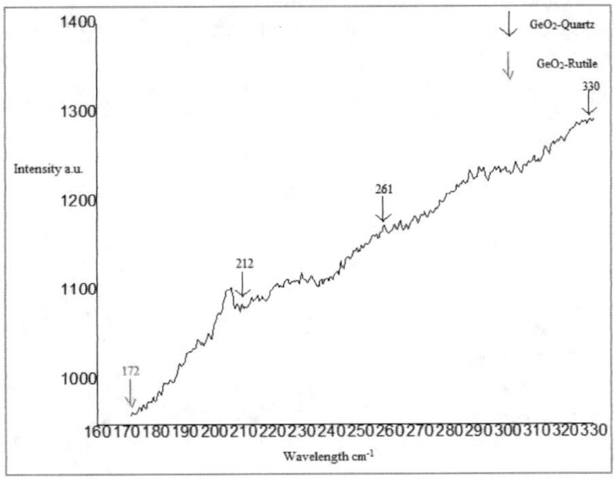

Figure 5. Low frequency range Raman spectra

4. Conclusions

In this paper, we have reported the formation of GeO_2 nanoparticles through microbial route using bacteria and low magnetic field at normal temperature and pressure. We also discussed the experimental results on formation of different phases, and phase transformations in GeO_2.

Investigation of phase transformation of GeO_2 nanocrystals have been done with the help of XRD and Raman spectra. The results demonstrated that most Ge atoms were transformed into GeO_2 after the oxidation of Germanium under microbial and magnetic environment. During the formation of GeO_2 from germanium, the polymorphs of GeO_2 were found with dominance of α-quartz as compared to rutile and vitreous structure. Raman scattering spectroscopy of biologically fabricated germanium dioxide successfully detected and identified all the Raman active modes which confirm the formation of α-quartz, rutile and vitreous phases. It was seen that Raman spectroscopy is better for phase transition analysis than XRD method.

References

[1] Ravindra M, Weeks R A and Kinser D L 1987 *Phys. Rev.* B **36** 6132
[2] Sahnoun M, Daul C, Khenata R and Baltache H 2005 *Eur. Phys. J.* B **45** 455
[3] Scott J F 1970 Raman spectra of GeO_2 *Phy. Rev.* B **1(8)** 1
[4] Laubengayer AW and Morton D S 1932 *J. Am. Chem. Soc.* **54** 2303
[5] Erwin P 1964 Crystal chemistry of tetrahedral structures (New York: Gordon and Breach)
[6] Smith G S and Isaacs P B 1964 *Acta Cryst.* **17** 842.
[7] Mernagh T P and Liu L-g 1997 *Phys. Chem. Miner.* **24** 7
[8] Krishna B G, Rao M J, Nalinikant B, Golhani D K and Zaidi S A H 2016 *Proc. IEEE Int. Con. (TENCON)* 2956.
[9] Krishna B G, Hegadi R, Rao M G and Soni S 2015 *Proc. IEEE Elect. Comp. Engg.* (Dhaka, Bangladesh)
[10] Ming L C and Manghnani M H 1983 *Phy. Earth Plan. Int.* **33** 26.
[11] Wu W et al. 2011 *J. Nanomat.* **2011** 841701.

[12] Madon M 1991 *Phys. Chem. Miner.* **18** 7.

[13] Atuchin V V, Gavrilova T A, Gromilov S A, Kostrovsky V G, Pokrovsky L D, Troitskaia I B, Vemuri R S, Carbajal-Franco G and Ramana C V 2009 *Crys. Grow. Desi.* **9**(4) 1829.

[14] Sanson A, Pokrovski G S, Giarola M and Mariotto G 2015 *EPL* **109** 26007.

Polarization mechanism and Shockley Read Hall recombination on Quantum Efficiency of InGaN/GaN Blue LED

Neha Anchal[a)] and Bijay Kumar Sahoo[b)]

Department of Physics, National Institute of Technology, Raipur-492010
[a)]nanchal.phd2018.phy@nitrr.ac.in; [b)]bksahoo.phy@nitrr.ac.in

Abstract.In_xGa_{1-x} N/GaN blue LEDs faces significant efficiency droop issue. The causes of efficiency droop are Shockley Read Hall recombination (SRH), Auger recombination (AR), carrier delocalization (CD) and electron leakage (EL). The SRH, Auger and electron leakage are functions of carrier concentration and temperature. InGaN/GaN superlattice has polarization electric field at interface. In this work we explore effect of built in polarization (BIP) on efficiency droop. It is shown that polarization field enhances SRH coefficient resulting in more droop in internal quantum efficiency of blue LED. Thus, for improvement in efficiency, polarization field required to be minimized which requires growth of the material in m-plane instead of c-plane.

1. Introduction

$In_xGa_{(1-x)}N$/GaN based blue light emitting diode (LEDs) are promising devices whose wavelength ranges from 420 to 440nm [1]. Traffic light, full colour display, optical storage, disinfection of medical equipments, and phototherapy are some application areas of $In_xGa_{(1-x)}N$/GaN blue LEDs [2, 3].However, the blue LED is suffering from internal quantum efficiency (IQE) droop. The causes of the efficiency droop are Shockley Read Hall recombination (SRH), Auger recombination (AR), carrier delocalization (ED) and electron leakage (EL) [4]. The SRH, Auger and electron leakage are functions of carrier concentration and temperature. Out of these causes, SRH is most important efficiency droop mechanism [5, 6]. SRH is part of non radiative recombination process. In SRH recombination, the electron in transition between bands passes through new energy states created with the band gap by defect in crystal lattice. The process is named after William Shockley, William Thornton Read and Robert N. Hall. Effect of SRH depends on energy band gap of material which is a function of temperature. A blue LED device must be balanced from electrical, optical and thermal point of view. $In_xGa_{(1-x)}N$/GaN has unique property known as built in polarization (BIP) at hetero interfaces [7]. BIP in $In_xGa_{(1-x)}N$/GaN is generated due to spontaneous polarization (SP) and piezoelectric polarization (PZ) of constituent semiconductors InGaN and GaN. BIP generates an electric field (~10^6V/cm) called as polarization electric (PE) field which modifies other properties of material [7]. In this work, we have investigated the role of BIP field on SRH mechanism and efficiency droop of$In_xGa_{(1-x)}N$/GaN LEDs.

2. Theoretical Model

The figure of InGaN/GaN blue LED is shown in figure 1. Its active region consists of InGaN/GaN superlattice (SL) or multi quantum well (MQW). GaN layer acts as barrier whereas InGaN acts as well. When InGaN layer is growth on GaN strains arises in InGaN layer. At interfaces due to lattice mismatch piezoelectric polarization charges are generated. Again both constituent semiconductor InGaN and GaN have spontaneous polarization. Thus, total polarization in any layer (InGaN or GaN) of the SL is $P^{total} = P^{sp} + e_{31}\varepsilon_1 + e_{32}\varepsilon_2 + e_{33}\varepsilon_3$.$P^{sp}$denotes spontaneous polarization, e_{31}, e_{32}, e_{33} are piezoelectric coefficients of InGaN, $\varepsilon_1, \varepsilon_2, \varepsilon_3$ are strain tensors [7]. In this type of strain $\varepsilon_1 = \varepsilon_2$, $e_{31} = e_{32}$ and $\varepsilon_3 = -2(C_{13}/C_{33})\varepsilon_1$, where C_{13}, C_{33} are elastic constants of InGaN. The values of parameters are taken from ref [7].PE field in well and barrier are given by equation [8] $E_w = L_b(P_b - P_w)/(\varepsilon_b L_w + \varepsilon_w L_b)$ and $E_b = L_w(P_w - P_b)/(\varepsilon_w L_b + \varepsilon_b L_w)$; where L_b, L_w are thickness of barrier and well, P_b, P_w are barrier and well polarization, $\varepsilon_b, \varepsilon_w$ are dielectric constants of the barrier and well. This BIP electric field interacts with electrons and holes resulting in various phenomena controlling efficiency.

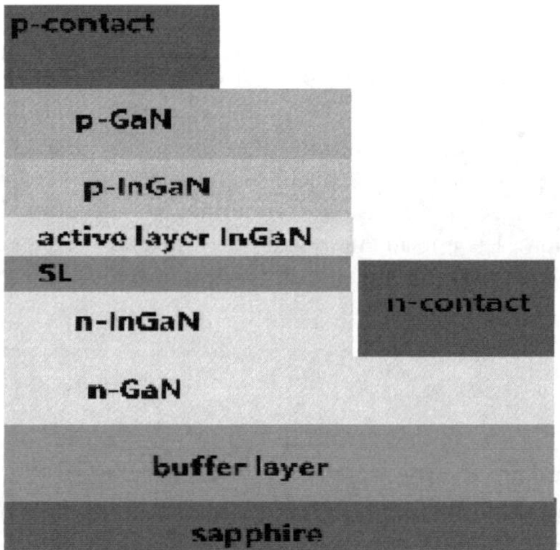

Figure 1 $In_xGa_{(1-x)}N/GaN$ blue LEDs

Efficiency of an LED is product of optical extraction efficiency (η_{OQE}) and internal quantum efficiency (η_{IQE}) $\eta = \eta_{IQE} \times \eta_{OQE}$ [9]. η_{IQE} defines rate of generation of photons in quantum well (QW) to number of carriers injected into the LED. The η_{OQE} is the fraction of photons emitted from the LED to photons produce in the QW. In other words, η_{OQE} stands for photons lost inside the LED. It is reported that the main cause of efficiency droop is not photon extraction. η_{IQE} is ratio of radiative recombination current to total input current to LED: $\eta_{IQE} = I_{rad}/I$. The total current in the QW is divided into two parts current that produces photons and current that is lost: $I = I_{rad} + I_{lost}$. Thus, $\eta_{IQE} = I_{rad}/(I_{rad} + I_{lost})$. This equation clarifies that efficiency droop can only occur when I_{lost} increases more than I_{rad}. The mechanisms responsible for I_{lost} are SRH, AR, ED and EL effect. It means $I_{lost} = I_{SRH} + I_{Aug} + I_{ed} + I_{leak}$.The current that is entering into QW is written as

$I_{QW} = I_{rad} + I_{SRH} + I_{Auger} + I_{ed}$. The total current becomes $I = I_{QW} + I_{Leak}$. The currents are defined as $I_{rad} = Bn^2, I_{SRH} = An, I_{Aug} = Cn^3, I_{ed} = k(n-n_0)^m$ and $I_{leak} = a\ (I_{QW})^p$.Where *A, B, C, k*(~1), n_0($n_0 = 5\times10^{18}\ cm^{-3}$), $m(\geq 2)$, *a, p* are constants and *n* is carrier density. Thus, if V_{QW} is volume of active region of all QWs then $I_{QW} = eV_{QW}[An + Bn^2 + Cn^3 + k(n-n_0)^m]$ where *e* is electronic charge. The internal quantum efficiency will become $\eta_{IQE} = (\eta_{inj} Bn^2)/[An + Bn^2 + Cn^3 + k(n-n_0)^2]$ [4, 9]. Here $\eta_{inj} = I_{QW}/I$, the ratio of carriers that recombine inside the QW represents injection efficiency. The radiative recombination coefficient is estimated from $B = B_0|F_{eh}|^2$; here $F_{eh} = \int \psi_e(z)\psi_h(z)dz$ is overlap integral of electron and hole wave functions. Similarly, $A = A_0|F_{eh}|^2$ and $C = C_0|F_{eh}|^2$ [6]. BIP electric field separates electron and holes away from each other, so overlap integral decreases resulting in small recombination. Thus, the coefficients *A*, *B* and *C* are influenced by polarization field. In this work, we estimated internal quantum efficiency with respect to SRH coefficient *A* under BIP field; and other coefficients (*B*,C *k* and *m*) are constants.

3. Result and Discussion

First strain is computed. Fig.2 shows lattice constant of GaN and lattice constant of InGaN as a function of In content x. From fig. it can be seen that as x increases lattice mismatch increases.

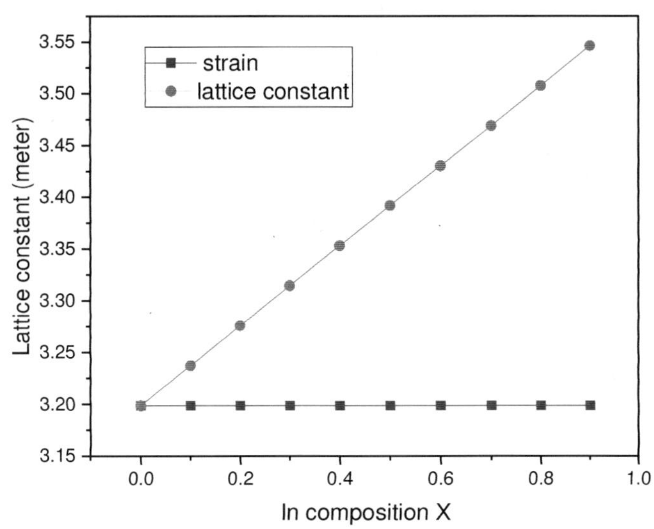

Fig.2 lattice constant with In content

First we have estimated polarization electric field *E* in both barrier and well. Figure 3 shows polarization electric field as a function of *In*content x. Next we have computed value of A taking BIP field into account. BIP field changes the band structures of InGaN. So, the wave functions of electrons and holes are modified. Three values of A are found: $A = 10^7\ /s$, $A = 10^8\ /s$ and $A = 10^9\ /s$.

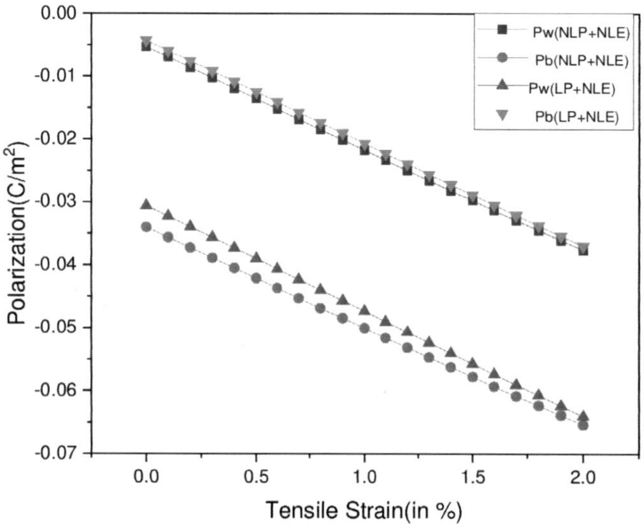

Fig. 3 polarization dependenc on strain

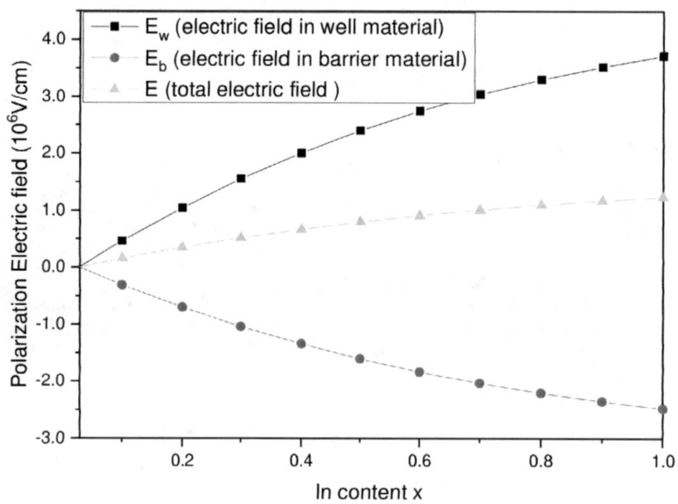

Fig. 4 .Polarization electric field in barrier and well as a function of x

Then, we have calculated internal quantum efficiency as a function of current density with A as a parameter at room temperature under the polarization field effect taking B and C are constants. In this calculation, we used data $B = 2\times10^{-11} cm^3 / s, C = 0$ [9].

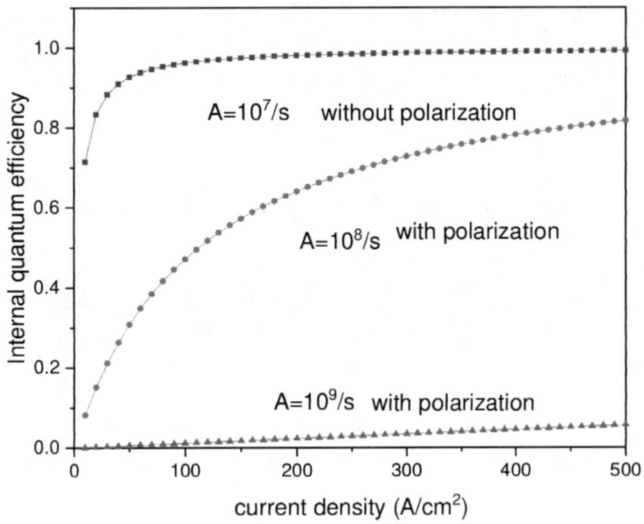

Figure 5.Internal quantum efficiency of InGaN/GaN blue LED.

Fig. 5 shows the graph between IQE and current density for with and without polarization mechanism. From fig 5 it can be observed that for no polarization effect, IQE becomes saturation after current density $60A/cm^2$.In this case IQE increases with current density for $A = 10^7 / s$. Red line shows IQE under the polarization effect for value $A = 10^8 / s$ as shown in fig. 5.For value $A = 10^8 / s$ IQE increases after current density $200A/cm^2$.The blue line in fig. 5 shows IQE for $A = 10^9 / s$ under the effect of polarization. In this case,IQE increases with current density. From fig.5 it is clear that as polarization increases SRH mechanism becomes strong due to which A value increases and it reduces the IQE. In the presence of polarization field this IQE is smaller than the no polarization case. From fig. 5 it is clear that material shows high efficiency inno polarization effect as compare to IQE with polarization.

4. Conclusion

In this work, we have investigated the role of polarization field on SRH mechanism and efficiency droop of$In_xGa_{(1-x)}N/GaN$ blue LEDs. It is found that as polarization field increases, A value increases due to which SRH mechanism become stronger to stronger and it reduces the IQE. Thus, device shows high IQE in no polarization case as compare to with polarization case motivating to growth SL in m-plane instead of c-plane to minimize polarization.

Acknowledgement

Author (NA) acknowledges with thanks to CSIR, New Delhi Govt. of India for financial support through awarding Junior Research Fellowship Sanction Letter No: 09/1116(009)/2018-EMR-I

References

[1] Yen-Kuang Kuo, Tsun-Hsin Wang, and Jih-Yuan Chang; *Appl. Phys. Lett.* 100, 031112 (2012)

[2] T. C. Wen, S. J. Chang, L. W. Wu, Y. K. Su, W. C. Lai, C. H. Kuo, C. H. Chen, J. K. Sheu, and J. F. Chen; *IEEE Trans Electron Devices,* 49, 1093 (2002)

[3] M. Lee, M. Yang, K. Song, and S. Park, *ACS Photonics*, DOI: 10.1021/acsphotonics.7b01453

[4] T. Jeong, Hyung-Jo Park, Jin-Woo Ju, H. Oh, J. Baek, J.Ha, G. Ryu, and H. Ryu, *IEEE Photonics Tech Letters,* 26, 649 (2014)

[5] G. Verzellesi, D. Saguatti, D. Saguatti, M. Meneghini, F. Bertazzi, M. Goano, G. Meneghessoand E. Zanoni, *J. Appl. Phys.* **114**(2013) 071101.

[6] F.Nippert, M.Mazraehno, M. Davies, M. Hoffmann, H. Lugauer, T. Kure, M. Kneissl, A. Hoffmann and M.Wagner,*Appl. Phys.* Lett. **113** (2018) 071107.

[7] O. Ambacher, J. Majewski, C. Miskys, A. Link, M. Hermann, M. Eickhoff, M. Stutzmann, F. Bernardini,V. Fiorentini, V. Tilak, B. Schaff and L. F. Eastman,*J. Phys.: Condens. Matter***14,** (2002) 3399.

[8] M.Leroux, N. Grandjean, M. Laugt, and J. Massies, *Phys. Rev. B*, **58**,(1998) 371

[9] Joachim Piperk, *Phys. Status Solidi A***207**, (2010), 2217.

IOP Conf. Series: Materials Science and Engineering **798** (2020) 012018 doi:10.1088/1757-899X/798/1/012018

Effect of temperature on electron leakage and efficiency droop of $Al_xGa_{1-x}N$ UV LED

Priyanka Sahare[a] and Bijay Kumar Sahoo[b]

Department of Physics, National Institute of Technology, Raipur-492010
[a]psahare.phd2018.phy@ nitrr.ac.in;[b] bksahoo.phy@nitrr.ac.in

Abstract. Al_xGa_{1-x} N UV LED is a promising light source for water purification, diagnosis and sterilization of medical equipment's, document authentication, security checking and general lightings. The device suffers from efficiency droop due to a number of factors like Shockley Read Hall recombination (SRH), Auger recombination (AR), electron delocalization (ED) and electron leakage (EL).In this work, effect of temperature on electron leakage and efficiency droop of AlGaN UV LED is investigated. It is found that electron leakage takes place at high injection current and high temperature. Electron leakage can be reduced by minimizing self-heating effect in the AlGaN UV LED.

1. Introduction

$Al_xGa_{(1-x)}N/GaN$ based UV LEDs are used in purificationof water, disinfection of medical equipment, UV curing, verification of document, phototherapy and medical analysis, security purpose. UV light wavelength ranges from 200 -400 nm [1]. It has maximum optical quantum efficiency (OQE) which ranges from 20 to 60% and optical power from 0.69 to 2.1mW. A high efficiency peak is observed at low current densities ~ 10 A/cm^2. Due to their various aspects, it is widely used; however it suffers from a problem named as efficiency droop that occurs at higher injected current. The efficiency of LEDs measured through radiative and non-radiative recombination. The non-radiative recombination processes are Shockley Read Hall recombination (SRH), Auger recombination (AR), electron delocalization (ED) and electron leakage (EL). These factors play important role in efficiency droop mechanism [4]. The factors SRH, AR, ED and EL depend on carrier concentration and temperature. However, out of these mechanismsEL is most effected one. In UV LEDs, electrons leaked from the active region to p-side of the device. Many reasons are responsible which results EL. They are thermionic emission, ballistic and quasi-ballistic transport of electrons, poor hole-transport inside the active region and polarization phenomenon of nitride semiconductors [5, 6, 7]. It is found that temperature is one important parameter for electron leaking resulting in droop of internal quantum efficiency (IQE). It is observed that rise in temperature enhances hole/electron ratio in multi quantum well (MQW) [8]. Rise in temperature of active region is named as self-heating. Due to self-heating of active regions, process like thermionic emission, ballistic and quasi-ballistic transport of electrons, poor hole-transport inside the active region takes place. It exhibitsthat increase in temperature, EL is increased.It has been found that in polar c – plane and non-polar m – plane based LEDs electron spilled and results reduction in efficiency when electron blocking layer (EBL) is not implemented [7]. EBL is used as a barrier for electrons. The problem of EL is resolved by growth of electron blocking layer over MQW region of LEDs device [9]. However, experiments reported that EBL is not able to stop the electron leakage and still EL remains in LEDs. All the factors (A, B, C, k and p) responsible for droop are functions of temperature of active region. Here we investigate effect of temperature on electron leakage and efficiency droop of $Al_xGa_{(1-x)}N/GaN$ UV LEDs UV. Taking other factors like SRH, Aug and ED coefficients (A, B, k and p) are constants.

2. Theoretical Model

Efficiency of LED can be represented as a product of optical extraction efficiency (η_{OQE})and internal quantum efficiency (η_{IQE}): $\eta = \eta_{IQE} \times \eta_{OQE}$ [9]. η_{IQE} is rate of generation of photons in quantum well (QW) to number of carriers injected into the LED. The η_{OQE} is the ratio of photons emitted from the LED to photons produce inside QW. In other words, η_{OQE} stands for loss of photons inside the LED. It is reported that the primary reason of efficiency droop is not photon extraction. Therefore in efficiency estimation we only consider IQE. η_{IQE} is ratio of radiative recombination current to total current input to LED: $\eta_{IQE} = I_{rad} / I$. The total current in the QW is divided into two parts current that produces radiative recombination process and current that is lost: $I = I_{rad} + I_{lost}$. Thus, $\eta_{IQE} = I_{rad} /(I_{rad} + I_{lost})$. This equation determines that efficiency droop can only occur when I_{lost} increases more than I_{rad}. The mechanisms responsible for $I_{lost}$are SRH, AR, ED and EL effect. It means $I_{lost} = I_{SRH} + I_{Aug} + I_{ed} + I_{leak}$. Thus, current that is entering into QW is presented as $I_{QW} = I_{rad} + I_{SRH} + I_{Auger} + I_{ed}$. The total current becomes $I = I_{QW} + I_{Leak}$. The currents are defined as $I_{rad} = Bn^2, I_{SRH} = An, I_{Aug} = Cn^3, I_{ed} = k(n - n_0)^m$ and $I_{leak} = a\ (I_{QW})^p$. Where *A, B, C, k*(~1), n_0($n_0 = 5\times10^{18}\ cm^{-3}$), $m(\geq 2)$, *a, m* are constants and *n* is carrier density and *p* is leakage parameter. Thus, if V_{QW} is volume of active region of all QWs then $I_{QW} = eV_{QW}[An + Bn^2 + Cn^3 + k(n - n_0)^m]$ where *e* is electronic charge. ThenIQE will become $\eta_{IQE} = (\eta_{inj} Bn^2)/[An + Bn^2 + Cn^3 + k(n - n_0)^m + \alpha I_{QW}^p]$ [4, 9].

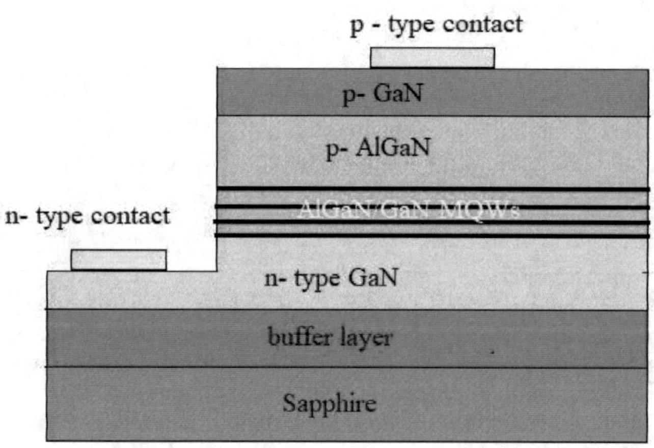

Figure 1. A picture of AlGaN/GaN UV LED

3. Result and Discussion

First leakage current I_{leak} is computed as a function of injection current I at various temperature. Following data are used in the calculation of leakage current: $A = 10^7 / s$, $B = 2\times10^{-11} cm^3 / s$, $C = 0$ and $p = 0.5$ [9]. We use these values for calculation of leakage current density which is a function of temperature. Figure 2 shows leakage current density with injection current density at temperature 250K, 300K, 350K, 400K, and 450K. From fig.2, it is observed that at 250K , current leakage is 16 A/cm^2; whereas at 450 K, current leakage is 22 A/cm^2. This shows that at higher temperature, high amount of current leakage takes place. It is to be pointed out that as temperature increases, the value of n_o is changed which in turn modifies value of m. To illustrate it further, leakage current is studied for another value of m.

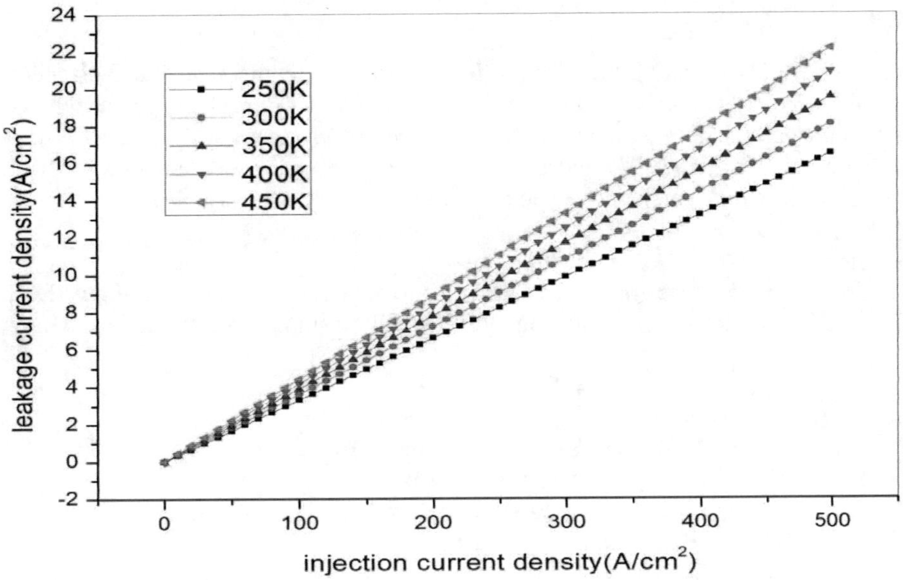

Figure 2. Leakage current density as a function of temperature

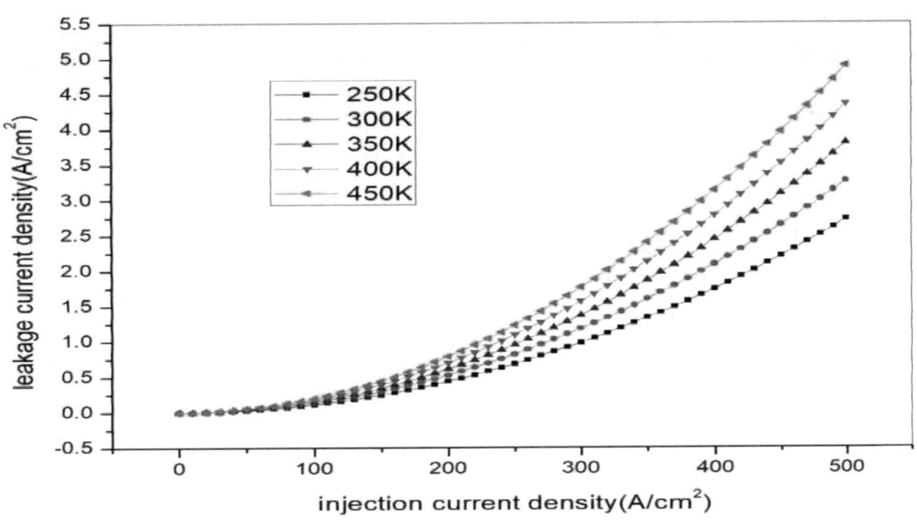

Figure 3. Leakage current density as a function of injection current at various temperatures

Figure 3 shows leakage current as a function of injection current *I* at various temperature with same constant values but *p =1*. In fig. 3 we use data $A = 10^7 / s$, $B = 2\times10^{-11} cm^3 / s$, $C = 0$ and $p = 1$ [9]. We observed that at 250K, 300K, 350K, 400K and 450K, leakage current density increases non - linearly. Leakage current enhances I_{loss}. As per our discussion, high I_{loss}reduces IQE of the LED. Thus, it is concluded that as temperature increases internal quantum efficiency decreases due to increase of lost current.

4. Conclusion

In this work, we have investigated the effect of temperature in EL mechanism on efficiency droop of $Al_xGa_{(1-x)}N/GaN$ UV LEDs. It is obtained that at high temperature leakage current is high. We conclude that under effect of temperature, more number of electrons is leaked to p-region without recombining with holes in the active region. To improve the performance, temperature of active region required to be reduced which can be achieved by minimizing self-heating.

Acknowledgments
Author (PS) acknowledges with thanks to Department of Science and Technology (DST) Govt. of India for financial support through awarding WOS-A Fellowship; project File No. SR/WOS-A/PM-97/2018

References

[1] Li, Zhang Y, Xu S, Bi W, Zhang , Kuo 2017*Materials* **10**1221.
[2] Joachim Piperk, 2010 *Phys. Status Solidi* A**207** 2217.
[3] H. Zhang, E. J. Miller, and E. T. Yu, 2004 *Appl. Phys. Lett.***84** 23.
[4] Meyaard , Qifeng Shen 2011 *Appl. Phys. Lett.* **99** 041112.
[5] X. Ni,1 X. Li, J. Lee,1 S. Liu, V. Avrutin, Ü. Özgür, H. Morkoç and A. Matulionis 2010 *J. Appl. Phys.***108** 033112.
[6] J.Xie, X. Ni 2008 *Appl. Phys. Lett.***93** 121107.
[7] S. S. Sahu, B. K. Sahoo 2019 *Thin Solid Films* **684** 59.
[8] Joachim Piperk, S. Li 2010*Opt. Quantum Electron* **42**, 89 .
[9] J. Lee, X. Li 2009 *Appl. Phys. Lett.***95** 201113.
[10] H. Zhang, E. J. Miller, and E. T. Yu 2004 *Appl. Phys. Lett.*, 84 23.

IOP Conf. Series: Materials Science and Engineering **798** (2020) 012019 doi:10.1088/1757-899X/798/1/012019

Studies on Reflectance-Temperature characteristics ofAlGaN semiconductor at three communication windows

Sakti Prasan Mohanty[1],C. S. Mishra[2], S. K. Sahoo[1†], G. Palai[2*]

[1]Department of Physics,Kalinga Institute of Industrial Technology(KIIT),Deemed to be University,Bhubaneswar -751024, India
[2]Gandhi Institute for Technological Advancement (GITA), Bhubaneswar -752054, India

Corresponding authors: [†]physushant@gmail.com, [*]gpalai28@gmail.com

Abstract. This paper reports the temperature variation of reflectance with respect to AlGaN wave guide structure.Plane wave expansion technique is employed for simulation to exhibit the reflectance pertaining to AlGaN wave guide structure.The simulation results show the approximate linear variation of reflectance with respect to temperature in the aforementioned AlGaN photonic wave guide structure at threecommunication windows (850 nm, 1310 nm, 1550 nm). Such characteristic study yields with accuracy,the realization of temperaturedependence of reflectance.

Keywords: Plane wave expansion, AlGaN, Reflectance, Temperature

1. Introduction

In recent years, integrated photonics finds potential applications in non-linear photonic devices. The direct band gap Group-III Nitride semiconductors are promising non-linear materials for the development of optoelectronic devices. Being a wide band-gap semiconductor with significant optical properties, Aluminum Gallium Nitride (AlGaN) suitably falls into this category and it is a good candidate for optical communication and sensing applications [1]. To pertain to the wave-guide structures for communication purposes, different quantities like refractive index and temperature, exhibit significant roles in experiencing the optical features [2-3]. With respect to the reflectance that stands as a function of refractive index in wave guide structures, this study realizes the temperature impact on this wave guide structure by the use of reflected signals. Earlier works have linked up to simulation analysis for light waves reflected from polymer waveguide structures taking account of temperature realization which includes three polymer waveguide structures on Silicon substrate by considering one wave length of 1550 nm at temperatures ranging from 300 K to 400 K[4] . Literature study also finds GaN grating structure or waveguide structure at 700 nm wavelength of light taking into account absorption loss and transmitted energy post incidence of light wave[5]. Related works have also taken care of reflected energy dependence of temperature in connection with refractive index and thickness variations [6]. This current paper delves into the refractive index change with room temperature and above upto 550 K at an energy gap of 1.5 eV for AlGaN and leads to the characteristic analysis of shifting of reflectance at three communication windows [7]. The three communication windows we have opted for in this paper for the clarification of temperature dependence of reflectance are 850 nm, 1310 nm and 1550 nm. Consequently plane wave expansion (PWE) method has been incorporated to explore the reflectance loss in aforementioned photonic waveguide structure. Temperature variation affects different parameters, most significantly refractive index, and structure of material and thickness of waveguide. The thickness of the waveguide structure for different temperatures, ranging from 300K to 550K is kept constant at 92 nm. The variation of refractive index with temperature at the three wave lengths (850 nm, 1310 nm, 1550 nm) is given in Table-1. This current paper is lined up as follows; proposed AlGaN waveguide structure, simulation results, analysis and result discussion are presented in section-2. Ultimately conclusions are presented in section-3.

2. Structure, Analysis and Discussion

To realize this Reflectance-Temperature characteristic, we propose an experimental set-up to investigate this variation. The experimental arrangement is shown in figure 1.

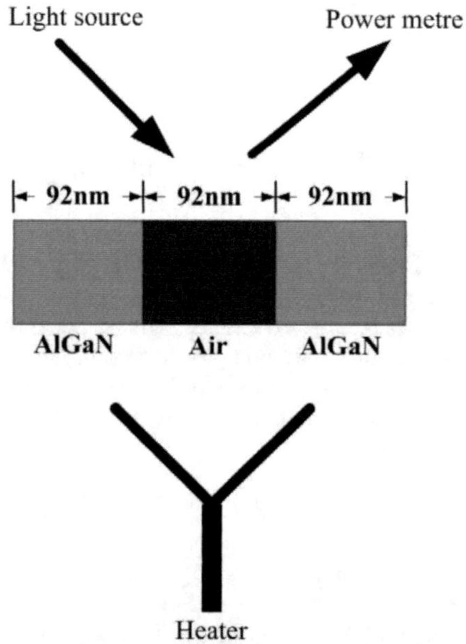

Figure 1: Experimental arrangement for Reflectance measurement from AlGaN waveguide structure

We have chosen the AlGaN waveguide structure of thickness 92 nm with air of same width in between. Light of wave length 850 nm first falls on this wave guide and a part of it gets reflected which is measured at the power meter. Consequently, other two light waves of wave lengths 1310 nm and 1550 nm fall and get reflected from the wave guide to repeat the measurements. A heater is used to achieve different temperature ranges.

Table 1: Refractive Indices at three communication windows with temperatures from 300 K to 550 K

Temperature in Kelvin	R.I. for 850 nm	R.I. for 1310 nm	R.I. for 1550 nm
300	2.846	2.787	2.768
350	2.849	2.789	2.77
400	2.852	2.793	2.774
450	2.859	2.799	2.78
500	2.861	2.802	2.783
550	2.864	2.804	2.785

Using data from Table-1 and employing PWE technique, simulation is done to find out reflectance of such waveguide with different temperatures varying from 300 K to 550K. The simulation results for temperature 300 K for 850 nm, 1310 nm, and 1550 nm are shown in figures 2(a)-(c) respectively. For other temperatures (350 K, 400 K, 450 K, 500 K, 550K) simulation is done as well, but results have not been shown here.

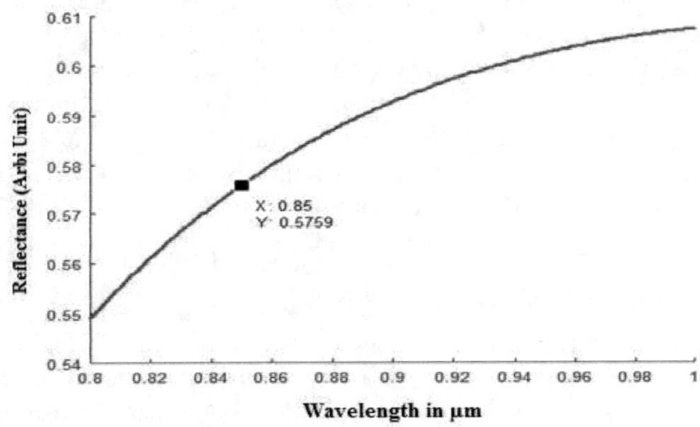

Figure 2(a): Reflectance variation with wavelength at temperature 300 K for wave length 850 nm

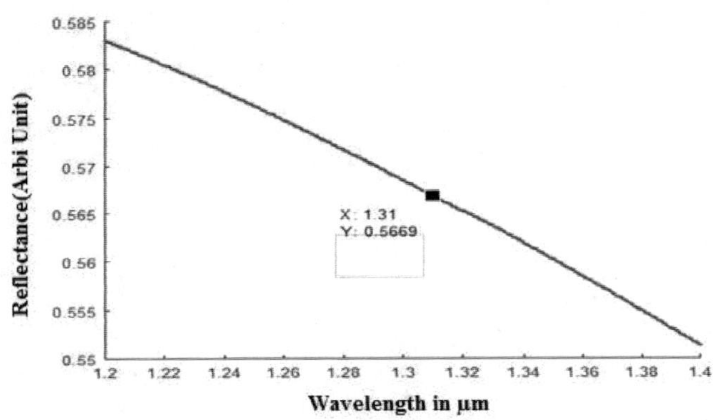

Figure 2(b):Reflectance variation with wavelength at temperature 300 K for wave length 1310 nm

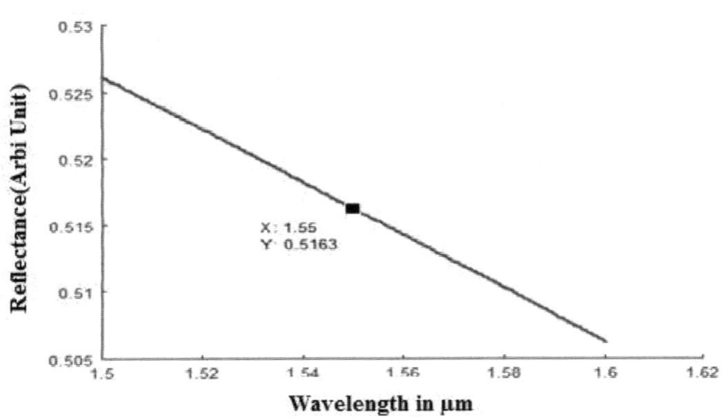

Figure 2(c):Reflectance variation with wavelength at temperature 300 K for wave length 1550 nm.

IOP Conf. Series: Materials Science and Engineering **798** (2020) 012019 doi:10.1088/1757-899X/798/1/012019

The figures 2(a)-(c) represent the variations between Reflectance (Arbi. Unit)along the vertical axis against thewave length(μm) along horizontal axis for 850 nm, 1310 nm and 1550 nm respectively. From these plots, it is observed that the variation of reflectance has an approximate linearity with these wavelengths. We have taken the measurement of reflectance at the three wave lengths that means 850 nm, 1310 nm, and 1550 nm in this paper.

Referring to these figures 3(a)-(c), the reflectances are 0.5759, 0.5669, and 0.5163 (Arbi. Unit) corresponding to wavelengths 850 nm, 1310 nm and 1550 nm respectively at 300K temperature. Similarly the reflectances for other temperatures are also calculated for the aforementioned waveguide structure. Attractive results have been observed that the reflectance for 850 nm increases linearly from 0.5500 to 0.5759 at temperature 300K-550K. Where in same waveguide structure, the reflectance decreases linearly with same temperature ranges. Using above values of reflectance for different temperatures at 850 nm, 1310 nm, 1550 nm, we plot their variations.

Finally three graphs have been plotted figures 3(a)-(c) with Temperature along horizontal axis and Reflectance along vertical axis corresponding to three wave lengths mentioned above.

For the wave length 850nm, the reflectance ranges from 0.5759 to 0.5775(Arbi. Unit), for the wave length 1310nm, the reflectance ranges from 0.5669 to 0.5721(Arbi. Unit) and for the wave length 1550nm, the reflectance ranges from 0.5163 to 0.5221(Arbi. Unit), corresponding to 300K-550K temperature ranges.

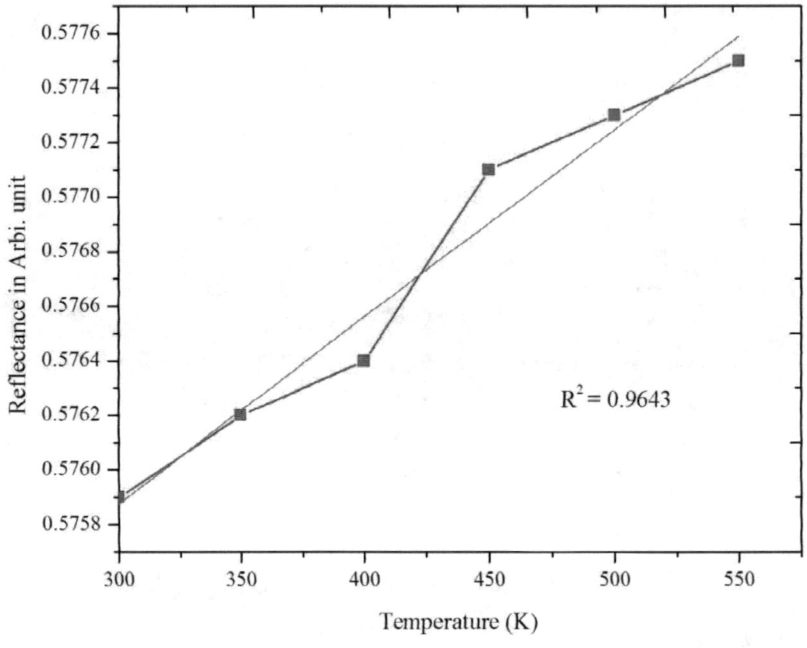

Figure 3(a): Reflectance vs. Temperature for wave length 850 nm

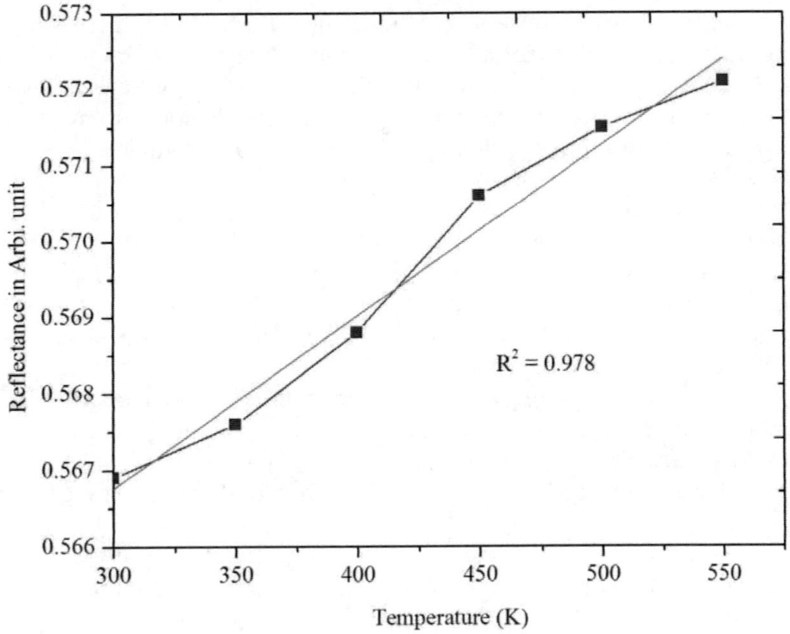

Figure 3(b): Reflectance vs. Temperature for wave length 1310 nm

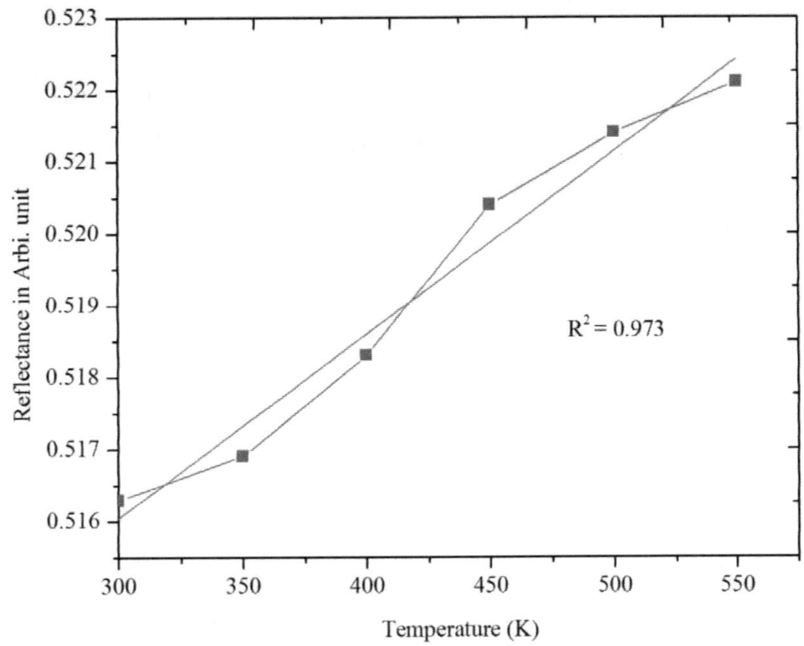

Figure 3(c): Reflectance vs. Temperature for wave length 1550 nm

3. Conclusions

With the help of simulation studies, the variation of Reflectance with Temperature is thoroughly investigated in the aforementioned AlGaN waveguide structure. The simulation is done with plane wave expansion (PWE) method whose results depict that the reflectancehas increasing linearity with temperature, ranging from 300 K-550 K for all the three communication windows (850 nm,1310 nm, 1550 nm) with AlGaNwaveguide structure of thickness 92nm. The trending linearity of approximately 96% is nicely verified with these results leading to an accurate demonstration of temperature in AlGaN wave guide structure at the aforementioned three communication windows.

References

[1] Eda GoldenbergAli Kemal Okyay2014Plasma- Optical characteristics of nanocrystalline $Al_xGa_{1-x}N$ thin films deposited by hollow cathode assistedatomic layer depositionJ. Vac. Sci. Technol. A **32** 031508

[2] N. Antoine-Vincent 2003Determination of the refractive indices of AlN, GaN, and $Al_xGa_{1-x}N$ grown on (111)Si substrates J. Appl. Phys. **93** 5222

[3] Natali F, Semond F, Massies J, Byrne D, Laüigt S, Tottereau O and Vennégués P 2003Correlation between threading dislocation density and the refractive index of AlN grown by molecular-beamepitaxy on Si(111)Appl. Phys. Lett. **82** 1386

[4] Goldhahn R, Scheiner J and Shokhovets S 2000Refractive index and gap energy of cubic $In_xGa_{1-x}N$,Appl. Phys. Lett. **76** 291

[5] Mishra C S and Palai G 2015Temperature and Pressure effect on GaN waveguide at 428.71 Terahertz frequency for Sensing application Optik – International Journal for Light and Electron Optics**126** 4685-4687

[6] Mishra C S and Palai G 2015Simulation studies for reflected light of polymer waveguide for realization of temperature Optik – International Journal for Light and Electron Optics**126** 3656–3658

[7] Tisch U, Meyler B, Katz O, Finkman E and Salzman J 2001Dependence of the refractive index of $Al_xGa_{1-x}N$ on temperature and composition atelevated temperatures J.Appl. Phys. **89** 2676

IOP Conf. Series: Materials Science and Engineering **798** (2020) 012020 doi:10.1088/1757-899X/798/1/012020

Influence of defect density and layer thickness of absorption layer on the performance of tin based perovskite solar cell

Priyanka Roy, Numeshwar Kumar Sinha, Sanjay Tiwari and Ayush Khare*

National Institute of Technology, Raipur

*Corresponding author's e-mail address: proy.phd2018.phy@nitrr.ac.in

Abstract. The tremendous progress exhibited by Perovskite solar cells (PSCs) within the decade has made PSCs the fastest growing PV cell to date. The PSCs has exhibited increment in power conversion efficiency (PCE from 3.2% to 25.2% within a decade. In PSCs, usually $MAPbX_3$ (X being Cl,Br and I) layer is used as an absorber layer. The incorporation of toxic Lead (Pb) is a matter of concern by some of the researchers. In order to deal with this Tin (Sn) is being used as an alternative of Pb. However, the performance of Sn based PSC still lags behind Pb based PSC. In this work, we demonstrate an efficient Sn based PSC with J_{Sc} of 32.6398 mA/cm^2, V_{oc} of 0.8946V, FF of 80% and PCE of 23.4%. The results exhibits the potential of Sn based PSCs which is not only efficient but also is environmental friendly. We study the impact of defect density and layer thickness of absorber layer over the performance of Sn based PSC. An analysis on the effect of doping in Electron Transport Layer (ETL) and Hole Transport Layer (HTL) over the performance of Sn based PSC is also conducted.

Key Words: Perovskite solar cell, Tin based perovskite solar cell, SCAPS-1D, Defect Density, Absorber layer thickness;

1. Introduction

The PSCs have exhibited their enormous potential by rapid advancement in performance within a decade. The PSCs have shown increment from 3.2% to 25.2% in PCE within a decade [1]. However, although these cells exhibit appreciable performance but still the incorporation of toxic Lead (Pb) is the matter of concern [2]. It is a well-known fact that toxicity due to lead can cause severe health risks and also has a deplorable impact on our ecosystem. This is one of the factors which is creating a negative impact over users and avoiding its market acceptance of the final product. $MAPbX_3$ (where X is Cl, Br or I) is the widely adopted absorber layer in the PSCs . Tin (Sn) have been used by several researchers as an alternative of Pb which is nontoxic. Unfortunately, Sn based PSCs suffers from drawbacks: (i) efficiency as high as Pb based PSCs are not yet achieved, (ii) instability in ambient conditions, (iii) degrades easily due to oxidation of Sn^{2+} into Sn^{4+} . All such demerits tend to restrict the development of Sn based PSCs. Till date the most efficient tin based PSC has attained PCE of 9.6%[3].It is seen that Tin based perovskite when deposited over TiO_2 has exhibited poor surface coverage [4]. Moreover, the issue of ion immigration takes place when TiO_2 is used as ETL. In order to deal with this here we use ZnO-NR as ETL. In this work, we demonstrate an efficient Sn based PSCs, and efficiency as high as 23.4% has been achieved. This work suggests the fact that tin based

PSC also has the ability to exhibit performance comparable to lead based PSC. An optimum range of absorber layer thickness, absorber defect density and doping level of ETL & HTL is also investigated.

2. Simulation Setup

SCAPS-1D (Solar Cell Capacitance Simulator) version 3.3.07 has been used to simulate the PSC. This software was specifically developed for CdTe and $CuInSe_2$ family. The several extensions improved its capabilities enabling the analysis of other types of solar cell. As compared to other softwares, SCAPS-1D has in-built operation window and differentiated models for defects, recombination and grating. This software is based upon general semiconductor physics, like poisons equation, drift diffusion equation and continuity equation. The basic equations are generally elucidated in 1D in a steady state condition. SCAPS analyses the physics behind the model and examines the electric field distribution, recombination profiles, individual current densities and carrier transport mechanism. This software enables to add upto 7 semiconducting layer with various material and defect properties[5].

2.1 Device Structure

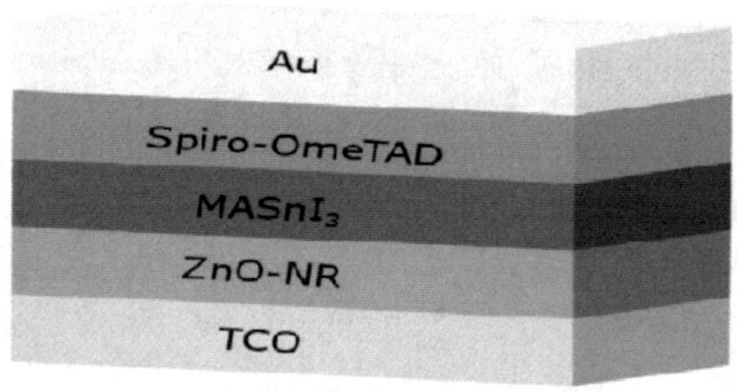

Figure 1. Schematic of device architecture of simulated Sn based PSCs: (a) Au/Spiro-OmeTAD/ MASnI3/ZnO-NR/FTO

We adopted planar architecture structure the schematic shown in Figure 1. The simulated PSC has five layers: **(i)** Gold (Au) as metal based cathode **(ii)** Hole transport layer (HTL) as p-type 2,2,7,7'-tetrakis-(N,N-dimethoxyphenyl-amine)-9,9'-spirobifluorene (Spiro-OmeTAD) **(iii)** absorber layer as Methylammonium lead trihalide ($MASnI_3$) **(iv)** ETL as Zinc oxide nanorod (Zno-NR) **(v)** Transparent Conductive Oxide(TCO) as Flourine doped Tin Oxide (FTO). The material perovskite as an absorber layer exhibits ambipolar behaviour with high mobility for both electron and hole. Its function is to absorb the incident radiation leading to the creation of the charge carriers. The function of ETL is to extract and transfer electron and block hole. Whereas, the function of HTL is to extract and transfer holes and block electrons.

2.2 Parameters

The parameters selected for simulation are taken from literature. Table-1 summarizes all the parameters used during simulation and Table-2 enclosed the data for front and back contact. The $MASnI_3$ based solar cell is simulated at AM1.5G at 300K. The absorber layer, HTL and ETL defect density was taken as 10^{14} cm^{-3} , 10^{15} cm^{-3} and10^{15} cm^{-3} respectively. The parameter values of various layers are extracted from research work carefully [6][7][8].

IOP Conf. Series: Materials Science and Engineering **798** (2020) 012020 doi:10.1088/1757-899X/798/1/012020

Table 1 The parameters used for simulation

Parameters	Spiro-omeTAD	$MASnI_3$	ZnO-NR	SnO_2: F
Thickness (μm)	0.5	0.6	0.5	0.25
Band gap(eV)	3.17	1.2	3.27	3.5
Electron affinity(eV)	2.1	4.17	4.3	4.4
Dielectric Permittivity	3	10	9	9
CB effective density of states (1/cm3)	2.5E+18	2.1E+18	2E+18	2.2E+18
VB effective density of states (1/cm3)	1.8E+19	1.9E+19	1.8E+20	1.8E+19
Electron thermal velocity(cm/S)	1E+7	3E+7	1E+7	1E+7
Hole thermal velocity(cm/S)	1E+7	3E+7	1E+7	1E+7
Electron mobility(cm2/VS)	2E-4	1.6E+0	1E+2	2E+3
Hole mobility(cm2/VS)	2E-4	2E-1	2.5E+1	1E+2
Shallow uniform donor density ND (1/cm3)	0	0	1E+18	2E+19
Shallow uniform acceptor density NA(1/cm3)	1E+20	6E+18	0	0

Table 2 Parameters for Back and Front contact

Parameters	Back Contact	Front Contact
Surface recombination velocity of electrons (cm/s)	1.00E+7	1.000E+5
Surface recombination velocity of holes (cm/s)	1.000E+5	1.000E+7
Metal work function (eV)	5.1444	4.43
Majority carrier barrier height relative to Ef (eV)	-0.0444	-0.0600
Majority carrier barrier height relative to Ev (eV)	-0.0000	-0.0028

3. Result and Discussion

The best performance is attained at thickness of 0.4μm with, PCE=23.4%, J_{Sc}= 32.6398 mA/cm^2, V_{oc} = 0.8946V and FF = 80%. The J-V curve of the simulated Sn based PSC is shown in Figure 2.

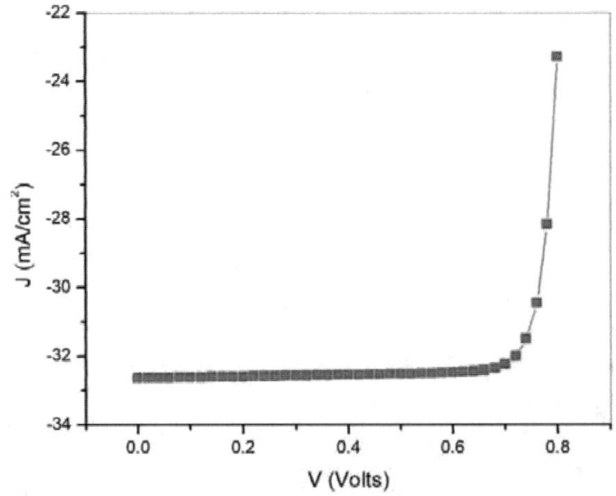

Figure 2. J-V curve of simulated Sn based PSC.

3.1 Impact of Defect Density and Thickness of Absorber Layer on the Performance

In order to obtain better PV performance the absorber layer plays a crucial role. The generation of photo excited charge carriers usually occurs at the absorber layer.The film quality and morphology of this layer has a direct impact over the PV performance of the PSC. The performance parameters of the PSC are highly dependent over the absorption layer thickness and defect density. In this work, we vary the absorber defect density (N_t) from range 1E^{12} to 1E^{16}. It is seen that the performance parameters reduces to a great extend as the N_t increases as shown in Figure 3. At N_t=1E^{12}, the performance parameters are: PCE= 30.17%, V_{oc}=1.357V , J_{sc}= 39.9123 mA/cm^2 , FF= 55.7% and when N_t=1E^{16} the performance parameters are: PCE= 2.56%, V_{oc}=0.6796V , J_{sc}= 5.0988 mA/cm^2 , FF= 73.84%. It is seen that PCE, J_{sc} and V_{oc} decreases with increases in N_t. Whereas the FF initially increases till 1E^{14} reaches to a maximum value and then decreases. The decrement in the performance with increase in defect density is basically due to the enhancement in the recombination process which leads to annihiliation of the charge carriers. At low defect density, the carrier diffusion length is high and recombination process is low thereby contributing to better PV performance.

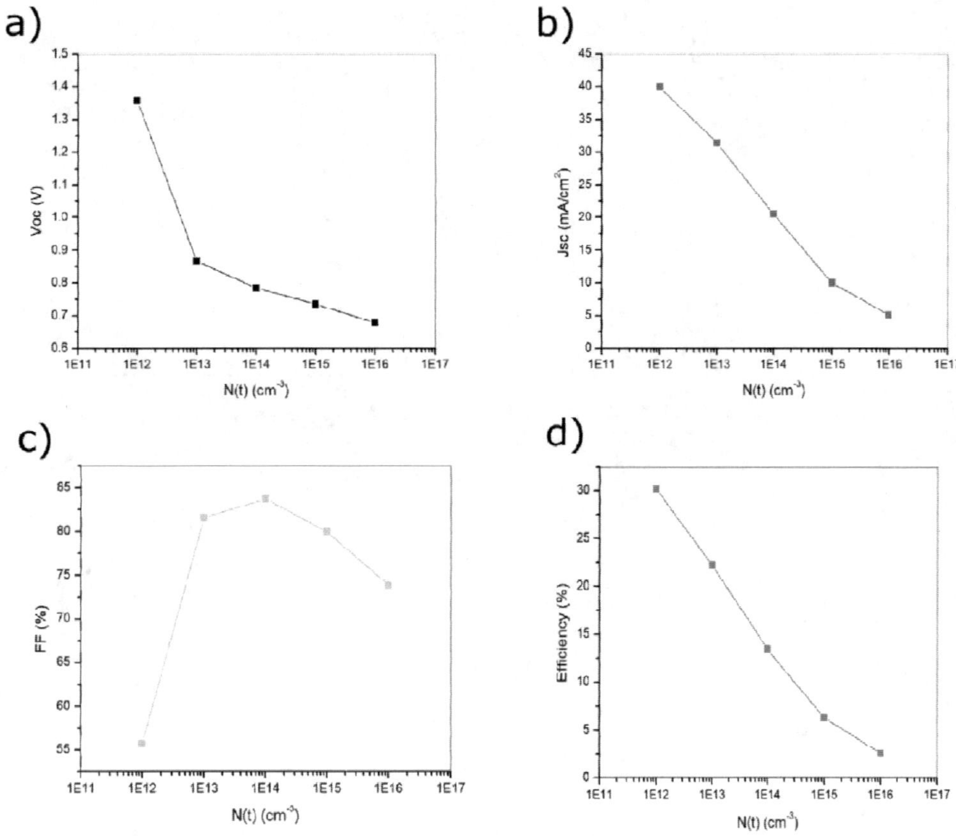

Figure 3. Variation of performance parameters of Sn based PSC with change in absorber defect density: a) Variation of Voc, b) variation of Jsc, c) variation of FF and d) variation of Efficiency as function of absorber layer defect density (N_t).

The performance of PSC depends upon the thickness of the absorber layer to a great extent as the photo excited charge carrier generation occurs at this layer. The absorber layer thickness has direct influence on the PV performance of the device as it affects the diffusion lengths of the charge carriers. If the absorber layer is too thick, then the charge carriers might recombine before reaching the charge collecting electrodes and too thin absorber layer will not be able to absorb the incident radiation properly, thereby reducing the overall performance of the device. In this study we vary the absorber layer thickness at range 0.1-1μm, keeping the thickness of ETL and HTL fixed. The result can be seen from Figure 4 that performance parameters are highly thickness dependent. The V_{oc} decreases with increase in thickness of the absorber layer. The J_{sc} and PCE increases initially and reaches maximum at t=0.4μm, and then slightly decreases.

IOP Conf. Series: Materials Science and Engineering **798** (2020) 012020 doi:10.1088/1757-899X/798/1/012020

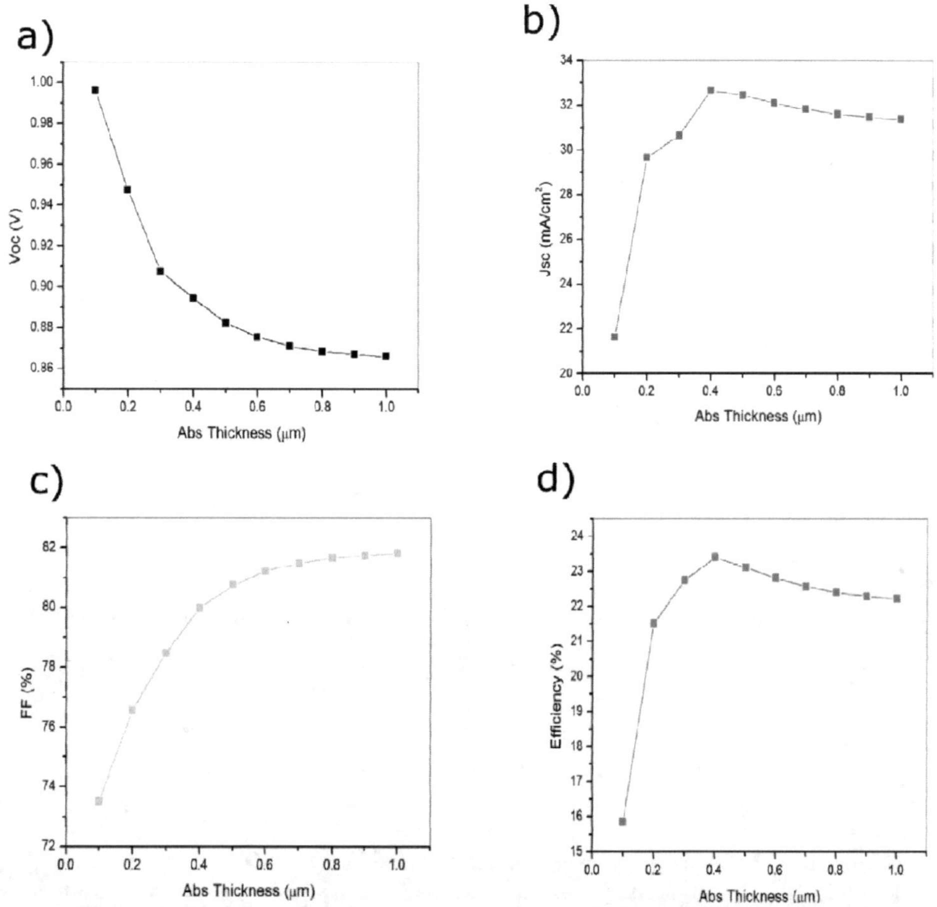

Figure 4. Variation of performance parameters of Sn based PSC with change in absorber layer thickness: a) Variation of Voc, b) variation of Jsc, c) variation of FF and d) variation of Efficiency as function of absorber's thickness.

3.2 Impact of Doping Level in HTL and ETL

The PV performance of solar cell is highly affected by the electrical behavior of the layer, which can be altered by addition of a dopant. The doping of HTL and ETL improves the PV performance by raising the interface electric field. In order to understand the impact of doping effect over HTL and ETL, we vary the acceptor carrier density (N_A) and donor carrier density (N_D) from $1E^{14}$ to $1E^{19}cm^{-3}$. Firstly, analysing the doping effect on HTL, we vary the acceptor carrier density (N_A). The variation in performance parameters with carrier concentration is shown in Figure 5. It is seen that, V_{oc} is constant till $1E^{16}$, and the increases after $1E^{16}$. Similarly J_{sc}, FF and PCE increases abruptly from $1E^{16}$ and value gets constant after $1E^{18}$ signifying the optimum range to be $1E^{16}$ to $1E^{18}$.

IOP Conf. Series: Materials Science and Engineering **798** (2020) 012020 doi:10.1088/1757-899X/798/1/012020

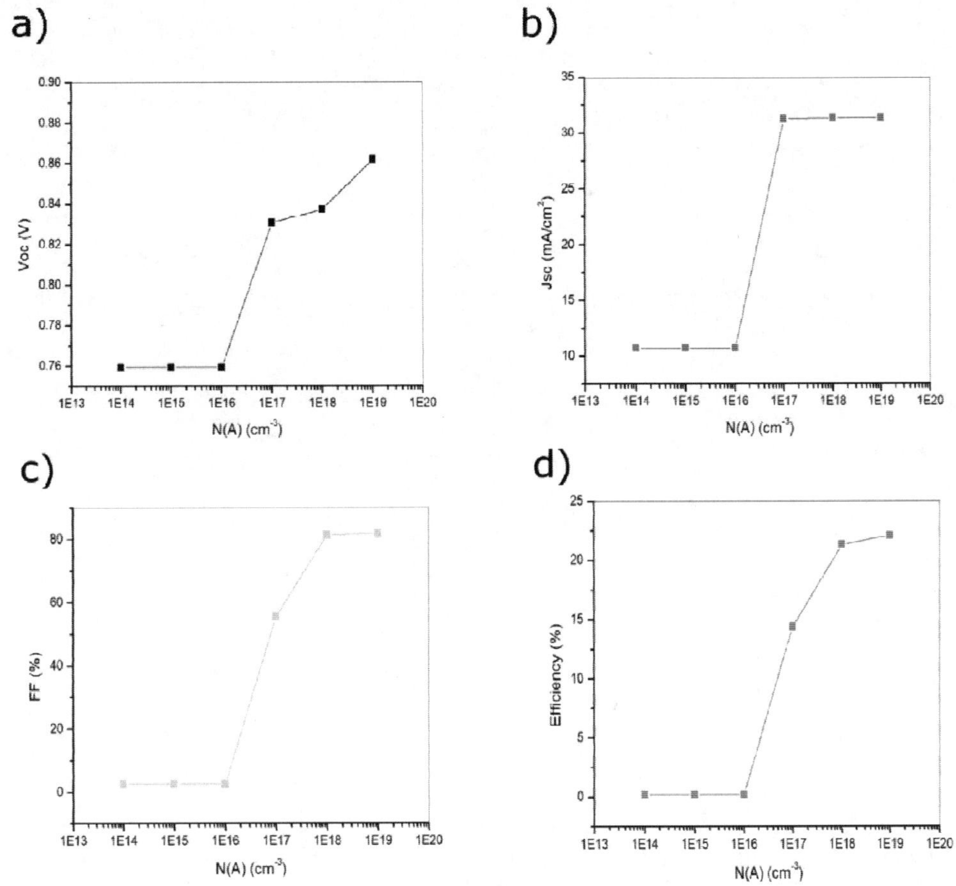

Figure 5. Variation of performance parameters of Sn based PSC with change in acceptor carrier concentration (N_A): Variation of performance parameters of Sn based PSC with change in absorber acceptor carrier concentration: a) Variation of Voc, b) variation of Jsc, c) variation of FF and d) variation of Efficiency as function of absorber's acceptor carrier concentration

It is seen that the V_{oc} doesn't depends upon N_D whereas J_{sc} and PCE increases from $1E^{16}$to $1E^{18}$ and then gets constant. The optimum range for carrier concentration is $1E^{16}$ to$1E^{18}$. The impact of donor carrier concentration (N_D) over the performance of Sn based PSC is shown in Figure 6 [6].

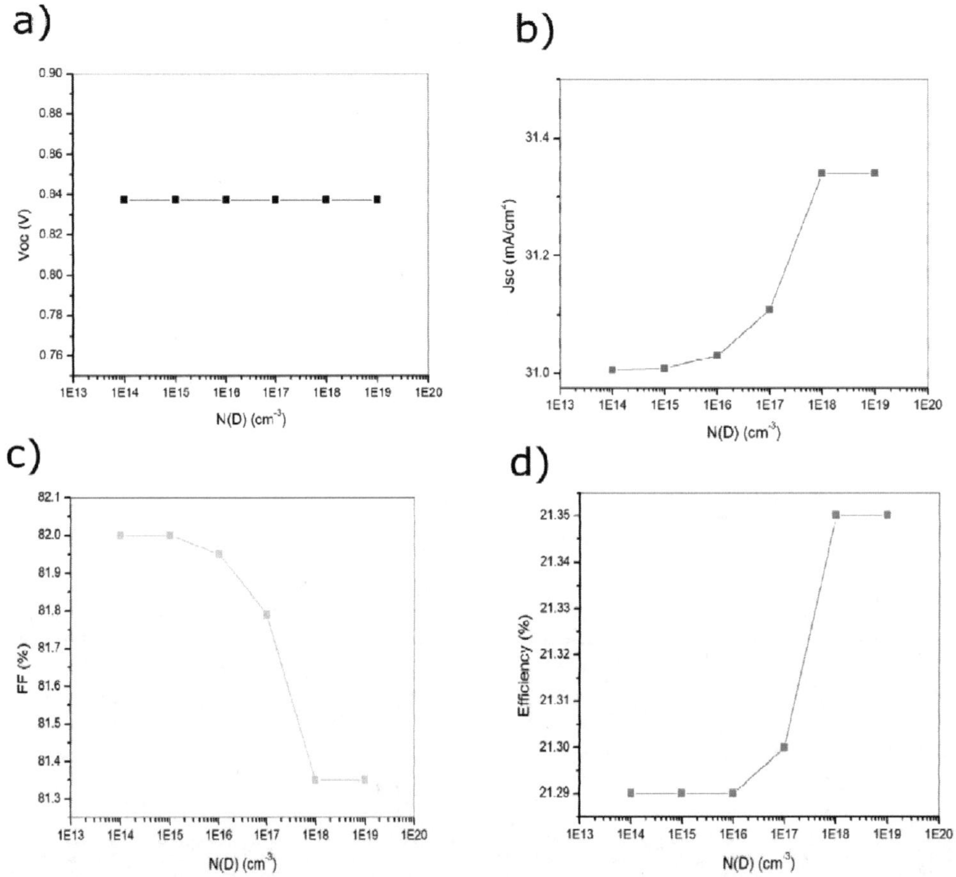

Figure 6. Variation of performance parameters of Sn based PSC with change in donor carrier concentration (N_D): Variation of performance parameters of Sn based PSC with change in absorber donor carrier concentration: a) Variation of Voc, b) variation of Jsc, c) variation of FF and d) variation of Efficiency as function of absorber's donor carrier concentration.

4. Conclusions

Table 3 Optimized values for Sn based PSC.

Optimized Parameters	ETL	$MASnI_3$	HTL
$N_t(cm^{-3})$		$1E^{12}$	
$N_A\,(cm^{-3})$			$1E^{16}$-$1E^{18}$
$N_D(cm^{-3})$	$1E^{16}$- $1E^{18}$		
Absorber thickness (μm)		0.4μm	

In this work, we demonstrated the potential of Sn based PSC with a performance being comparable to

Pb based PSC by attaining a high PCE of 23.4%. The impact of defect density and thickness of absorber layer is studied. It is seen that defect density of absorber layer must be as low as possible and optimum layer thickness should 0.4μm to attain high performance. The optimum doping concentration of ETL and HTL required for better PV performance was studied. The obtained results are summarized in Table-3. However, practically the device suffers from issue like poor device stability due to easy oxidation from Sn^{2+} to Sn^{4+}. This issue can be solved by improving the fabricating process which can deposit high quality defect free films and using proper device encapsulation.

Acknowledgement
The authors would like to thank Professor Marc Burgelman, Department of Electronics and Information Systems, University of Gent for the development of the SCAPS software package and allowing its use.

References

[1] NREL 2019 PV Research Cell Record Efficiency Chart
[2] Lyu M, Yun J H, Chen P, Hao M and Wang L 2017 *Adv. Energy Mater.* **7** 1602512
[3] Shao S, Liu J, Portale G, Fang H H, Blake G R, ten Brink G H, Koster L J A and Loi M A 2018 *Adv. Energy Mater.* **8** 1702019
[4] Hao F, Stoumpos C C, Guo P, Zhou N, Marks T J, Chang R P H and Kanatzidis M G 2015 *J. Am. Chem. Soc.* **137** 11445–52
[5] Marc Burgelman, Koen Decock, Alex Niemegeers, Johan Verschraefen S D 2018 SCAPS manual
[6] Du H J, Wang W C and Zhu J Z 2016 *Chinese Phys.* B **25** 108802
[7] Mandadapu U, Vedanayakam S V and Thyagarajan K 2017 Simulation and Analysis of Lead based Perovskite Solar Cell using SCAPS-1D Indian J. Sci. Technol. **10** 1–8
[8] Anwar F, Mahbub R, Satter S S and Ullah S M 2017 *Int. J. Photoenergy* **2017** 1

IOP Conf. Series: Materials Science and Engineering **798** (2020) 012021 doi:10.1088/1757-899X/798/1/012021

Synthesis and characterization of CdTe thin film on FTO by electrodeposition technique for CdTe/CdS solar cells

Lekha Verma and Ayush Khare

Department of Physics, Thin Film Research Laboratory, National Institute of Technology, Raipur, 492010, India

*Corresponding author's e-mail address: lekhaverma.85@gmail.com

Abstract. In this work, the synthesis of the CdTe thin film (absorber layer) on fluorine-doped Tin oxide (FTO) substrate by electrodeposition technique of CdTe/CdS solar cell has been presented. As a result of the superior efficiency in energy conversion by absorbing energy in a broad range of the solar spectrum, thin film CdS/CdTe solar cells have been paid attention worldwide as a potential candidate for photovoltaics (PV). The electrodeposition (ED) technique is suitable for the synthesis of CdTe material due to its ease, low cost, scalability and manufacturability. Usually, the quality of these deposited thin films depends on several growth parameters, and the type of conducting glass substrate plays a crucial role. CdTe material grown up on the top of the FTO coated glass substrate having dimensions 75mm* 25mm*1mm and was characterized for its optical properties by UV-Vis spectrometer (UV-Vis), Photoluminescence (PL) and Fourier transform infrared spectrometer (FTIR). XRD result confirms the cubic crystalline character of the deposited material. The band gap of CdTe film estimated using Tauc's plot is found to be 1.5eV. It is expected that this article will enrich the features of CdTe/CdS devices for solar energy conversion and stimulate further innovative research interest on this interesting topic.

1. Introduction

Cadmium Telluride (CdTe) is an II-VI group compound semiconductor. Due to its band gap of 1.5 eV (direct) and high photoluminescence quantum efficiency, it is having various imaging applications. CdTe can be used in several optoelectronic devices such as gas sensors, photo catalysis, biological detection, solar devices, photo detectors, UV sensors, nonlinear optical materials, various luminescence devices, etc. [1]. CdTe photovoltaic (PV) has the lowest emissions of coal, the lowest consumption of water, and the shortest energy reimbursement time for any current PV technology. Efficient photonic devices based on thin films can be produced using a variety of thin film manufacturing technologies such as radio frequency magnetron sputtering, chemical vapour deposition, laser ablation, electrodeposition (ED), and closed space sublimation [2-5]. ED is a simple and economically cheap method for the preparation for the CdTe absorption layer of the solar cell. The traditional ED process requires the use of a 3-electrode device [cathode (working electrode), anode (counter electrode), and the reference electrode] and mainly the researches has been done were based on the three-electrode method [6]. In the 3-electrode system, the use of the reference electrode is due to changes in the concentration of electrolytes throughout the deposition. Impurities such as Ag^+ and

K^+ ions which present in the reference electrode, will leak into the electrolyte deposition and significantly decrease the efficiency of CdTe-based solar cells [7]. Thus, the simple 2-electrode system is used in this study, instead of the traditional 3 electrode system. To examine its suitability for the electronic and PV applications, this article reports a comprehensive study on CdTe layers developed by the 2-electrode ED process.

2. Materials and methods

Before the deposition, the fluorine tin oxide (FTO) coated glass substrates were washed carefully with diluted HCL and acetone. After that the slides were thoroughly washed by deionised water several times. CdTe thin films have been deposited by ED technique. CdTe electroplating deposition was carried out by the combination of 2 electrodes: (a) cathode (glass / FTO) (b) anode (carbon electrode) immersed in an electrolyte. This method is usually performed in an aqueous electrolyte containing high concentration (0.5–2.0M) of Cd ions from cadmium sulfate ($CdSO_4$) and very low concentrations (~1.0–5.0mM) of Te from tellurium dioxide (TeO_2) in aqueous solutions. Initially, the electro-purification process was carried out for 50h to remove the impurities present in the solution. An appropriate amount of diluted TeO_2 solution was added to adjust the concentration level of Te, and 1000 ppm of $CdCl_2$ was added into the solution. At the beginning when precipitation started, stirring was done using magnetic stirrer for 20min at a temperature of 85^0C and pH of the solution was maintained at 2±0.02. The deposition was made in static condition. After the deposition, the deposited films were rinsed with double distilled water and dried out at room temperature. Characterization of the CdTe films was using specific techniques to determine their structural, optical, and luminescence properties. The deposition process of CdTe is governed by the following chemical reactions [8]:

$$TeO_3^{2-} + Cd^{2+} + 3H_2O + 6e \longrightarrow CdTe + 6OH^-$$

The electrodeposited CdTe is shown in Figure 1. The deposited film is almost uniform and blackish in color.

Figure 1. CdTe film electrodeposited on FTO.

3. Results and Discussions

3.1 UV-Visible (UV-Vis) spectrum

For the development of PV devices, the study of optical parameters including optical absorption, transmission, and the band gap energy (E_g) is very important. Using Shimadzu 1800 UV visible spectrometer, optical absorption measurement was performed. Spectral absorption study provides the information about the nature of the band gap (i.e. direct or indirect) of semiconducting materials [9]. Optical absorption spectrum of ED-CdTe film is shown in figure 2(a) and the E_g value of this film is estimated using Tauc's plot as shown in figure2(b). The enhancement in absorption helps to increase the amount of photo-generated charge carriers in the CdTe film, which gives the beneficial impact on the performance of the solar cell. The observed average band gap value is found to be E_g=1.5eV which is consistent with the literature [10].

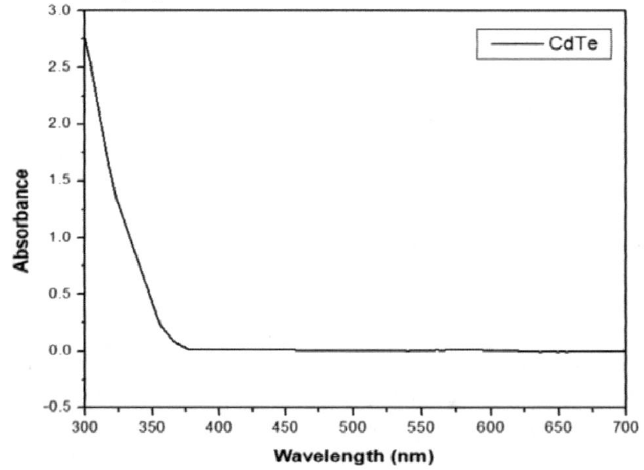

Figure 2. (a) Absorption spectrumof ED deposited CdTe thin film.

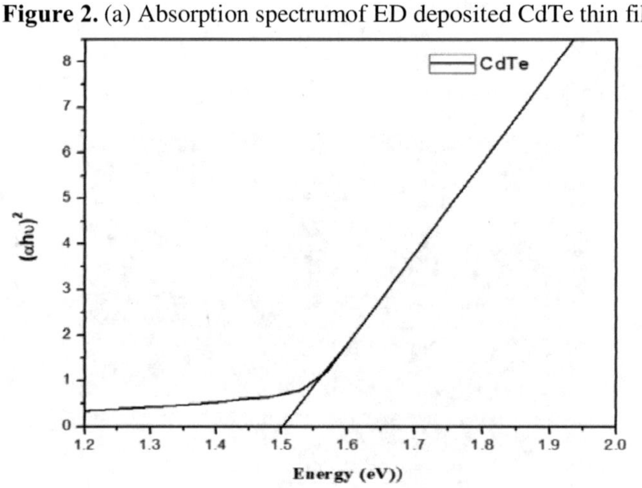

Figure 2. (b) Tauc's plot of ED deposited CdTe thin film.

3.2 Photoluminescence (PL) studies

PL is a reliable method to examine the transitions of electrons from higher energy levels to lower energy states. Figure 3 represents the room temperature PL spectrum of electrodeposited CdTe layer in the wavelength range of 750-850nm. The wavelength of the laser beam used as the source of excitation (He-Ne) was 635nm (1.96eV). The PL spectrum of the film exhibit two different bands, centered at 797nm (E_g = 1.55eV) and 840nm (E_g = 1.47eV), which is very close to the previous studies [11]. Such spectral lines are caused by the excitons in the shallow acceptors or donors [12].

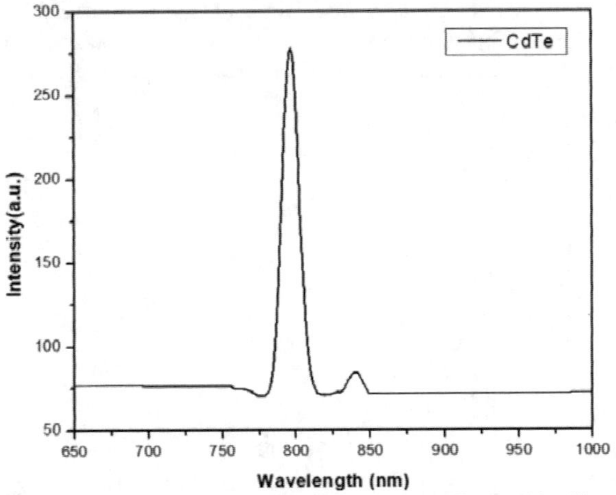

Figure 3. PL spectrum for ED deposited CdTe thin film.

3.3 Fourier transforms infrared spectroscopy (FTIR) analysis

FTIR is an analytical technique to identify the nature of the material as well as it gives information about the impurities present in the deposited film. Bruker Alpha FTIR spectrometer was used to record IR absorption spectrum. Figure 4 presents the FTIR transmission spectrum of electrodeposited CdTe thin film. The spectrum consists two intense bands at: 1642cm^{-1} due to the C=O stretching vibration and 3398cm^{-1} due to the C–O–H stretching vibrations of the carboxylic acid group [13]. The absorption band around 1123 cm^{-1} is assigned to the presence of C=O stretching vibrations [14].

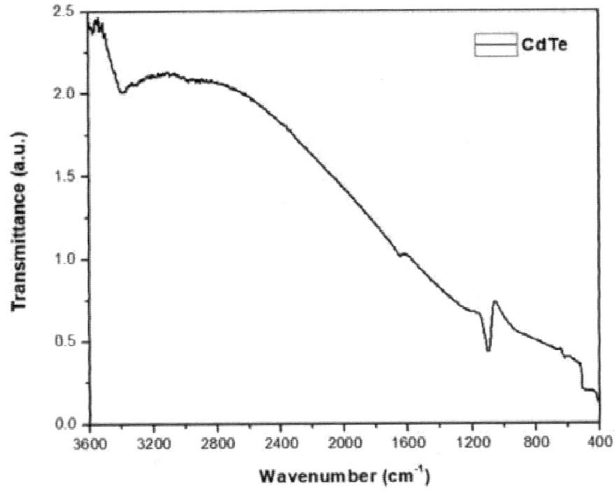

Figure 4. FTIR spectrum for ED deposited CdTe thin film.

3.4 X-Ray Diffraction (XRD)

XRD is an effective analytical method to determine the phase and determine the crystalline size of deposited materials [15]. The XRD spectrum for ED-CdTe film grown on FTO substrate is displayed in figure 5, which was recorded in 2θ range of 0°-60° using Cu Kα radiation of 1.5406 A° wavelength. The diffraction peaks (111), (002), (022) and (113) validated to the cubic phase of the CdTe thin film from JCPDS data card no.98-062-0533. The respective peaks are corresponding to the diffraction angles 23.5, 27.2, 38.9 and 46.0. Thus the CdTe film exhibit cubic crystalline nature.

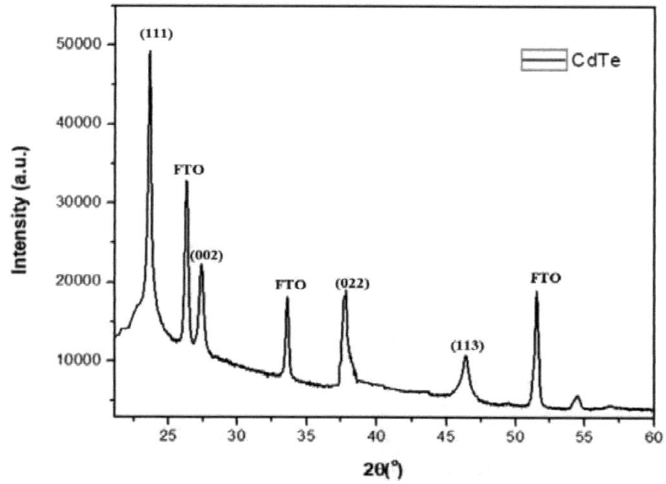

Figure 5. XRD spectrum for ED deposited CdTe thin film.

With the help of following equations (1), (2), and (3) crystallite size (D), dislocation density (δ) and strain (ε) of CdTe material are calculated as shown in table 1.

$$D = \frac{0.9\lambda}{\beta cos\theta} \quad ...(1)$$

$$\delta = \frac{1}{D^2} \quad ...(2)$$

$$\varepsilon = \frac{\beta cos\theta}{4} \quad ...(3)$$

Where,
D= crystalline size
λ= wavelength of x-ray radiation
β= full width of half maxima (FWHM)
θ= Braggs angle

Table 1. XRD parameters of ED-CdTe film.

Parameters	**CdTe**
Average crystalline size (D)	44.78nm
Dislocation density (δ)	49.9 lin/m^2
Microstrain (ε)	0.15 lin^{-2}m^{-2}

4. Conclusions

In the present work CdTe layer has been successfully electrodeposited using 2-electrode system. Optical and structural properties of CdTe thin film were investigated by using different characterization techniques. The optical band-gap of deposited CdTe film is found as 1.51 eV. PL spectrum of CdTe layer consist two characteristics PL peaks situated at 797nm and 840nm. In the FTIR spectrum C=O and C-O-H stretching vibrations are found to dominate. XRD analysis shows that the CdTe film is cubic crystalline in nature with preferred orientations in (111), (002), (022), and (113) directions.

Acknowledgements
The authors are thankful to the Science and Engineering Research Board (SERB)project no. EMR/2017/000431, New Delhi, India for their financial assistance.

References

[1] Bera S R and Saha S 2018 *Materials Today: Proceedings* **5** 3476–85
[2] Kulkarnia R, Rondiyaa S, Pawbakea A, Waykara R, Jadhavara A, Jadkara V, Bhordea A, Datec A, Pathand H and Jadkard S 2017 *Energy Procedia* **110**188–95.
[3] Almeida D B, Rodriguez E, Moreira R S, Agouram S, Barbosa L C, Jimenez E and Cesar C L 2009 *European Conference on Biomedical Optics **06*** 7371-10.
[4] Ferekides C S, Marinskiy D, Viswanathan V, Tetali B, Palekis V, Selvaraj P and Morel D L 2000 *Thin Solid Films* **361** 520-526
[5] Dharmadasa I M, Madugu M L, Olusola O I, Echendu O K, Fauzi F, Diso D G, Weerasinghe A R, Druffel T, Dharmadasa R, Lavery B, Jasinski J B, Krentsel T A and Sumanasekera G 2017 *Coatings* **7** 1-17
[6] Echendu O K, Okeoma K B, Oriaku C I and Dharmadasa I M 2016 *Adv. Mater. Sci. Eng.* **2016** 1-8
[7] Salim H I , Patel, V, Abbas A, Walls J M and Dharmadasa I 2015 *J. Mater. Sci.: Mater. Electron.* **26** 3119-28
[8] Wanga J, Zhoub X, Lva P, Yangc L, Dinga D, Niua J, Liua L, Lia X, Fua W and Yanga H 2016 *Electrochim. Acta* **202** 32–38
[9] Sinha T, Lilhare D and Khare A 2019 *Rare Metals*https://doi.org/10.1007/s12598-019-01331-1
[10] Deivanaki S, Jayamurugan P, Mariappan R and Ponnuswamy V 2010 *Chalcogenide Letters* **7** 159 –63
[11] Vatavu S, Zhao H, Caraman I, Petru Gaşin P and Ferekides C 2009 *Thin Solid Films* **517** 2195–2201
[12] Tetyorkin V V, Sukach A V, Stariy S V and Boiko V A 2012 *Semicond. Phys Quan. Elect. Optoelectron* **15** 340-344
[13] Oluwafemi O S, Flowers J, Adeyemi O O, Bhagyaraj S, Ncapayi V, Zikalala N and Gwebu E T 2019 *Nano-struct. Nano-objects***17** 223-28
[14] Rodzik L, Łańcucka J L, Szuwarzyński M, Szczubiałka K and Nowakowska M 2017 *Nanotechnology* **28** 1-13
[15] Kumar S G and Rao K K 2014 *Energy Environ. Sci.* **7** 1-45

Effect of annealing on structural, morphological and optical properties of CdTe/(Cd-Zn)S thin films

Devjyoti Lilhare* and Ayush Khare

Thin Film Laboratory, Department of Physics, National Institute of Technology
G E Road, Raipur – 492 010 India

*Corresponding author's e-mail address dev17jyoti@gmail.com

Abstract. Here, we present the results of an investigation on structural, morphological and optical properties of CdTe/(Cd-Zn)S films deposited on FTO substrate through electrodeposition technique for solar cell application. The deposited films were heat-treated in the presence of $CdCl_2$ for the recrystallization. Various techniques, such as X-ray diffraction (XRD), field emission scanning electron microscopy (FESEM), high-resolution transmission electron microscopy (HRTEM), optical studies (absorption, transmission, band-gap and photoluminescence studies) were employed to examine the structural, morphological and optical properties. XRD spectra show the preferential orientation (111) in CdTe films. Uniform and bounded grains are observed upon annealing of the films. The changes in CdTe film were attributed to nucleation of the CdTe grains during the $CdCl_2$ treatment. The band-gap (E_g) value for CdTe estimated from absorption spectra and corresponding Tauc's plots comes to be ~1.50eV. Summary of results suggests that the annealing of different component films is helpful in improving solar cell performance.

Keywords: (CdTe)/(Cd-Zn)S/FTO; CBD; electrodeposition; annealing; solar cell

1. Introduction

In the future, solar power generation will serve as a source of energy. Due to the less material usage and lower cost manufacturing processes, thin film solar cells have emerged to become the future of solar energy technology. In photovoltaic (PV) technology, Cadmium telluride (CdTe) has the potential to become a leading energy producer in industry of solar energy in the coming decades. CdTe is positioned to possess a high absorbance coefficient. It can be easily deposited by high growth-rate processes, and has an optimum band-gap (~1.50 eV). Thus, it has a high potential as a thin-film solar cell absorber layer on an industrial level [1]. During last many years, CdTe/(Cd-Zn)S thin film heterostructure has attracted researchers for intense research in PV community [2-19].

In CdTe/(Cd-Zn)S solar cells, (Cd-Zn)S film (100 nm) absorbs 36% of the incident photon with energy higher than the band-gap energy [20]. CdTe has the direct optimum band-gap with over 5×10^5/cm absorption coefficient defining that a large amount of the energetic photons with higher energy than the energy of band-gap can be absorbed within a few μm of absorber layer of CdTe [21] and only a thin layer of the material is needed for total absorption of the solar spectrum (<2 μm). These characteristics make it an ideal material for PV conversion.

Various established techniques, such as spray pyrolysis, electrodeposition (ED), sputtering, close spaced sublimation(CSS) and chemical bath deposition (CBD) have been used to grow (Cd-Zn)S layers [22-26]. There are two different categories to divide these techniques; (i) low-temperature and (ii) high-temperature. In CBD and ED techniques, produces (Cd-Zn)S layers below the 100 °C temperatures and for device fabrication as-deposited layers are not suitable. So, in way to get better the structural, optical and electrical properties of these layers, they must go through the heat treatment at approximately 400 °C for 20 min in air before the CdTe deposition over the glass/FTO/(Cd-Zn)S substrates. Heat treatment in air generally advances the crystallinity, and enhances the properties of the materials. (Cd-Zn)S used as a window layer material with the wide band-gap (E_g= 2.46 eV) has been deposited by CBD technique and the CdTe used as absorber layer material with narrow band-gap (E_g~1.50 eV) was grown by the electrodeposition technique. CBD is a popular method for depositing (Cd-Zn)S films [27]. Xu et al. reported the efficiency of the CdTe/(Cd-Zn)S solar cell increased from 15.42% to 17.71% [28]. The chemically deposited (Cd-Zn)S films have a tendency to form the mixed phases (cubic and hexagonal), and therefore they come with poor crystalline quality. So, for the enhancement of optical and electronic properties annealing treatment becomes necessary to the chemically deposited (Cd-Zn)S films. Annealing is the effective method to enhance the properties of the films [29]. $CdCl_2$ treatment is also very effective to activate the cells. Following treatment with $CdCl_2$, film recrystallizes into the hexagonal phase, resulting in improved crystallinity and a lesser defects density [30]. $CdCl_2$ treatment can also improve the doping (n-type) of (Cd-Zn)S films, resulting in a higher built-in capacity in the CdTe film, leading to increased photovoltage [31]. CdTe–(Cd-Zn)S inter-diffusion is also affected by the $CdCl_2$ treatment [32]. Because of the large lattice mismatch at the interface of CdTe/(Cd-Zn)S, good solar cells are achieved by regulating S and Te interdiffusion [33, 34].

In this work, CdTe/(Cd-Zn)S films were deposited onto a FTO substrate by electrodeposition technique. Structural, optical and morphological properties were investigated using different techniques. Annealing effect has been at the focus of the work. The objective of our work is to differentiate the effect of annealing before and after $CdCl_2$ treatment films and to analyze the CdTe/(Cd-Zn)S/FTO film to enhance the knowledge about fabrication of thin film solar cell.

2. Materials and method

2.1. *Materials*

(Cd-Zn)S films are synthesized using cadmium acetate [$Cd(CH_3COO)_2$, 99.9% pure], zinc acetate [$Zn(CH_3COO)_2$, 99.9% pure] triethanolamine (TEA, 99.0% pure), thiourea (99% pure) and ammonia solution (NH_3OH, 25.0%) provided by Merck India. For CdTe electrolytic solution, cadmium sulphate ($CdSO_4$, 98% extra pure) from Loba Chemie, Tellurium Oxide (TeO_2, 99%) from Sigma-Aldrich, and Cadmium Chloride ($CdCl_2$, monohydrated pure) from Merck India, are used. All the chemicals are of analytical status and are used as received. Distilled water is used as a solvent for the preparation of different solutions.

2.2. *Film deposition*

We have worked on two types of films: annealed before $CdCl_2$ and annealed after $CdCl_2$ treatment. The CdTe/(Cd-Zn)S solar cell fabrication was as followed: a (Cd-Zn)S window layer was deposited on an FTO substrate using CBD technique. In the first film, (Cd-Zn)S is heat treated (annealed without $CdCl_2$) in air for 20 min at 400°C and dipped in to the electrolytic solution prepared by adding 1M solution of $CdSO_4$, a smaller amount (low concentration) of TeO_2, and 1000ppm level of $CdCl_2$. The pH was adjusted to 2.00 ± 0.02 through using NH_4OH or H_2SO_4 of high purity. In the second film, (Cd-Zn)S was directly placed into the electrolytic solution without any annealing. After the dip coating of $CdCl_2$ it was annealed for 20 min at 400°C. In both the cases, (Cd-Zn)S and CdTe [deposited on (Cd-Zn)S)] were used as window and absorber layer, respectively. In this deposition, 2-

electrode system was employed in the process of electroplating: (i) cathode: Glass/FTO and (ii) anode: high purity carbon electrode. Both the electrodes were immersed into an electrolytic solution. Full device structure was completed after forming back metal contact [glass/FTO/(Cd-Zn)S /CdTe/back metal contact] shown in figure 1.

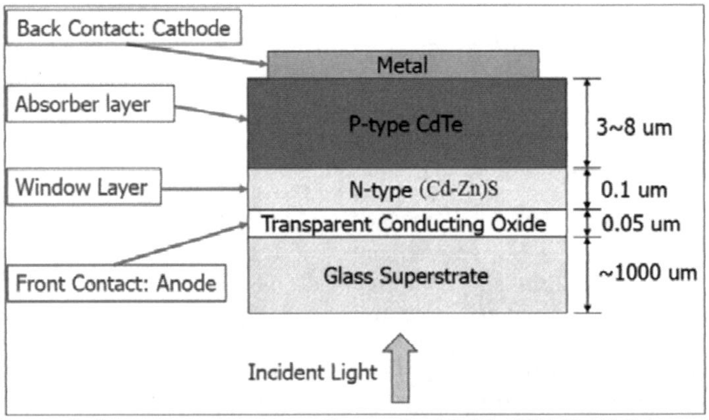

Figure 1. Superstrate structure of a standard CdTe/(Cd-Zn)S Solar cell

2.3. Measuring instruments

In the way of identification of the structural classification and crystallographic orientation of the CdTe/(Cd-Zn)S films, PANalytical X'pert Multifunctional Powder X-ray Diffractometer (2θ= 20° to 80° and λ= 1.54 Å) at NIT Raipur is used. The applied operating voltage is 40 kV and current at 30 mA. The topographical study is done by using SIGMA Field Emission Scanning Microscope (CARL ZEISS EVO) at NIT Raipur, India. HRTEM studies are done using JEOL JEM-2100F/HR from IIT Kharagpur. The optical studies (absorption and transmission spectra) are done by using Shimadzu Make UV-1800 Spectrophotometer in the 600 nm to 1000 nm wavelength range at NIT Raipur. PL emission spectra are recorded using Shimadzu RF-5301 PC Spectrofluorophotometer at an excitation wavelength of 632 nm at NIT Raipur.

3. Results and discussion

3.1. Structural characterization

3.1.1 XRD studies

XRD is the first tool to identify the phases of prepared materials, their crystallinity and crystallite size within the layers. The typical XRD patterns of two different films (annealed before $CdCl_2$ and annealed after $CdCl_2$ treatment) are shown in figure 2.The XRD patterns reveal that the deposited films are polycrystalline in nature. The most intense peak in both the cases at (111) orientation is observed at 2θ= 24.11 which is identified as the zincblende structure. Another peaks of CdTe at (220) and (311) orientation (2θ= 39.40 and 2θ= 46.55, respectively) is observed only in before annealing case. These peaks are not observed in the film after annealing. The crystal structure of the film did not change upon annealing but another peak of cubic phase was missing in other film. The highly preferential growth along the particular crystallographic orientation of CdTe is (111) and other two peaks, (220) and (311) are extremely weak along the (111) plane. This entails that CdTe film became polycrystalline upon annealing. Similar trend is observed in the paper reported by Lee et al. [35]. They observed that annealing and $CdCl_2$ treatment made steadier zincblende structure of CdTe films and made CdTe films polycrystalline forms.

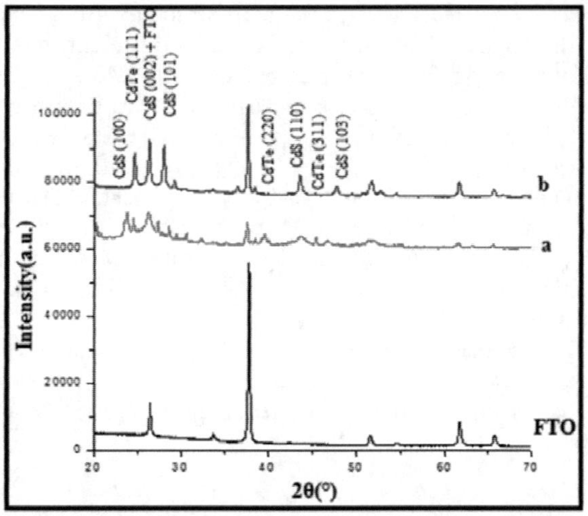

Figure 2. XRD Spectra of CdTe/(Cd-Zn)S film: (a) annealed before $CdCl_2$ treatment (b) annealed after $CdCl_2$ treatment

Table 1. XRD data of CdTe/(Cd-Zn)S films deposited on FTO substrate

Sample	2θ (°)	d values (Å)	Intensity	FWHM (in radian)	hkl	D (in nm)
Annealed Before $CdCl_2$ treatment	23.862	3.72611	11779	0.027	(111)c	54.79
	39.538	2.27747	5152	0.0224	(220)c	68.31
	46.958	1.93342	3078	0.0169	(311)c	93.38
Annealed after $CdCl_2$ treatment	24.11	3.70611	13804	0.0177	(111)c	83.79
	39.538	--	--		No peak	--
	46.958	--	--		No peak	--

In CdTe layers, the peak intensities, occurrence of peaks and angles can be accredited to the random grains orientation. In Table 1, the obtained XRD data and the structural parameters of CdTe/(Cd-Zn)S thin films are shown. The crystallite sizes (D) are determined using the Scherrer's formula [36]:

$$D = \frac{k\lambda}{\beta\cos\theta} \tag{1}$$

where, k = 0.94, λ= 1.54Å (wavelength of the X-rays used), β is FWHM of the diffraction peak (in radian) and θ is the Bragg's angle.

Generally, it is considered that CdTe growth follows the structure zone model as reported by Luschitz et al. [37, 38] in which the structure of a grain was seen as a succession from needle like structure to a huge fine defined grain structure demonstrated at low to high temperature [38]. Similarly, the thickness of the film is also an affecting factor for the grain size [39]; hence, thicker films may have a bigger apparent size of grain. So, it is clear that the formation of the grain boundary in CdTe solar cells can influence the significant choice of deposition technique and conditions.

The recrystallization effect also affects the CdTe grain structure. But it occurs primarily for low temperature deposition, where the as-deposited (before $CdCl_2$ treatment) grain size is small and after $CdCl_2$ treatment, the size of the grain is considerably enhanced. [40-43]. In case of deposition at higher temperature, the grain size of as-deposited film is nearly similar to the treated grain size because of high temperature [44-46]. It may be possible due to the (Cd-Zn)S interface recrystallization [47]. Similar outcome can be seen in the article reported by Quadros et al. [48]. They reported that during the deposition at low temperature, in case of post-$CdCl_2$ treatment, CdTe recrystallizes to a much bigger grain size, but no change is observed in case of post-annealing at high temperature deposition. This situation creates a difficulty while considering the structures of grains in CdTe solar cells. Sometimes, varied deposition methods become extremely difficult due to their conflicting results that it may be shift with respect to one another.

In the present case, it is concluded from Table 1 that the film annealed before $CdCl_2$ treatment has increasing grain size in spite of after $CdCl_2$ treated film. However, film after $CdCl_2$ treatment has higher intensity in (111) orientation so that grain size is also large as compared to the film before $CdCl_2$ heat treatment.

3.1.2 FESEM studies

FESEM is a useful tool to check the uniformity of large area material coverage. SEM is also widely used to determine the topographical image, shape and size of the grains and available defects/voids. For the improvement in thin film solar cells, deep knowledge of surface topography is very important. FESEM images of two different types of deposited films: annealed before $CdCl_2$ treatment and annealed after $CdCl_2$ treatment is shown in figure 3. In both the films, rose flower like structures are observed. This indicates the hydrophilic nature of CdTe/(Cd-Zn)S film. An overlapped layered growth structure and uniformity in nanoparticles are also observed that may be accredited to the chemically deposited (Cd-Zn)S films [49]. In figure 3(a), it is clearly evidenced that grains are agglomerated leading to large crystalline grains [50]. The effect of annealing on (Cd-Zn)S film ($CdCl_2$ heat treatment for 20 min at 400°C in air) showed a different result where crystallites are combined and bounded together to form large crystals and the clusters became large grains of CdTe. Grain sizes differ in few microns size, and these layers demonstrated similarity to the structure of CdTe grown by high temperature methods [51]. In duration of $CdCl_2$ treatment, a liquid surface is created at the grain boundary and Cl is easily moved along the grain boundary. Nevertheless, diffusion of Cl is sluggish in the crystalline material grains. For each CdTe grain that existing between the back electrical contact and (Cd-Zn)S, the Cl-diffused layer is therefore merely a thin skin [52].

$CdCl_2$ Heat treatment improved the coalescence of nanocrystals from smaller to larger to get better material layers. After $CdCl_2$ treatment, the grain growth appears collectively with gaps in between grains. This suggested that $CdCl_2$ treated CdTe film which is produced from Cl precursor has an improved excellence for PV devices because there is no noticeable gaps between the grains, and for less scattering it has larger grain sizes at the boundaries and with high electron mobility [51].

In figure 3(b) also has a rose flower like compact structure but most of the nanoparticles observed distinct and non-uniform around the rose like structure. This result indicates that annealing before $CdCl_2$ treatment enhances the size of grain in CdTe/(Cd-Zn)S films with recrystallization and occurrence of $CdCl_2$ encourage a sintering mechanism inside the CdTe film to increase average grain size and decrease intergrain pore size [53].

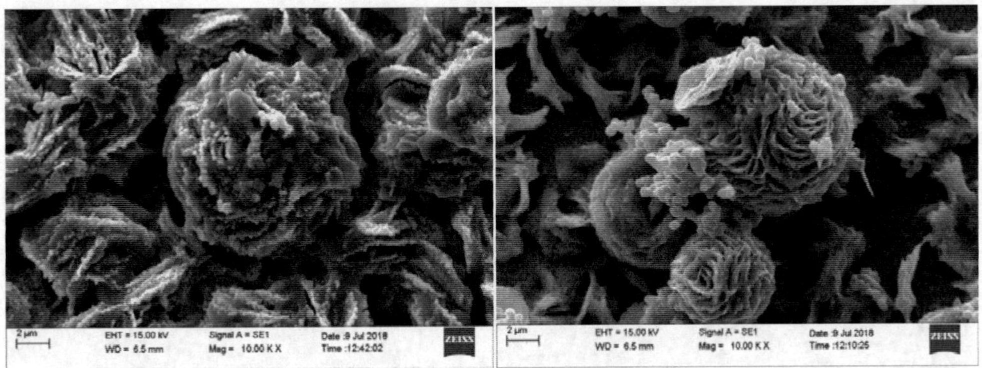

Figure 3. FESEM images of CdTe/(Cd-Zn)S film: (a) annealed before $CdCl_2$ treatment (b) annealed after $CdCl_2$ treatment.

3.1.3 HRTEM studies

HRTEM is a unique way to access the information at the nanometer (nm) scale. It is also useful for morphological study which provides the shape, size and nanocrystalline behavior. In figure 4 HRTEM images of two types of deposited films (annealed before and after $CdCl_2$ treatment) are shown. In both the cases micrographs demonstrated the layers consisting of small grains of tens of nm dimensions. A columnar shape is observed in a lot of these grains, which is usual to the substrate of (FTO/(Cd-Zn)S). For fabrication of high efficiency solar cells, there is a need to create the large CdTe grains which is enlarging from (Cd-Zn)S layer to the top surface [50].

The prepared film was treated at 400℃ for 2 min that is the optimum temperature and time for obtaining the highest efficiency. Figure 4(a) shows that the HRTEM image has a perfect lattice. There are no stacking faults. No epitaxial co-ordination is observed between the (Cd-Zn)S and CdTe layers. Several types of stacking faults powerfully affect the electrostatic potential within the CdTe grains and act as hole traps, and acquire high stacking fault energies of up to 235 mJ/m^2 [54]. In figure 4(b), the HRTEM analysis suggests that CdTe grains are epitaxial related to the (Cd-Zn)S grains [also confirmed by high resolution lattice imaging] [55] and they are full of faults and columnar. Selected area electron diffraction (SAED) image shows that the CdTe film has cubic phase, whereas the (Cd-Zn)S film has a combination of cubic and hexagonal phases. The HRTEM images demonstrate thermal grooving on the CdTe surface close to the grain boundaries and independent reformation of the (Cd-Zn)S and CdTe films as a result of the reduction of interfacial energy.

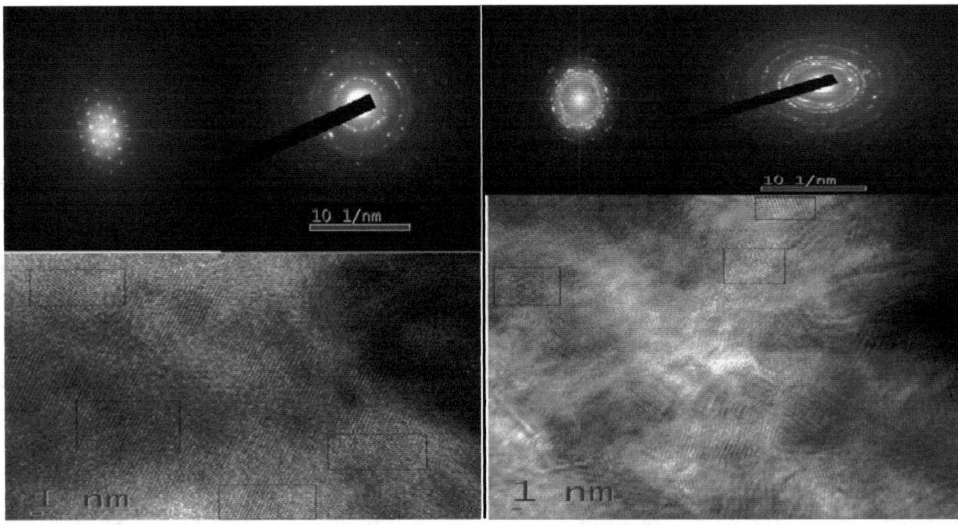

Figure 4. HRTEM images of CdTe/(Cd-Zn)S film: (a) annealed before $CdCl_2$ treatment (b) annealed after $CdCl_2$ treatment.

Figure 5 demonstrates the EDX spectrum of CdTe/(Cd-Zn)S film. EDX is a useful tool to identify the available elements present into the composition, product imperfections and defects. It also helps to investigate the failure and identification of cause, catalyst quality, poisoning and elemental distribution.

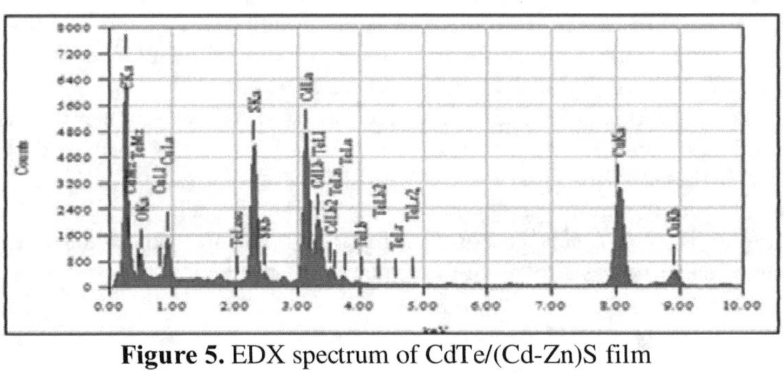

Figure 5. EDX spectrum of CdTe/(Cd-Zn)S film

3.2 Optical Characterizations

3.2.1 Absorbance and transmittance

To develop the PV device, the optical transmission, absorption and the energy band-gap (Eg) of the materials are the essential factors to understand and determine. These parameters analyze the frequency and intensity of the emerging rays. With the help of the optical studies, the colour of an object can be easily identified as it depends on the processes of light emission, transmission and reflection by the object according to the sensitivity of the human eye to the different colours.

Optical absorption spectral studies of CdTe/(Cd-Zn)S/FTO films were performed in the wavelength range of 600 nm - 1000 nm as shown in figure 6. Both the films show high absorbance and less transmittance as the CdTe nature for making better solar devices. But in the absorbance spectra of annealed before $CdCl_2$ treatment (figure 6(a)) showed little high absorbance than annealed after $CdCl_2$ treatment (figure 6(b)). The absorption edges are observed to slight shift to lower wavelength side in case of after $CdCl_2$ treatment. The optical absorption properties of CdTe improved when grown through process of annealing before $CdCl_2$ treatment as compared to after $CdCl_2$ treatment.

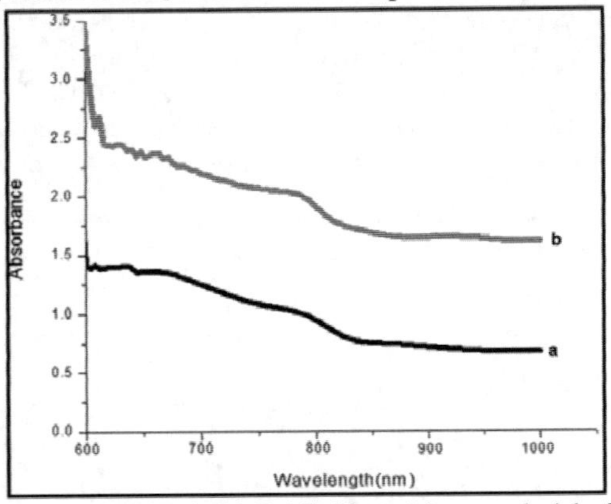

Figure 6. Absorption spectra of CdTe/(Cd-Zn)S film: (a) annealed before $CdCl_2$ treatment (b) annealed after $CdCl_2$ treatment

In order to determine the energy band-gap of the deposited films, absorption spectra were recorded. With the help of the absorption spectra, energy band-gaps can be easily determined by plotting $(ah\nu)^2$ versus hν (Tauc's plot) by the extrapolation of the linear portion of the curve [56]. The optical direct band-gap values can be determined by the following relation:

$$\alpha(\nu)h\nu = c(h\nu - E_g)^m \quad (2)$$

Where, α(ν)= absorption coefficient, E_g= energy band-gap of the material, hν= photon energy, c= light velocity and m= constant value (determined type of optical transition). When the value m= 1/2 and 3/2 it is allowed for direct transition, and when m= 2 and 3 it is for allowed indirect transitions. In the present case, absorption coefficient (α(ν)) is in the order of 10^4 cm^{-1} for all the direct band-gap semiconductors [57].

The energy band-gap values are estimated for both the deposited films: annealing before $CdCl_2$ and after $CdCl_2$ treated films (figure 7). In energy band-gap value of annealed before $CdCl_2$ treatment (figure 7(a)) is observed 1.51 eV while in annealed after $CdCl_2$ treatment (figure 7(b)), the energy band-gap value is 1.49 eV. The slight difference in two films may be attributed to the absorption spectra because there is slight shift in absorption edge. The annealing before $CdCl_2$ treated film is thinner than the other film so that thin layers generally demonstrate slightly increased band-gap values due to quantum effects and unfinished surface coverage [50].

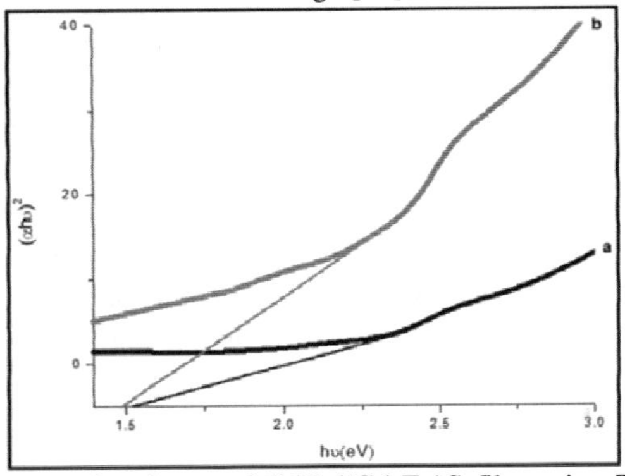

Figure 7. Estimation of energy band-gap of CdTe/(Cd-Zn)S film using Tauc's plot: (a) annealed before $CdCl_2$ treatment (b) annealed after $CdCl_2$ treatment

For transmission of the light, the scattering and absorption are the two depending factors [58]. The transmittance of a film is complementary to its absorption. The transmittance spectra of CdTe/(Cd-Zn)S/FTO films were recorded in the wavelength range of 600 nm - 1000 nm shown in figure 8. It is observed that in accordance with the high absorption, transmission is less in both the cases. For making solar cell devices, there is a need of high absorption quality and less transmittivity. So, in that way, the film annealed before $CdCl_2$ treatment film is useful for the better productivity of solar cell devices.

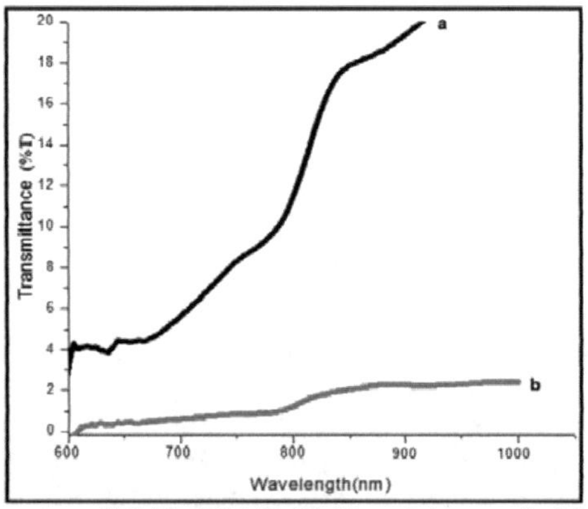

Figure 8. Transmittance spectra of CdTe/(Cd-Zn)S film: (a) annealed before $CdCl_2$ treatment (b) annealed after $CdCl_2$ treatment

3.2.2 Photoluminescence studies

PL study is a precise study to investigate the relative energies of the states (ground and excited states), concentrations of the emitting species, and electronic transitions. The electrons release photons with energy corresponding to the material's energy band-gap in band-to-band transitions. Additionally, Photons emit to defect levels from the conduction band (CB), and then defect levels to the valence band (VB). Therefore, peaks can occur under the value of the energy band-gap, but the results can be affected because of the different transitions.

PL spectra of two types of deposited techniques (annealed before $CdCl_2$ treatment and annealed after $CdCl_2$ treatment) are shown in figure 9. The PL spectra observed at an excitation value of 632 nm (1.96 eV). In both the cases band located at ~1.50 eV that can be due to the emission of the defects between the energy levels [59–61]. The existence of defects and particle size distribution may lead to the broadening of the PL spectrum. Broad bands are initiated from the defects in the crystal. The peaks intensities and their FWHM showed nearly identical to the quality of material. Broad peak is observed with high intensity in annealed before $CdCl_2$ treated film (figure 9(a)) demonstrated high crystalline quality, whereas, annealed after $CdCl_2$ treated film has very low intensity (figure 9(b)). This signifies the existence of a huge number of shallow acceptor and donor type defects in the material. Thus, transitions occur from donor to acceptor that produces the photons less than the energy band-gap. In some conditions, where emissions are greater than the energy band-gap values that may be reason of presence of nano-size particles attributing to the quantum confinement effects [62]. The PL spectra of the CdTe/(Cd-Zn)S showed that, although the intensity changed significantly from the center of the grain to the grain boundary but the peak position remained at the same position (~1.50 eV), analogous to the band-gap of the CdTe. The spectra observed from before $CdCl_2$ treatment have large grain and have higher intensities with a peak position at ~1.50 eV and similarly same peak position is observed in after $CdCl_2$ treatment with small grains and low intensity, demonstrating the predominant contribution from lower gap alloys [63].

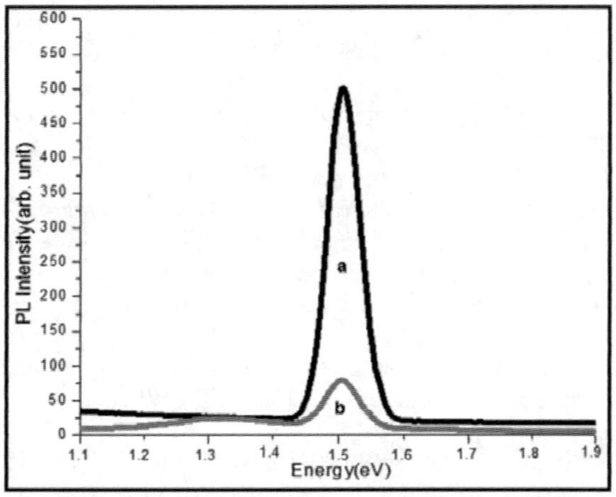

Figure 9. PL intensity of CdTe/(Cd-Zn)S film: (a) annealed before $CdCl_2$ treatment (b) annealed after $CdCl_2$ treatment

4. Conclusions

(Cd-Zn)S/CdTe films were deposited onto a FTO substrate by electrodeposition technique. Two types of films were deposited: annealed before $CdCl_2$ treatment and annealed after $CdCl_2$ treatment. Structural, optical and morphological properties were investigated. In XRD spectra, the most intense peak at (111) orientation is observed and identified as the zincblende structure. Annealed before $CdCl_2$ treatment film showed better crystallinity and has large grain sizes. A rose flower like structure is observed in both the films but in annealed before $CdCl_2$ treatment film observed better uniformity in crystal and grains. In HRTEM, perfect lattice is observed for the annealed before $CdCl_2$ treatment film. No epitaxial co-ordination is observed between the (Cd-Zn)S and CdTe layers. Conversely, annealed after $CdCl_2$ treatment film observed full of faults and epitaxial that may affect the electrostatic potential within the CdTe grains. Annealed before $CdCl_2$ treatment film has higher absorption quality and less transmittivity which is very suitable for the fabrication of the solar cell devices. The PL spectra of the CdTe/(Cd-Zn)S showed that, although the intensity changed significantly but the peak position remained at the same position (~1.50 eV) which is analogous to the energy band-gap of the CdTe.

References

[1] Hussain K M A, Mahmood Z H, Syed I M, Begum T, Faruqe T and Parvin J 2015 *Am. J. Mater. Sci. appl.* **2** 91
[2] Hossain M S, Rahman K S, Karim M R, Aijaz M O, Dar M A, Shar M A, Misran H and Amin N 2019 *Sol. Energy* **180**, 559
[3] Kasim U, Narayanan H and Anthony O 2008 *Leonardo J. Sci.* **12** 111
[4] Kartopu G, Clayton A J, Brooks W S M, Hodgson S D, Barrioz V, Maertens A, Lamb D A and Irvine S J C 2014 *Prog. Photovolt: Res. Appl.* **22** 18
[5] Brooks W S M, Irvine S J C, Barrioz V and Clayton A J 2012 *Sol. Energy Mater. Sol. Cells* **101** 26
[6] Castillo R H, Acosta M, Riech I, Rodriguez G S, Gamboa J M, Acosta C and Zambrano M 2017 *Optik* **148** 95
[7] Osman M A and Abd-Elrahim A G 2018 *Optical Mater.* **77** 1
[8] Osman M A, Abd-Elrahim A G and Othman A A 2018 *Mater. Character.* **144** 247
[9] Prem Kumar T and Sankaranarayanan K 2009 *Chalcogenide Lett.* **6** 617

[10] Lee J H, Song W C, Yang K J and Yoo Y S 2001 *Thin Solid Films* **416** 184
[11] Zhou J, Wu X, Teeter G, To B, Yan Y, Dhere R G and Gessert T A 2004 *Physica Status Solidi B* **241** 775
[12] Werta S Z, Echendu O K, Egbo K O and Dejene F B 2019 *Thin Solid Films* **689** 137511
[13] Rugen-Hankey S L, Clayton A J, Barrioz V, Kartopu G, Irvine S J C, McGettrick J D and Hammond D 2015 *Sol. Energy Mater. Sol. Cells* **136** 213
[14] Yao H, Shen H, Zhu X, Jiao J, Li J and Wang W 2016 *Ceram. Int.* **42** 2466
[15] Kartopu G, Williams B L, Zardetto V, Gurlek A K, Clayton A J, Jones S, Kessels W M M, Creatore M and Irvine S J C 2019 *Sol. Energy Mater. Sol. Cells* **191** 78
[16] Barrioz V, Proskuryakov Y Y, Jones E W, Major J, Irvine S J C, Durose K and Lamb D A 2017 *Mater. Res. Soc. Symposium Proceedings* 1012 Y12
[17] Jones E W, Barrioz V, Irvine S J C and Lamb D 2009 *Thin Solid Films* **517** 2226
[18] Irvine S J C, Barrioz V, Lamb D, Jones E W and Rowlands-Jones R L 2008 *J. Cryst. Growth* **310** 5198
[19] Chavhan S D, Senthilarasu S and Lee S H 2008 Appl. Surf. Sci. **254** 4539
[20] Yamaguchi T, Yamamoto Y, Tanaka T, Demizu Y, Yoshida A, 1996 *Thin Solid Films* **281–282** 375
[21] Noufi R and Zwiebel K 2006 High-efficiency CdTe and CIGS thin-film solar cells: highlights and challenges *In: Proc. 4th World Conf. on Photovoltaic Energy Conversion* 317
[22] Basol B M 1988 *Sol. Cells* **23** 69
[23] Lokhande C D 1990 *Mater. Chem. Phys.* **26** 405
[24] Orton J W, Goldsmith B J, Chapmann J A and Powell M J 1982 *J. Appl. Phys.* **53** 1602
[25] Romeo N, Bosio A, Canevari Vand Seuret D1987 *Sol. Cells* **22** 23
[26] Schaffner J, Feldmeier E, Swirschuk A, Schimper H J, Klein A and Jaegermann W 2011 *Thin Solid Films* **519** 7556
[27] Rome N, Bosio A, Canevari V and Podesta A 2004 *Sol. Energy* **77** 795
[28] Xu X B, Wang X Y, Gu W P, Quan S and Zang Z 2017 *Superlattices Microstruct.* **109** 463
[29] Archbold M D, Halliday D P, Durose K, Hase T P A, Smyth-Boyle D, and Govender K January 3–7, 2005 Characterization of thin film cadmium sulfide grown using a modified chemical bath deposition process, *in: Proceedings of the Conference Record of the 31st IEEE Photovoltaic Specialists Conference, Coronado Springs Resort Lake Buena Vista, FL* 475
[30] Moutinho H R, Albin D, Yan Y, Dhere R G, Li X, Perkins C, Jiang C S, To B and Al-Jassim M M 2003 *Thin Solid Films* **436** 175
[31] Fritsche J, Kraft D, Thißen A, Mayer T, Klein A and Jaegermann W 2002 *Thin solid Films* **403–404** 252
[32] Wu X, Ribelin R, Dhere R G, Albin D S, Gessert T A, Asher S, Levi D H, Mason A, Moutinho H R and Sheldon P September 15–22, 2000 High-efficiency $Cd_2SnO_4/Zn_2SnO_4/Znx\ Cd_{1-x}S/(Cd\text{-}Zn)S/CdTe$ polycrystalline thin-film solar cells, *in: Proceedings of the Conference Record of the 28th IEEE Photovoltaic Specialists Conference, Anchorage* 470
[33] Jaegermann W, Klein A, Fritsche J, Kraft D and Spath B 2005 *Mater. Research Soc. Symposium Proceedings* **865** F6.1
[34] Fritsche J, Klein A and Jaegermann W 2005 *Adv. Eng. Mater.* **7** 914
[35] Lee J H, TextscLee H Y, Park Y K, Shin S H and Park K J 1998 *Jpn. J. Appl. Phys.* **37** 3357
[36] Sanap V B and Pawar B H 2009 *Chelco. Lett.* **6** 415
[37] Luschitz J, Siepchen B, Schaffner J, Lakus-Wollny K, Haindl G, Klein A and Jaegermann W 2009 *Thin Solid Films* **517** 2125
[38] Luschitz J, Lakus-Wollny K, Klein A and Jaegermann W 2007 *Thin Solid Films* **515** 5814
[39] McCandless B E, Moulton L V and Birkmire R W 1997 Prog. Photovolt. **5** 249
[40] Zoppi G, Durose K, Irvine S J C and Barrioz V 2006 *Semicond. Sci. Technol.* **21** 763

[41] Galloway S A and Durose K 1995 SEM/EBIC observations of CdTe/(Cd-Zn)S thin film solar cells Microscopy of Semiconducting Materials *(Bristol: Institute of Physics Publishing)* 709
[42] Paudel N R, Young M, Roland P J, Ellingson R J, Yan Y and Compaan A D 2014 *J. Appl.Phys.* **115** 064502
[43] Pookpanratana S, Liu X, Paudel N R,Weinhardt L, Bar M, Zhang Y, Ranasinghe A, Khan F, Blum M, Yang W, Compaan A D, and Heske C 2010 *Appl. Phys.Lett.* **97** 172109
[44] Moutinho H R, Al-Jassim M M, Abulfotuh F A, Levi D H, Dippo P C, Dhere R G and Kazmerski L L 1997 Studies of recrystallization of CdTe thin films after $CdCl_2$ treatment *Proc. 26th IEEE Photovoltaic Specialists Conf.* 431
[45] Ferekides C S, Balasubramanian U, Mamazza R, Viswanathan V, Zhao H and Morel D L 2004 *Sol. Energy* **77** 823
[46] Levi D H, Moutinho H R, Hasoon F A, Keyes B M, Ahrenkiel R K, Al-Jassim M, Kazmerski L L and Birkmire R W 1994 Micro through nanostructure investigations of polycrystalline CdTe: correlations with processing and electronic structures *Proc. 24th IEEE Photovoltaic Specialists Conf.* 127
[47] Major J D, Bowen L and Durose K 2012 *Solar cells Prog. Photovolt.* **20** 892
[48] Quadros J, Pinto A L, Moutinho H R, Dhere R G and Cruz L R 2008 *J. Mater. Sci.* **43** 573
[49] Lilhare D, Sinha T, Verma L and Khare A 2019 *Semicond. Sci. Technol.* **34** 125010
[50] Dharmadasa I, Madugu M L, Olusola, O I, Echendu O K, Fauzi F, Diso D, Weeradinghe, A, Dharmadasa R, Lavery B, Jasinski J B, Krentsel, T A and Sumanasekera G 2017 *Coatings* **7** 17
[51] Manaf N A A, Hussein I S, Madugu M L, Olusola O I and Dharmadasa I M 2015 *Energies* **8** 10883
[52] Marfaing Y 2001 *Thin Solid Films* **387** 123
[53] Roh J S and IM H B 1988 *Sol. Energy mater.* **23** 2267
[54] Yoo S H, Butler K T, Soon A, Abbas A, Walls J M and Walsh A 2014 *Appl. Phys. Lett.* **105** 062104
[55] Al-Jassim M M, Hasoon F S, Jones K M, Keyes B M, Matson R J and Moutinho H R 1993 Proc. 23rd IEEE PVSC 459
[56] Tauc J and Menth A 1972 *J. Non-Cryst Solids* **569** 8
[57] Hankare P P, Kokate A V, Asabe M R, Delekar S D and Chougule B K 2006 *Mater. Sci. Eng. B* **133** 37
[58] Kavitha N, Chandramohan R, Valanarasu S, Vijayan T A, Rex Rosario S and Kathalingam A 2015 *J. Mater. Sci: Mater. Electron.* **27** 2574
[59] Abken A E, Halliday D P and Durose K 2009 *J. Appl. Phys.* **105** 064515
[60] Ahmad-Bitar R N 2000 *Renew. Energy* **19** 579
[61] Vigil O 1 997*J. Vac. Sci. Technol. A* **15** 2282
[62] Glinka Y D, Lin S H, Hwang L P, Chen Y T and Tolk N H 2001 *Phys. Rev. B.* **64** 085421.
[63] Dhere R, Rose D, Albin D, Asher S, Al-Jassim M, Cheong H, Swartzlander A, Moutinho H, Coutts T, and Sheldon P October 3, 1997 *Presented at the 26th IEEE Photovoltaic Specialists Conference, September 29Ð Anaheim, California*

IOP Conf. Series: Materials Science and Engineering **798** (2020) 012023 doi:10.1088/1757-899X/798/1/012023

Effect of precursor ratios on the structural, morphological and optical properties of CBD-CdS films

Tarkeshwar Sinha and Ayush Khare

Department of Physics, National Institute of Technology,Raipur - 492010 India

*Corresponding author's e-mail address: ts22701@gmail.com

Abstract. Cadmium sulfide (CdS) films were deposited on fluorine doped tin oxide (FTO) glass substrates by chemical bath deposition (CBD). X-ray diffraction (XRD), scanning electron microscopy (SEM), atomic force microscopy (AFM), photoluminescence (PL), and Fourier transform infrared (FT-IR) analyses were employed to compare the structural, morphological, and optical properties of deposited films. The XRD studies confirmed the cubic crystalline nature of the deposited films, and the intensities of the peaks were found to vary for different precursor ratios. The results of SEM study show that the films were smooth and uniform. The RMS surface roughness values of the CdS films were found to be in the range of 25.42nm to 35.33nm. Energy band-gap values of the CdS films were found in the range of 2.32eV to 2.39eV. Two distinct bands (centred at 489nm and 556nm) were found in the PL emission spectra. S-C-N,Cd-S, and C-S vibrations were observed to dominate in the FT-IR spectra.

1. Introduction

The Cadmium sulfide (CdS) is one of the most popular II-VI group compound semiconductor due to its use in wide range of applications such as, light emitting diodes, optoelectronic devices, thin film solar cells, photo-degradation, and piezo-electronic devices [1-2]. CdS can be synthesized in different size and shape such as triangle, cube, sheet, tube, wire, film, rod, powder, fiber, etc. Multiple deposition techniques have been developed to deposit CdS films such as thermal evaporation, spray pyrolysis, electrodeposition, chemical bath deposition (CBD), molecular beam epitaxy [3-4]. CBD technique is easy to perform among these deposition techniques and can be implemented using low cost experimental set-up. The quality and applicability of CdS films is affected by various deposition parameters including deposition time, complexing agent, pH variation, bath temperature, and annealing treatments[5–6]. However, the effect of variations in cadmium (Cd) and sufur (S) ratio has not been explored well in the literature. Therefore, we are reporting the rigorous study of variations in structural, morphological, and optical properties of CdS films due to changes in precursor ratio.

2. Experimental details

Fluorine doped tin oxide (FTO) coated glass slides were used as substrates. Cadmium acetate and thiourea were used as precursor materials. Amonia (NH_3) has been used as a complexing agent as it helps to maintain the pH of the solution too. Chromic acid, acetone, and double distilled water were used for the cleaning of all the required items before the film deposition. A hot plate magnetic stirrer was used to homogenize the solution. Three different solutions were prepared with different Cd/S ratios (a) 1:1, (b) 1:2, and (c) 1:3. Then, all the three solutions (with an FTO slide dipped inside) were placed in the water bath for 1h at 70 °C. After the completion of deposition process, all the three CdS films were lifted carefully and washed ultrasonically in distilled water. X-ray diffraction (XRD) technique was used for structural analysis of the films. For the study of morphological behaviour of the films, scanning electron microscopy (SEM) and atomic force microscopy (AFM) characterization techniques were employed. UV-visible spectroscopy was used for the optical analysis of the films. Photoluminescence (PL) was used to study the luminescence behaviour of the films and Fourier transform infrared (FT-IR) spectroscopy was employed to analyse their infrared absorption spectra.

3. Results and discussions

3.1 Structural analysis

The XRD diffractograms of CdS films deposited for different precursor ratio is shown in figure 1. Film deposited for Cd/S ratio 1:2 and 1:3 have three characteristic XRD peaks (111), (200) and (220). But the film deposited for Cd/S ratio 1:1 has only (111) and (200) characteristic peaks[7]. Thus, all three films have cubic crystalline nature, but the intensity of peaks is different for different film. The film deposited for Cd/S ratio 1:1 and 1:3 are poorly crystallized as compared to the film deposited for CdS ratio of 1:2. It suggests that the crystalline behaviour of CBD-CdS films depends on the precursor ratio. The crystalline sizes of films deposited for Cd/S ratio 1:1, 1:2, and 1:3 are calculated by Scherrer formula, which are found as 60.28nm, 65.34nm, and 52.47nm, respectively. Hence, the crystallite size estimation also suggests that the film deposited for Cd/S ratio of 1:2 is more crystalline than the other two films.

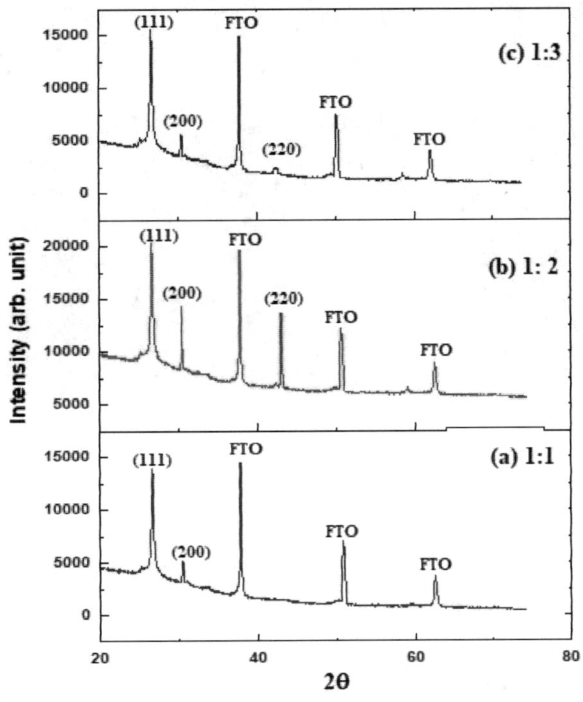

Figure 1. XRD patterns of CdS films deposited for Cd/S ratio (a) 1:1, (b) 1:2, and (c) 1:3.

IOP Conf. Series: Materials Science and Engineering **798** (2020) 012023 doi:10.1088/1757-899X/798/1/012023

3.2 Surface morphology

The surface morphology of the CdS films deposited for different Cd/S ratio is shown in figure 2. As shown in SEM micrographs (figure 2) of the CdS films, one can easily notice that the precursor ratio affects the morphology of the films significantly. The film deposited for Cd/S ratio of 1:2 shows the better morphological behaviour as compared to other two films. The RMS surface roughness values of films deposited for Cd/S ratio 1:1, 1:2, and 1:3 are found as 30.36 nm, 25.42nm, and 35.33nm, respectively. These values have been calculated from the AFM analysis using the instrument's software and the corresponding AFM images are shown in figure 3.The smoothness and low roughness of the film surface suggests that the film is suitable for optoelectronic applications[8-9].

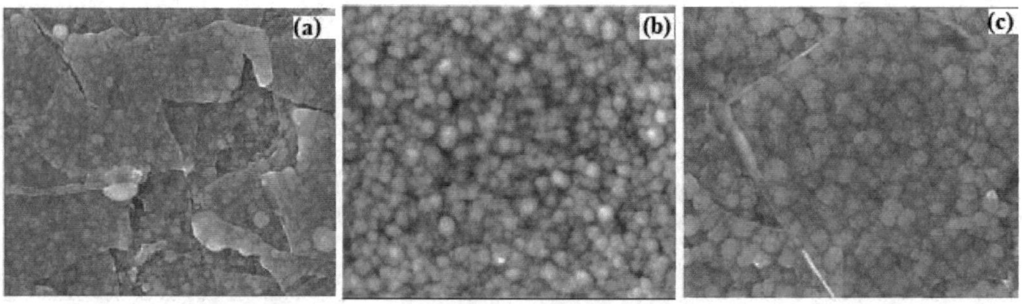

Figure 2. SEM images of CdS films deposited for Cd/S ratio (a) 1:1, (b) 1:2, and (c) 1:3.

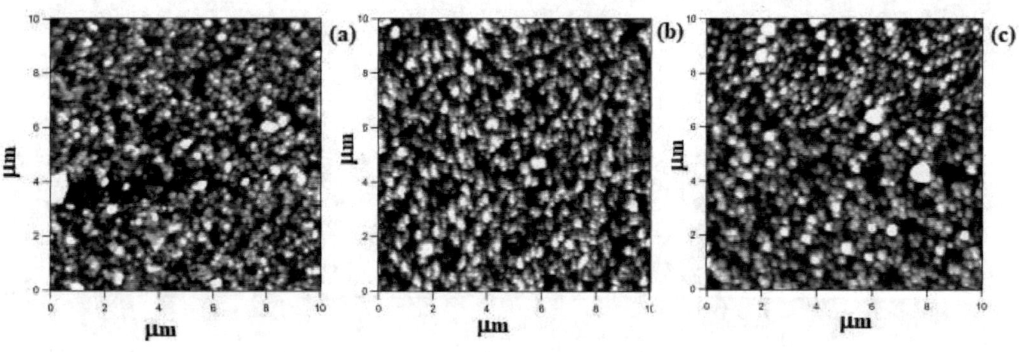

Figure 3. AFM images of CdS films deposited for Cd/S ratio (a) 1:1, (b) 1:2, and (c) 1:3.

3.3 Optical analysis

The optical transmission spectra of CdS films deposited for different Cd/S ratio is shown in figure 4.It shows that the transmittance values of the films varied from 60% to 90%, specifically in the longer wavelength region. Due to its better morphology and smoothness on the surface, the film deposited for Cd/S ratio of 1:2 exhibits better transmission behaviour as compared to other two films. The optical band-gap energy of the CdS films has been estimated from the Tauc's plot as shown in figure5. The band-gap energy of the films deposited for Cd/S ratio of 1:1, 1:2, and 1:3 are found as 2.32 eV, 2.36 eV and 2.39 eV, respectively. The optical transparency and band-gap values matched pretty well with the literature [10].

IOP Conf. Series: Materials Science and Engineering **798** (2020) 012023 doi:10.1088/1757-899X/798/1/012023

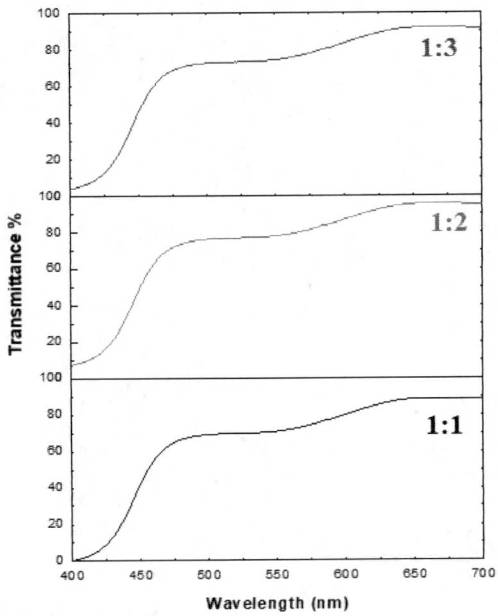

Figure 4. Transmittance spectra of CdS films deposited for Cd/S ratio 1:1, 1:2, and 1:3.

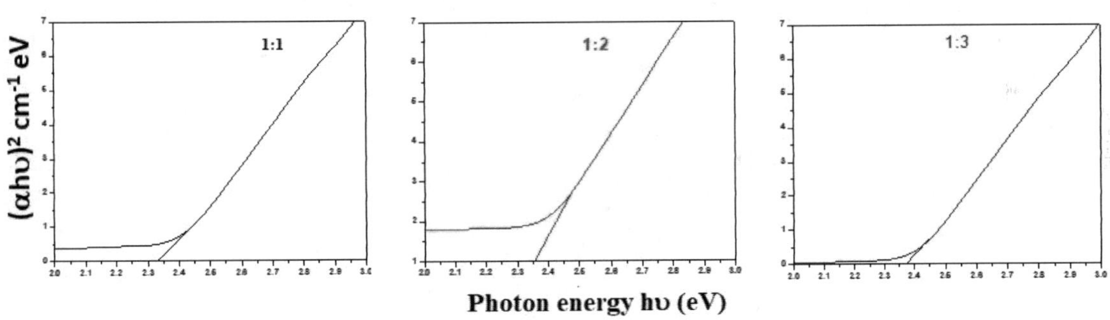

Figure 5. Tauc's plots of CdS films deposited for Cd/S ratio 1:1, 1:2, and 1:3.

3.4 PL study

Figure 6 presents the PL spectra of CdS films deposited for different Cd/S ratios. PL spectroscopy has been performed to study the luminescence spectra ($\square_{Ex}$=442 nm) recorded in the wavelength range from 400nm (3.1eV) to 750nm (1.65eV). The structures of PL emission spectra of all three films exhibit the similar curve shapes, but the intensity of film deposited for Cd/S ratio of 1:2 is maximum suggesting that it has better PL behaviour as compared to other two films. Two distinct bands have been obtained in the PL spectra. The first stronger peak centred at 2.53 eV (489nm) is associated with the band to band transition and the second one centred at 2.23 eV (556nm) is attributed to radiative recombination of electrons and holes through surface/defect states [11].

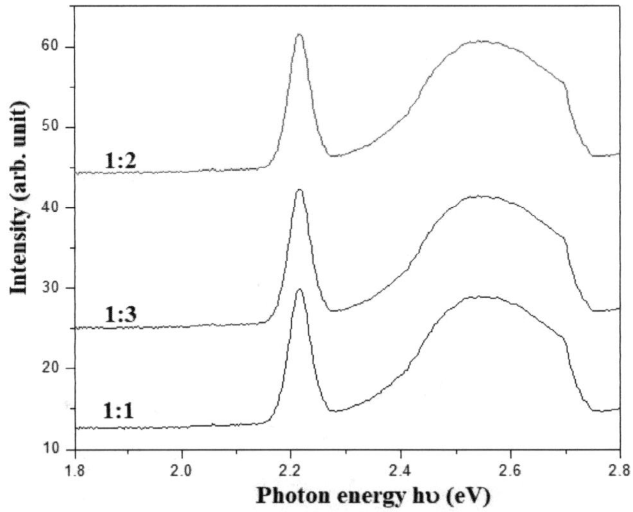

Figure 6. PL spectra of CdS films deposited for Cd/S ratio 1:1, 1:2, and 1:3.

3.5 FT-IR study

The characteristic peaks exhibited by room temperature FT-IR spectra of deposited films are presented in figure 7. The absorption bands at 1448 and 1971 cm^{-1} are associated with the triple-bonded S–C–N vibrations. Band situated at 1642 cm^{-1} is associated with the bending vibrations of CdS. Another broad band at 1309 cm^{-1} is related with the C-S stretching bands. Some other weak bands observed between 684 and 917 cm^{-1} are associated with the stretching of C–O bond, C-S bond, O–H bond, and Cd-S bond of CdS. These findings are in good agreement with the previous studies [12-14].

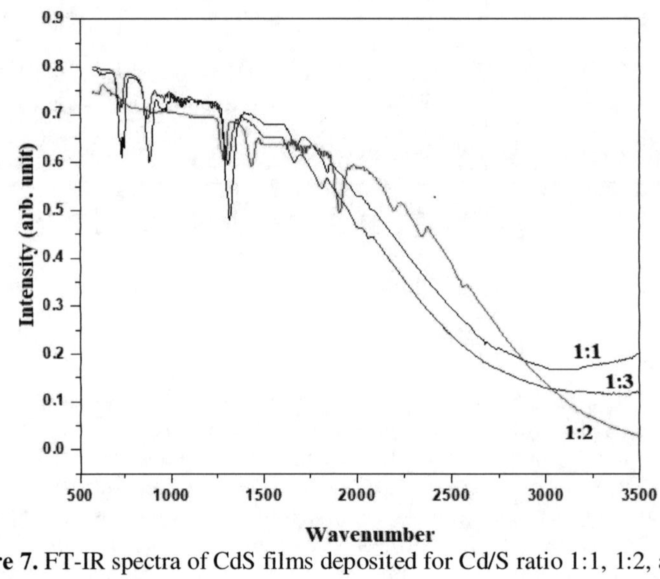

Figure 7. FT-IR spectra of CdS films deposited for Cd/S ratio 1:1, 1:2, and 1:3.

4. Conclusions

Three CdS films with different Cd/S ratio have been deposited by CBD technique. Films showed cubic crystalline nature and the film deposited for Cd/S ratio of 1:2 exhibits best crystallinity. The SEM analysis showed that the morphological nature of films depends significantly on the Cd/S ratio and it varied with the change in the precursor ratio. AFM analysis confirms that the films deposited for 1:1 and 1:3 are rougher than the film deposited for Cd/S ratio of 1:2. Two bands (situated at 489nm and 556nm) are found in PL spectra of the CdS films. Stretching bands of C-S, O–H, C–O, S-C-N, and Cd-S bonds of CdS films are found in the FT-IR spectra.

References

[1] Moreno-Regino VD, Castañeda-de-la-Hoya FM, Torres-Castanedo CG, Márquez-Marín J, Castanedo-Pérez R, Torres-Delgado G and Zelaya-Ángel O 2019 *Res. Phys.* **13** 102238
[2] Moualkia H, Rekhila G, Mahdjoub A and Trari M 2017 *J. Mater. Sci.: Mater. Electron.* **28** 19105-12
[3] Isik M, Gullu H H, Delice S, Parlak M and Gasanly N M 2019 *Mater. Sci Semicond. Process.* **93** 148-52
[4] Nobari N, Behboudnia M and Maleki R 2017 *Mater. Sci. Eng. B* **224** 181-89
[5] Sinha T, Lilhare D and Khare A 2019 *Rare Metals* https://doi.org/10.1007/s12598-019-01331-1
[6] Carreón-Moncada I, González L A, Rodríguez-Galicia J L and Rendón-Angeles J C 2016 *Thin Solid Films* **599** 166-73
[7] Patil S B and Singh A K 2010 *Appl. Surf. Sci.* **256** 2884-89
[8] Mukherjee A, Fu M, Ghosh P and Mitra P 2015 *Mater. Lett.* **141** 39-42
[9] Kumar P, Saxena N, Gupta V, Singh F and Agarwal A 2014 *J. Appl. Phys.* **116** 043517
[10] Priya N S, Kamala S S, Anbarasu V, Azhagan S A and Saravanakumar R 2018 *Mater. Lett.* **220** 161-64
[11] Wang C, Wang H M and Fang Z Y 2009 *J. Alloys Compd.* **486** 702-5
[12] Islam M A, Haque F, Rahman K S, Dhar N, Hossain M S, Sulaiman Y and Amin N 2015 *Optik* **126** 177-80
[13] Chander S and Dhaka M S 2017 *Thin Solid Films* 638 179-88
[14] Maticiuc N, Katerski A, Danilson M, Krunks M and Hiie J 2017 *Sol. Energy Mater. Sol. Cells* **160** 211-16

IOP Conf. Series: Materials Science and Engineering **798** (2020) 012024 doi:10.1088/1757-899X/798/1/012024

Characterization of Photovoltaic Property of a $CH_3NH_3Sn_{1-x}Ge_xI_3$ Lead-Free Perovskite Solar Cell

A Verma[1], A K Diwakar[1],R P Patel[2], A Verma[3]

[1]Department of Physics, Kalinga University, Atal Nagar, Raipur (C.G.) India-492101
[2]Department of Pure and Applied Physics, Guru GhasidasVishwavidyalaya,Koni, Bilaspur, (C.G.), India-495009
[3]SoS in Physics and Astrophysics, Pt. Ravishankar Shukla University, Raipur (C. G.), India-492010
Corresponding authere mail ID: alokeverma1785@gmail.com

Abstract: Using nano blending of lead-free perovskite organometalliccompound ($MABO_3$) is finding by blending of methylammonium and SnI_3. The structure of testing matterial and PV property is characterized on this work. The perovskite structure confirms by XRD analysis and it is havingan orthorhombic crystal phase structure, which can show the perovskite structure. UVPS and UVVIS analysisof the band position of valance at 5.70 eV in competition usingin absence of air and the bandgap of optical is 2.0 eV. The Spin coating of the methylammonium iodide and lead-free iodide (SnI_3) on a TiO_2 film yields by mixed solution process, 1.80-nm-diameter $CH_3NH_3Sn_{1-x}Ge_xI_3$ dots on the TiO_2 surface.

Keywords:$CH_3NH_3Sn_{1-x}Ge_xI_3$,Perovskite, UVPS, UVVIS, SnI_3 (lead- free iodide).

1. Introduction

Last few years, in this field lots of work and developing many new types of solar cell material. Lead is a very harmful element for our environment. So we work on this field and remove to lead in solar cells and developing new type solar cell synthesis from Sn and Ge. It's a fourth-generation solar cell full fulfills our requirement of energy and more efficient from the traditional solar cell [1]. Perovskites material having lots of physical property one of the most important properties of these materials easily shows the PV properties. Nanocrystals of semiconductor are the time-honored ample manners due to QCE and its endless optical transitions within the middle of the bands of electronic in the substance crystals become discrete in the nanosize crystals and in that, way the properties of the nanomaterial size are dependent on size [2]. The semiconductor of nanoparticles are having various optical properties like using in displays, BIS and PSCs. Therefore the application of solar cell like nanomaterials of semiconductor can be used as a photosensitive light-absorbing material in either a solid-state pn junction structure or PEC junction type [3]. The iodide- based redox electrolyte is investigated in the PV performance of $CH_3NH_3Sn_{1-x}Ge_xI_3$-sensitized perovskite solar cell.

2. Methods

4-Hydroxybutyric acid lactone solution with equimolar CH_3NH_3I and SnI_3doped Ge on a TiO_2surface using making of $CH_3NH_3Sn_{1-x}Ge_xI_3$ by direct deposition. CH_3NH_3I was synthesis fromusing methylamine 18.2 mL andHI 10 mL (55 wt % in water) in a erlenmeyer flask of250-mL at 0°C for 2 Hrs [4]. The precipitate was collected by evaporation at 80°C for 1 Hrs, which is followed by wash five times with DME and become dry at 100°C forover night in a air less dryer. The syntheses CH_3NH_3I powder was mixing with SnI_3doped Ge at a 1:1 ratio of mole in 4-Hydroxybutyric acid lactone at 80°C for twoHrs, that is used as a solution of coating on thelayered formation of $CH_3NH_3Sn_{1-x}Ge_xI_3$on the TiO_2. 43 wt.% concentrated solution of coting used, that is having CH_3NH_3I 2.2 g (13 mmol) and $SnI_3$6.0 g (13 mmol) in 4-Hydroxybutyric acid lactone 10 mL.FTO conductive glass was pretreated with 0.1 M $[CH_3CH_2OCOCH{=}C(O\text{-})CH_3]_2Ti(OCH(CH_3)_2)_2$in $CH_3(CH_2)_3OH$, wherein the TiO_2 was deposited as a past and one Hrs. heated at 550°C. Using IQ surface profiler α-step determination of the thickness of the annealed TiO_2 film. The coating solution of perovskite was spreay going on the TiO_2base (38.46 $\mu L/cm^2$) and 13 sec rotating at a speed of 2,400 rpm in the comprehensive atmosphere. $CH_3NH_3Sn_{1-x}Ge_xI_3$formed at the TiO_2 surface was 15 min desiccated at 100°C. Counter electrod of platinum was prepared by spreading a droplet of 7 mMChloroplatinic acid hydrate in $(CH_3)_2CHOH$ on an FTO substrate and 20 min heated at 400°C in presence of air. Sandwiching of$CH_3NH_3Sn_{1-x}Ge_xI_3$ and TiO_2 onworking electrode and counter electrode was using 25-μm-thick Surlyn. The redox electrolyte was readyby liquifying 0.9 M LiI, 0.45 MI_2, 0.5 M test-2,6-Di-tert-butylpyridine, and 0.05 M urea in $CH_3COOCH_2CH_3$, which was given the space of the vacuum-packed electrodes aforementioned to measurement [6]. PXRD profiles has been noted by diffractometer Rigaku D/MAX2200/PC using graphite-monochromatedCuKα radiation (λ = 1.5 Å). Here, over the 2θ range from 5° to 100° for 4 sec in each 0.02° step oncomprehensive temperature at data collected. Lattice parameter determines by the using of the TREOR software. XRD measurement of $CH_3NH_3Sn_{1-x}Ge_xI_3$powder was obtained by desiccating the solution at 100°C of the equal molar mixture CH_3NH_3I and SnI_3 doped Ge [6]. EDXRS combined with a FESEM using the investigation of the distribution of perovskite $CH_3NH_3Sn_{1-x}Ge_xI_3$ in the TiO_2 film.The energy of X-ray corresponding to Sn, Ti, , I and Gehas beennoted and cross-sectional area in the TiO_2 filmrepoted by SEM [3].Here, the confermention of the presence of element by XRD and SEM, and mapping are synchronized. HRTEM using the investigation of the SEM image at 200kV.

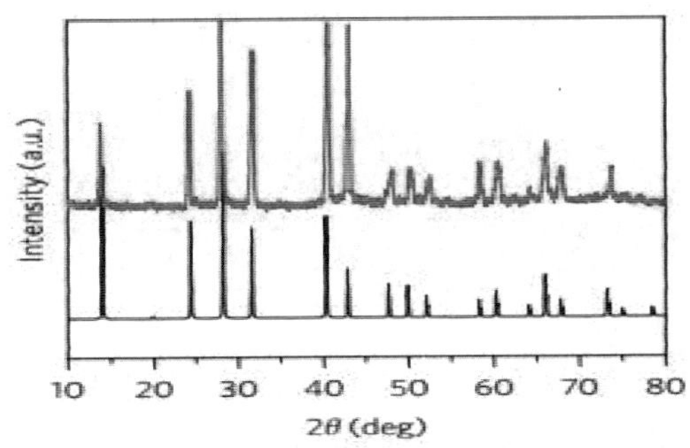

(a)

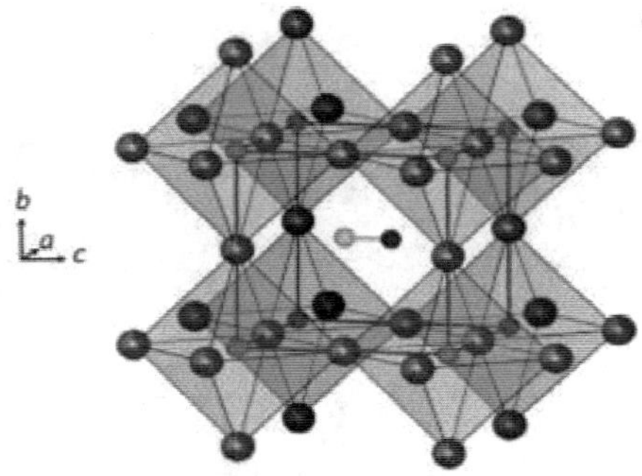

(b)

Figure 1 (a) XRD pattern and (b) crystal structure (b) of $CH_3NH_3Sn_{1-x}Ge_xI_3$.

3. Results and discussion

The XRD pattern shows thefigure 1a of the $CH_3NH_3Sn_{1-x}Ge_xI_3$. The unit cell value of a = 8.66 Å, b = 8.11 Å and c= 30.33 Å having the orthorhombic structure of $CH_3NH_3Sn_{1-x}Ge_xI_3$. TEM image shows the Figure 4 of the $CH_3NH_3Sn_{1-x}Ge_xI_3$deposited on TiO_2, where the $CH_3NH_3Sn_{1-x}Ge_xI_3$ spots are noticeablyacquired and lightly distributed on the TiO_2. The coating on the TiO_2 surface through the spin-coating method using the CH_3NH_3I and SnI_2doped Ge leads to making a spot of $CH_3NH_3Sn_{1-x}Ge_xI_3$. The Diameter size of 2.0 nm is the average size of the deposited $CH_3NH_3Sn_{1-x}Ge_xI_3$.

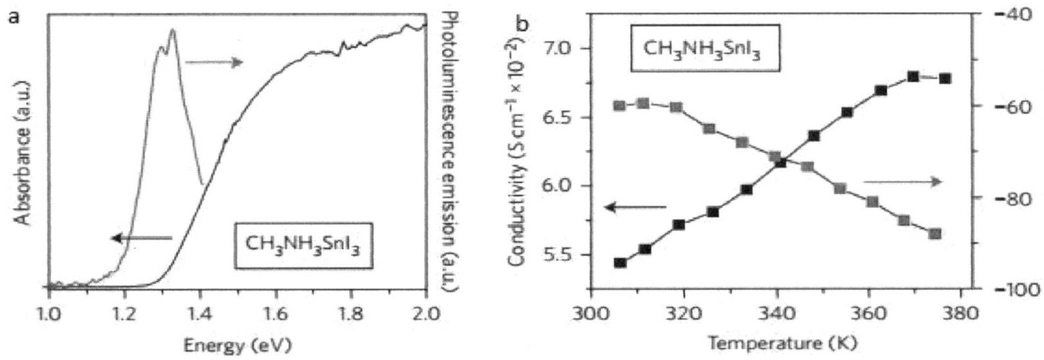

Figure 2 (a) OA and PL spectra, conductivity and Seebeck coefficient (b) temp function of sample of $CH_3NH3SnI_3$.

Figure 3 Cross-sectional SEM image (scale bar: 500 nm).

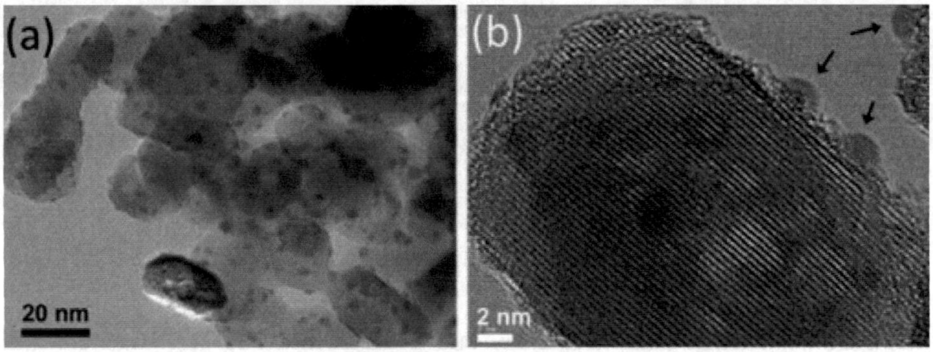

Figure 4 (a) wide view and (b) magnified image of $CH_3NH_3Sn_{1-x}Ge_xI_3/TiO_2$TEM micrographs

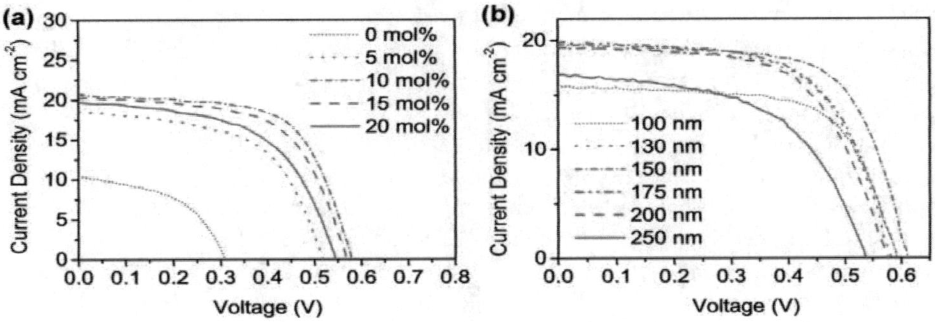

Figure 5 J – V curves of$CH_3NH_3Sn_{1-x}Ge_xI_3$ with (a) different SnI_3 molar ratios and (b) different perovskite layer thicknessesdevived based.

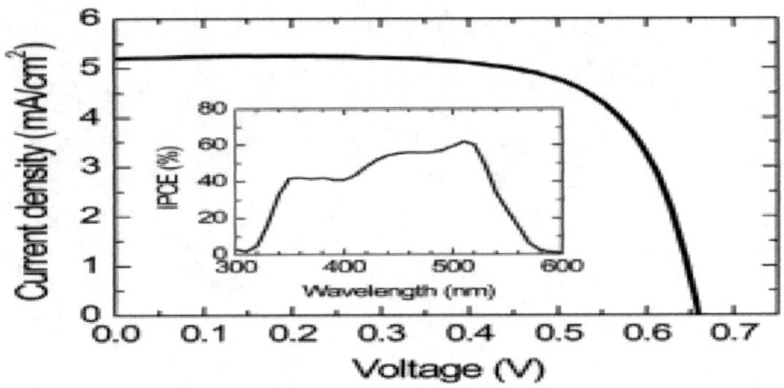

Figure 6 PID/V curve of the $CH_3NH_3Sn_{1-x}Ge_xI_3$-sensitized solar cell under the global standard spectrum one-sun light intensity.

Energy-dispersive X-ray spectroscopy mapping shows the Figure 3 and in the mesoporous TiO_2 film Sn, Ge and I are well-distributed 3-D. In Energy-dispersive X-ray spectroscopy, element analysis atomic percentages are found to be 2.0 % and 5.0% for Sn, Ge and I, respectively. These results are shown that the ratio of Sn, Ge to I is close to 1:3.Figure 6 shows the PID/V curve of the $CH_3NH_3Sn_{1-x}Ge_xI_3$-sensitized solar cell under AM 1.5 G one-sun light intensity. A PID of 5.2 mA/cm^2, 0.66 V, and 0.70FF are noted at the global standard spectrum one sun 100 mW/cm^2 illumination, leading to an overall CE of 2.4%.The electron excitation starts to come about at around 570 nm, which is steady with the probableEg of 2.2 eV.

4. Conclusions

Succesfullysynthesized of lead-free perovskite solar cell based on organic-inorganic hybridization of new nanocrystalline surface used TiO_2. $CH_3NH_3Sn_{1-x}Ge_xI_3/TiO_2$was notated to bean orthorhombic phase-type of perovskite and 2.2 eV optical bandgap and E_v position to be 5.6 eV based on UPS measurement. The position of conduction band 0.8 eV noted for $CH_3NH_3Sn_{1-x}Ge_xI_3$ was higher than that TiO_2, that is show the allowed photo excited electrons injection from $CH_3NH_3Sn_xGe_xI_3$ to TiO_2. An overall conversion efficiency of 3% noted and sensitized by $CH_3NH_3Sn_{1-x}Ge_xI_3/TiO_2$, under full sunn illumination.

Reference

[1] Verma A, Diwakar A K, & Patel R P 2019Int. J. Sci. Res. in Physics and Applied Sciences 7 24.
[2] Goesmann H and Feldmann C 2010 AngewandteChemie International Edition 49 1362.

[3] Im J, Chung J, Kim S et al.2012Nanoscale Res Lett 7 353.
[4] Boopathi K M, Ramesh M,Perumal P, Huang Y, Tsao C, Chen Y, Lee C and Chu C 2015 J. Mater. Chem. A 3 9257.
[5] SeokwooJ & Braun P 2003ChemInform 15 1256.
[6] Im J H, Lee C R, Lee J W, Park S W & Park NG 2014 Nanoscale 3 4088.
[7] Ke W, Kanatzidis M G 2019 Nat Commun 10 965.
[8] Noel N K, Stranks S D, Abate A, Wehrenfennig C, Guarnera S, Haghighirad A A, Sadhanala A,Eperon G E, Pathak S K, Johnston, et al. 2014 Environ. Sci. 7 3061.
[9] Hao F, Stoumpos C C, Cao D H, Chang R P H, Kanatzidis M G 2014 Nat. Photon. 8 489.
[10] Kumar M H,Dharani S, Leong W L, Boix P P, Prabhakar R R, Baikie T, Shi C, Ding H, Ramesh R, Asta M, et al. 2014 Adv. Mater. 267122.
[11] Zhao X G, Yang J H, Fu F, Yang D, Xu Q, Yu L, Wei S H, Zhang L 2017 J. Am. Chem. Soc. 139 2630.

IOP Conf. Series: Materials Science and Engineering **798** (2020) 012025 doi:10.1088/1757-899X/798/1/012025

Green and chemically synthesized ZnO nanoparticles: A comparative study

P K Upadhyay[1], Vikas Kumar Jain[2], A K Shrivastav[3], Shashank Sharma[4] and Ravi Sharma[5*]

[1]Department of Physics, Govt. Nagrik Kalyan Mahavidyalaya Ahiwara, Durg (C.G.) India

[2]Department of Chemistry, Govt. Engineering College, Sejbahar, Raipur (C.G.) India

[3] Department of Physics, Dr C. V. Raman University, Kargi road kota , Bilaspur (C.G.) India

[4]Mechanical Department, University College of Engineering, Osmania University, Hyderabad, India

[5]Department of Physics, Govt. Arts and Commerce Girls College Raipur (C.G.) India

*Corresponding author's e-mail address: rvsharma65@gmail.com

Abstract: Green synthesis of ZnO nanoparticles using plant extract is an alternative method to traditional chemical synthesis method. In the present work, we report the synthesis of zinc oxide nanoparticles (ZnO) synthesized by two methods, first by green synthesis by using zinc acetate and leaves extract of Ocimum Tenuiflorum (Tulsi leaves) and secondly by chemical route method using organic capping agent. The effect of synthesis methods on structural, optical, and photoluminescence property of ZnO nanoparticle were analyzed. Formation of zinc oxide nanoparticles has been confirmed by UV-Visible absorption spectroscopy, X-ray diffraction (XRD), Fourier transform infrared spectroscopy (FTIR), and Scanning Electron Microscope (SEM). The Scanning Electron Microscope reveals non spherical morphology of nanoparticles. The particle sizes of nanoparticles were calculated experimentally using XRD and SEM. The particle size measured by XRD pattern was found to be 63nm for chemically synthesized ZnO and 38nm for ZnO sample prepared by green synthesis. SEM images showed interesting flower like structure and agglomeration of nanoparticles. The blue-shift in absorption spectra was observed with reducing size of the nanoparticles. The FTIR spectra inferred that the stabilizing agent passivated the surface of the particles. The PL spectra showed blue shift for the sample prepared by green synthesis. The results suggest that ZnO sample synthesized be green route is better in comparison to the chemically synthesized samples.

Keywords: Photoluminescence, Optical Characterization, ZnO nanoparticles, FTIR, Green synthesis.

1.Introduction

Nanomaterials play a very important role in today's material world. The nanomaterials exhibit unique and different physical, chemical and biological properties when compared to their macro scaled counterparts [1]. Nanoparticles are extensively studied for their optical, electronic, catalytic, magnetic, antimicrobial properties. In the last few decades, inorganic nanoparticles, whose structures showed significantly new and improved physical, chemical, and biological properties and functionality due to their nanosize was much studied. Several types of inorganic metal oxides have been synthesized and remained in recent studies like TiO_2, CuO, and ZnO. Among the metal oxide, ZnO NPs are of maximum interest because they are inexpensive to produce, safe and can be prepared easily. Zinc oxide nanoparticles have drawn the attention of many researchers for their unique optical and chemical properties that can be easily tuned by changing the morphology. Zinc oxide nanoparticles belong to the class of metal oxides, which is characterized by photocatalytic and photo-oxidizing capacity against chemical and biological species [2].Zinc Oxide is a versatile semiconductor material and has attractive properties like wide band gap (3.2eV), large exciton binding energy (60 meV), high stability and environment friendly material. ZnO nanoparticles have been used in various cutting-edge applications like electronics, communication, sensor, cosmetics, environmental protection and biological applications like biological sensing, biological labeling, gene delivery, drug delivery and nanomedicines etc. [3–8].

There are many conventional physical and chemical methods that have been used for the synthesis of ZnO nanoparticles, such as, chemical vapor synthesis, laser ablation, solvothermal and sol-gel method [9-12]. However, these methods are usually expensive, labor-intensive, and are potentially hazardous to the environment and living organisms. The chemical method involves the use of toxic chemicals which can prove to be hazardous for the environment and the person handling it[13].In physical and chemical methods, additional capping and stabilizing agent are needed. Most of the synthesis techniques use organic solvents and reducing agents [14–17]. Thus, there is an obvious need for an alternative, cost-effective and at the same time safe and environmentally sound method for synthesizing nanoparticle. Therefore there is a need for "green synthesis" that offers numerous benefits of eco-friendliness, where toxic chemicals are not used for the synthesis process. In green synthesis method plants, bacteria, fungi, algae etc are used. This method involves a biological approach in which different types of microorganisms/ plant extracts are used to synthesize different metallic NPs and has many advantages over other chemical methods as they are greener, energy saving and cost-effective. Various researchers studied the biosynthesis of zinc oxide nanoparticles using agricultural wastes, plants and their parts and fungal cultures that has emerged as an alternative to chemical synthetic procedures because it does not require elaborate processes and provide natural capping agents[18–23].But, some problems that are often experienced in synthesizing metal nanoparticles by green synthesis are stability and aggregation, control of crystal growth, morphologies, sizes, and distribution, which are important issues and continue to be solved. In present study, we have synthesized samples of ZnO nanoparticles both by chemical (chem) and green method and compared the structural and optical properties of them. The structural characterization by XRD, SEM and optical characterization by UV-VIS spectrometry, FTIR was made. The synthesized nanoparticles produced by both methods have different size and morphology. Results revealed that the ZnO nanoparticles synthesized by green route shows enhanced structural and optical properties than chemically synthesized ZnO nanoparticles. The present work shows that, the green synthesis of ZnO sample have better results in comparison to the chemically synthesized samples.

2.Experimental

The samples of ZnO nanoparticles were synthesized by two methods.

(i) Green Synthesis

For the preparation of leaves extract of ocimumtenuiflorum leaves was first washed many times with water and dried in sunlight. 10 gm of dried leaves with 100 ml distilled water were taken in a 250 ml borosil beaker and was boiled for 10 min. until solution turns in reddish colour. The mixture solution was then cooled to room temperature. The leaves extract filtered utilizing filter paper 50 ml of ocimumtenuiflorum leaf extract was boiled to 60-70°C using magnetic stirrer heater. Then 5 gm of zinc nitrate was added to the leaf extract, when temperature reaches to 70°C it reduces to deep reddish paste. This paste was then dried at temperature 100- 130°C for 40-45 minutes. The dried light yellow powders of zinc oxide nanoparticles were obtained. The ZnO nanoparticles so obtained were preserved in the air-tight bottles for characterization purpose.

(ii) Chemical method

Zinc acetate (Zn $(CH_3COO)_2$. $2H_2O$), thiourea ($(NH_2)_2CS$) and ammonia (25% NH_3) was used as the starting materials, for the preparation of the ZnO samples. The samples were prepared by dissolving zinc acetate, thiourea and ammonia in deionised water, separately. The chemical bath technique was used. Zinc acetate, thio-urea and ammonia solutions were mixed at ratio of 1:1:1 and rapidly heated to 80ºC stirring it continuously. The prepared ZnO nanoparticles were washed with acetone, ethanol and deionized water. All the reagents of AR-grade were used without further purification.

The morphologies and sizes of as prepared ZnO nanoparticles were determined by X-ray diffraction using Cu Kα radiation (λ = 1.5418 Å). Rigaku Rotating Anode (H-3R) diffractometer was used for X-ray diffraction patterns. XRD data were collected over the range 20°– 70° at room temperature. For scanning electron microscope (SEM), JSM – 5,600 LV was used. Absorption spectra of the samples prepared was recorded with the help of Perkin Elmer Model Lamda 950 UV-VIS spectrophotometer. The FTIR spectra were recorded by the help of FTIR spectrometer (SHIMADZU) in the range of 500 cm^{-1} to 4,000 cm^{-1}. For PL measurements Perkin Elmer spectroflurophotometer Model LS 45 was used.

3. Results and Discussions

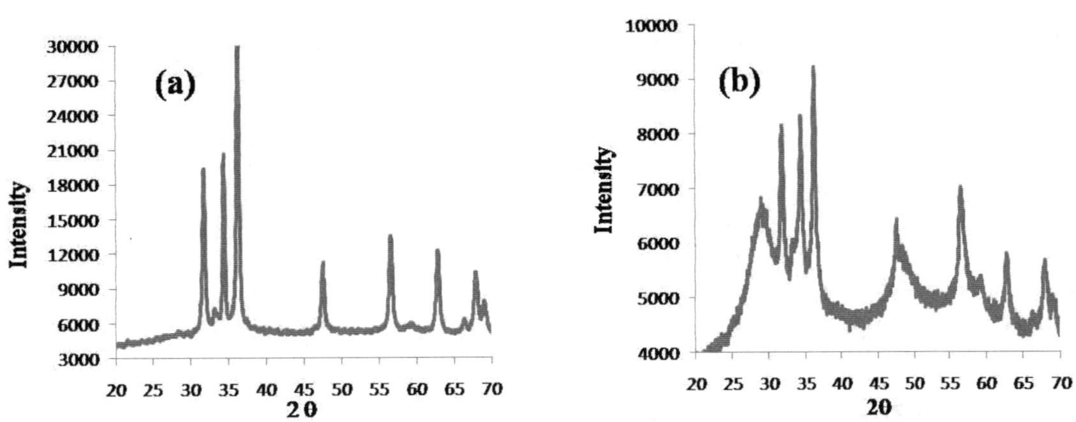

Figure.1 (a) XRD of nanoZnO by Chemical method (b) XRD spectra of nano ZnO by green method.

IOP Conf. Series: Materials Science and Engineering **798** (2020) 012025 doi:10.1088/1757-899X/798/1/012025

The crystal structure and purity of ZnO nanoparticles were characterized by X- ray diffraction (XRD) studies. Figure 1a shows the XRD pattern of the nano ZnO synthesized by chemical method. From the above figure it could be inferred that the synthesized ZnO nanoparticles are in pure phase, i.e. no impurities are seen in the XRD pattern. The peaks obtained at $2\theta^{o}$ values of 31.62, 34.33, 36.12, 47.33, 56.31, 62.64, 66.03, 67.64, and 68.73 can be attributed to the planes (100), (002), (101), (102), (110), (103), (200) and (112), (201) respectively. All the peaks are in good agreement with hexagonal wurtzite structure of ZnO well matched with standard data (JCPDS 36-1451). Broadening in the XRD peaks could be seen very clearly, which confirms the nanocrystalline nature of zinc oxide. Figure 1b shows XRD pattern of ZnO nanoparticles by green synthesis method. Broad diffraction peaks were observed for eco-friendly method with one extra peak attributed to plane (001) obtained at $2\theta^{o}$ value of 27.82, rest all other peak are slightly shifted towards lower angle side, which may be due to increasing lattice strain and inter planar spacing [24,25].The broadening occurs in all the peaks. This is due to the smaller particle size and large number of crystal planes. The intensity in the signal of their diffraction patterns is reduced due to broadening of peaks. The XRD result indicates green synthesized ZnO nanoparticles have enhanced crystalintity than chemical method.

For both the samples the average crystallite size was determined from the XRD pattern parameters using Debye Scherrer equation $D = k\lambda/\beta\cos\theta$, where D is the average crystallite size, k is Scherrer's constant equal to 0.90, β is the full-width at half-maximum (FWHM), and θ is the Bragg's angle [26]. The broadening in peaks due to crystal imperfection, distortion and strain was calculated using the formula $\varepsilon = \beta \tan\theta$ [27].

Table 1 XRD parameters

Samples	Lattice Parameter (Å)		d spacing	Lattice Strain	Crystallite Size
	a	c			
Chemical Method	3.264 Å	5.228 Å	2.4992	0.1782	63nm
Green Method	3.243 Å	5.209 Å	2.4763	0.1964	38nm

Table 1 shows the lattice parameter, crystallite size, d–spacing and lattice strain of ZnO nanaoparticles synthesized from chemical and green method. The crystallite size of the chemically synthesized ZnO was found to be 63nm, whereas, the size of ZnO synthesized by green method was found 38nm.

The Morphology of synthesized ZnO nanoparticles was investigated by scanning electron microscopy. Figure 2 (a and b) shows the SEM image of ZnO synthesized from chemical and green method respectively. The sample of ZnO synthesized the chemical method shows the distinctive and abundant flower-shaped structure, which were principally the ZnO nanoparticles. There are huge hexagonal arrays of nanoparticles assimilated to form flower-shaped bundles. It can be seen that many of the particles are agglomerated, this is due to the formation of ZnO nanoparticles during first few minute of the experiment and later these particles agglomerates and their size becomes larger. The average particle sizes obtained from SEM images was ~ 65 nm. The SEM patterns agree well with the XRD results in terms of both size and shape. In the figure 2(b), highly agglomeration of particles is clearly seen. The average size of these synthesized non spherical powder ZnO nanoparticles is ~ 40 nm. Many of the particles are in the range less than 30nm.The hexagonal structure is not clearly visible for the sample prepared by green synthesis. As seen from the SEM images, green ZnO nanoparticle has lower particle size than the chemically synthesized ZnO nanoparticles. The change in morphology observed is due to green synthesis and natural capping agent.

Figure 2. (a) SEM of ZnOnano by Chemical method, (b) SEM of nanoZnO by green method.

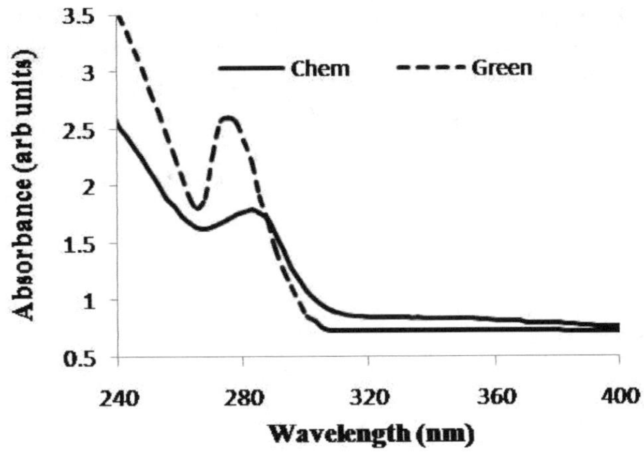

Figure. 3 Absorption spectra of samples synthesized by Chemical and Green method.

The UV–visible absorption spectra of ZnO nanoparticles synthesized by chemical method and by green synthesis as a function of wavelength are shown in Figure 3 In both the cases, the absorption band shows a blue shift due to the quantum confinement effect, as compared with the bulk. This optical phenomenon indicates that these nanoparticles have a quantum size effect [28]. Figure 3 shows the optical absorption spectra of the nanoZnO measured between 240 nm- 400 nm. It was found that the spectra are featureless and very less absorption occurs in the visible region. The absorption peak for the synthesize nanoZns by chemical method was recorded at 284nm. The excitonic absorption peak of ZnO nanoparticle synthesized by green method was found at 274 nm, which lies much below the band gap wavelength of 388 nm of bulk ZnO [29]. The peak at ~274 nm is due to inter band transition of electrons from the more inner shell to the uppermost shell. It is possible that, due to aggregation and agglomeration, particle size increases and material settled down at the bottom of container causing decrease in the absorbance [30]. This type of behavior is typical for many semiconductors due to internal electric fields within the crystal and inelastic scattering of charge carriers by phonons [31, 32].

Absorption coefficient (α) associated with the strong absorption region for both the samples of nanoZnO were calculated from absorbance (A) and the sample thickness (t) using relation $\alpha = 2.303A/t$. The Tauc relation → $(\alpha h\nu)^{1/n} \approx (h\nu - E_g)$, was used to calculate optical band gap of ZnO nanoparticles, where, α is the absorption coefficient, $h\nu$ is the energy of incident photons and the value of n depends upon the type the transition which may have 1/2, 2, 3/2 and 3 values corresponding to the allowed direct, allowed indirect, forbidden direct and forbidden indirect transitions, respectively. The characteristic absorption edge reflects the band gap of the material. The band gap calculated from the above data was ~4.36eV for the sample prepared by chemical method and the band gap for ZnO nanoparticles synthesized by green method was found to be ~4.52 eV. In both cases the optical band gap was found higher than standard value 3.54 eV. The increase in the band gap of the ZnO nanoparticles with the decrease in particle size may be due to a quantum confinement effect.

FTIR is an effective method to reveal the composition of products. The composition of the samples and the presence of various bonds were studied. Figure 4 (a) and 4(b) shows the FTIR spectra of the ZnO nanoparticles prepared by chemical and ZnO nanopaticles by green synthesis respectively. Both the figures show a smooth similar curve having little difference. The peak around 501 and 540 cm^{-1} are assigned to the Zn-O band in both the cases. In figure 4a bands around 1200 cm^{-1} and 1100 cm^{-1} are due to the characteristic frequency of inorganic ions. The absorption peak of the S-H vibration at 2,450–2,465 cm^{-1} is not observed in the IR spectrum of green synthesis (Figure 4b), indicating that the thiourea molecules were bound to the ZnO nanoparticle surface. A comparative strong peak was observed at 1086 cm^{-1}, which is because of C- O stretching of the capping agent. The broad peak in the higher energy region at 3439 cm^{-1} is due to OH stretching and the peak at 1625 cm^{-1} is due to OH bending of adsorbed moisture in the sample and all the other peaks are attributed to the characteristic of the material.

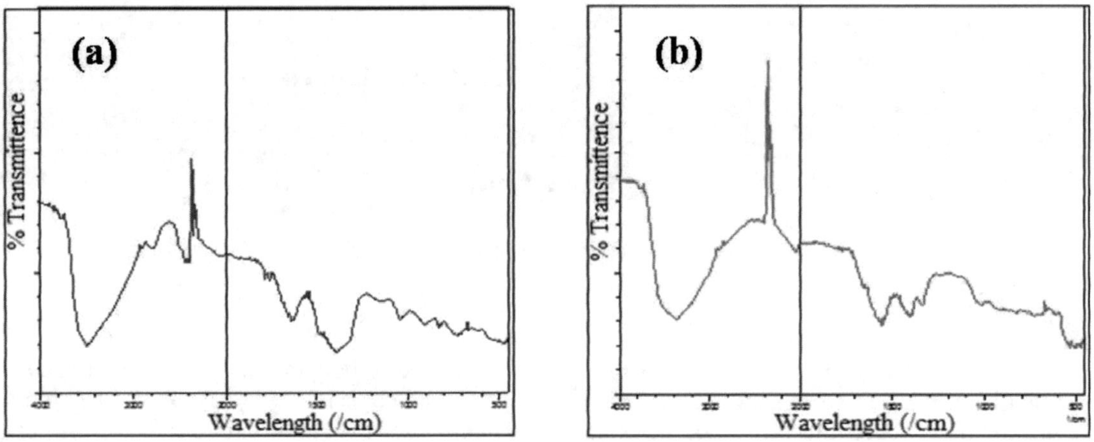

Figure 4 (a) FTIR of ZnO nano by Chemical method, (b) FTIR of nano ZnO by green method.

Figure 4(b) show characteristic peaks at 511, 591, 671, 712, 826, 921, 1,058, 1,146, 1,386, 1,460, 1,638, 1,787, 2,448, 2,801, 3,450 cm^{-1} and some other associated peaks. The bands around 501 and 640 cm^{-1} are assigned to the Zn-O band. Similar investigations reported earlier shows that metal oxides give absorption bands in the fingerprint regions below 1000 cm^{-1} arising from inter-atomic vibrations [33]. A strong intensity band at 1,058 cm^{-1} may be due to S-O-C stretching. The stretching vibrations assigned to the C-S linkage occur in the region of 800–600 cm^{-1} and the weak S-S stretching vibration falls between 600 - 400 cm^{-1}. The peak at 1600-1700cm^{-1}may be due to nitrogen–oxygen interaction. The distinct peaks at 1386 $cm^{-1}$and 1146 $cm^{-1}$are due to phosphorus–oxygen interaction, establishing the presence of covalently bonded phosphates on ZnO nanoparticles. The broad absorption peak at 3450 cm-1 can be attributed to the characteristic absorption of hydroxyl group.

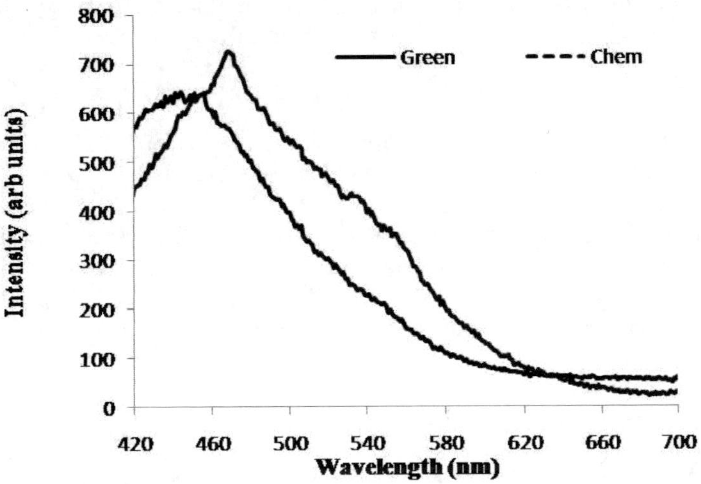

Figure 5. Photoluminescence spectra of samples by Chemical and Green synthesis.

Study of the photoluminescence (PL) is an interesting property of material that can provide valuable information on the purity and quality of the material. Luminescence studies provide information regarding defect states. PL studies may reveal the shift in defect states or the change in the density. Figure 5 shows the room temperature PL spectrum of the ZnO nanoparticles prepared by chemical method as well as by green synthesis method recorded in the range 420 nm-700 nm. The PL measurements were recorded at excitation wavelength of 370nm. Only one broad peak was found for both the samples. Stoke's shift in the peak was observed, the PL peak was obtained at 445 nm for green synthesis sample, whereas, the peak at 469 nm was found for chemically synthesized sample. This PL emission shifts towards smaller wavelength is due to the decrease in the size of the sample, which are in good agreement to the XRD and SEM results. This PL spectra may be because of the electron–hole recombination at a deep level emission in the band gap caused by intrinsic point defects and surface defects, e.g., oxygen vacancies, zinc interstitials, and the incorporation of hydroxyl groups in the crystal lattice during solution growth [34]. The broad band luminescence for both ZnO nanoparticles is related to transitions from donor states to deep acceptor states.

IOP Conf. Series: Materials Science and Engineering **798** (2020) 012025 doi:10.1088/1757-899X/798/1/012025

4. Conclusions

The nanoparticals of ZnO have been successfully synthesized by a simple chemical reaction using aqueous medium; in which thiourea was used as the capping agent and by green synthesis using leaves extract of OcimumTenuiflorum (Tulsi). The XRD pattern indicated the growth of the ZnO nanoparticles in hexagonal phase. The crystallite size measured by XRD pattern was found to be 63nm by for chemically synthesized ZnO and 38nm for ZnO sample prepared by green synthesis. SEM images showed interesting flower like structure and agglomeration of nanoparticles prepared by chemical synthesis. The SEM photograph also shows that the particle sizes are bigger for chemically synthesized samples. The blue-shift in absorption spectra was observed for the sample prepared by green synthesis. The solid-state theory based on the delocalized electron and hole within the confined volume explains the blue-shifted optical absorption spectra. The FTIR spectra inferred that the stabilizing agent passivated the surface of the particles. Prominent IR peaks were analyzed and assigned. The PL spectra showed blue shift for the sample prepared by green synthesis. Results revealed that the ZnO nanoparticles synthesized by green route shows enhanced structural and optical properties than chemically synthesized ZnO nanoparticles.

References

[1] Chaturvedi S, Dave P N and Shah N K 2012 *J. Saudi Chem. Soc.* **16** 307 -325

[2] Sharma D, Rajput J, Kaith BS, Kaur Mohinder and Sharma Sapna 2010 *Thin Solid Films* **519** 1224–1229

[3] Hussein J, El-Banna M, Razik T A and El-Nagga M E 2018 *Int. J. Biol. Macromol* **107** 748-754

[4] Wang L, Kang Y, Liu X, Zhang S, Huang W and Wang S 2012 Sens. Actuators B: Chem. **162** 237–243

[5] Cross SE, Innes B, Roberts MS, Tsuzuki T, Robertson TA and McCormick P 2007 Skin Pharmacol. Physiol. **20** 148–154

[6] Zhou J, Xua N and Wang Z L 2006 *Adv. Mater.* **18** 2432–2435

[7] Rasmussen J W, Martinez E, Louka P and Wingett D G 2012 *Exp. Opin. Drug. Deliv.* **7** 1063–1077

[8] Shaheen T I, El-Naggar M E, Abdelgawad A M, Hebeish A 2016 *Int. J. Biol. Macromol.* **83** 426-32

[9] Lobiak E V, Shlyakhova E V, Bulusheva L G, Plyusnin P E, Shubin Yu V and Okotrub A V *J. Alloys and Compounds* **621** 351-356

[10] Cho J M, Song J K and Park S M 2009 *Bulletin Korean Chem. Soc.* **30** 1616-1618

[11] Wang C, Shen E, Wang E, Gao L, Kang Z, Tian C, Lan Y and Zhang C 2006 *Current Appl. Phys.* **6** 499-502.

[12] Hasnidawani J N, Azlina HN, Norita H, Bonnia N N, Ratim S and Ali E S *Procedia Chemistry* **19** 211-216

[13] Kharissova OV, Rasika HV, Kharisov D B I, Perez B O and Jimenez Perez V M 2013 *Trends in Biotechnology* **31** 240-248

[14] Kumari B, Sharma S, Singh N, Verma A, Satsangi V R, Dass S and Shrivastav R 2014 *Int. J. Hydrogen Energy* **39** 18216–18229

[15] Ambika S and Sundrarajan M 2015 *J. Photochem. Photobiol. B Biol.* **149** 143–148

[16] Kavitha K, Paulpandi M, Ponraj T, Murugan K, Sumathi S 2016 *Karbala Int. J. Mod. Sci.* **2** 46–55

[17] Ravikumar S, Gokulakrishnan R and Boomi P 2012 *Asian Pacific J. Trop. Dis.* **2** 85–89

[18] Kumar B, Smita K, Cumbal L and Debut A 2014 *J. Saudi Chem. Soc.* **21** 605–609

[19] Haverkamp RG and Marshall AT 2009 *J. Nanoparticle Research* **11** 1453-1463

[20] Kumar B, Smita K, Cumbal L and Debut A *J. Saudi Chem. Soci* **18** 364-369

[21] Raliya R, Tarafdar J C 2013 *Agric. Res.* **2** 48–57

[22] Kadhim Q A, Alwan R M, Ali R A, Jassim A N 2016 *Nanosci. Nanotechnol. 6* 1–5

[23] Jamdagni P, Khatri P, Rana J S 2016 *J. King Saud Univ Sci.* **30** 168-175
[24] Kumar S, Asokan K, Singh RK, Chatterjee S, Kanjilal D, Ghosh A K 2014 *RSC Adv.* **4** 62123 - 62131
[25] Bindu P and Thomas S 2014 *J. Theor. Appl. Phys.* **8** 123 – 134
[26] Sharma R, Dhoble S J, Bisen D P, Brahme N and Chandra B P 2011 *Int. J. Nanoparticles* **4** 64 - 76
[27] Varol S F, Babür G, Çankaya G, Kölemen U 2014 *RSC Adv.* **4** 56645 -56653
[28] Huang M H, Mao S, Feick H, Yan H, Wu Y, Kind H, Weber E, Russo R and Yang P D 2001 *Science* **292** 1897–1899
[29] Kumbhakar P, Singh D, Tiwary C S, Mitra A K 2008 *Chalcogenide Letters* **5** 387-394
[30] Moss T S, Burrell G J and Ellis B 1973 *Semiconductor Opto-Electronics*, Butterworth & Co. Ltd.
[31] Honsi H M, Fayek S A, Al-Sayed S M, Roushdy M and Soliman M A 2006 *Vacuum* **81** 54-58
[32] Sawby A, Selim M S, Marzouk S Y, Mostafa M A and Hosny A 2010 *Physica B, Physics of Condensed Matter* **405** 3412-3420
[33] Kwon Y, Kim K, Lim C and Shim K 2002 *J. Ceramic Processes* **3** 146–149
[34] Xu S and Wang Z 2011 *Nano Research* **4** 1013- 1098

IOP Conf. Series: Materials Science and Engineering **798** (2020) 012026 doi:10.1088/1757-899X/798/1/012026

Influence of silver nanoparticle on Raman and photoluminescence of heat treated Ag-Er codoped tungsten tellurite glasses

Sunil Patidar[1], Ghizal F Ansari[1] and S K Mahajan[2*]

[1]Physics Department, Madhyanchal Professional University, Bhopal (M.P.) 462044, India
[2]Departments of Applied Physics, Samrat Ashok Technological Institute, Vidisha (M.P.) 464001, India
*Corresponding author's e-mail address: sachin_k_mahajan@rediffmail.com

Abstract. The glasses of composition TeO_2 -WO_3- Li_2O-Er_2O_3 with $AgNO_3$ (TWLEOAG) were prepared by the high-temperature melting method. Surface Plasmon resonance (SPR) bands are observed in the range 520-535 nm and 640-650nm of without erbium-doped glass by absorption spectrum. The TEM measurement reveals the presence of spherical AgNPs in this glass. The DSC has also been analyzed for the stability of transition temperature T_g and on set crystallization T_0 of this glass system. The Raman spectrum revealed that TWLEOAG glasses were mainly formed by the [TeO4], {TeO3}, [WO_4] and [WO_6] units and [TeO_4] units convert to [TeO_3] units with the addition of WO_3 and Li_2O content into tellurite glasses system. In particular TWLEOAG05 glass samples, heat-treated different time interval system reveal the enhancement in Raman peaks intensity with small shift in due to Ag nanoparticle. Also photoluminescence enhancement of red upconversion compare to green emission of TWLEOAG05 for different annealing duration has been observed under 808nm excitation. In this work optimum 0.5 mol% AgNP in TWEOAG glass would be useful for optical devices.

1. Introduction

In past decades tellurite based glasses with rare-earth ions have attracted much attention due to nonlinear properties, exhibit good infrared transmission and low phonon energy but could not improve the chemical and glass-forming ability of glass with addition of other oxide [1,2]. In most of tellurite glasses WO_3, Bi_2O_3, PbO also are mixed to provide better stability, to improve the glass-forming ability and optical properties [3] for their promising applications in optical devices and photonics. Further Li_2O or other alkali oxides are also mixed as a glass modifier, which changes its internal structure due to the formation of non-bridging oxygens [4]. When doping with rare-earth (RE) ions in such tellurite based glass host material may yield new material with modified in optical properties. It is also know that optical signal of RE ions will be enhance with codoping of other ions such as Yb^{3+} ions. Another way to embed nanoparticle (NPs) with rare-earth ions in glasses lead to enhance the optical emission. [5]. This enhancement may be progressed in the presence of the intense localized electromagnetic field in the interface of the surface of metallic Ag or Au nanoparticles (NPs) and dielectric medium. In such a situation, RE ions, photonic mode density would be enhanced tremendously by the interaction between the metal particle and rare-earth ions. Moreover, there is a limit of high concentration dopants beyond which tellurite glass displays luminescence quenching through energy transfer with Ag NPs in presence of alkali material. R.J. Amjad et al. [6] have reported the surface enhanced Raman and upconversion emission in Er^{3+}-doped zinc-tellurite glasses contain silver NPs, and an enhancement by about four times in Raman signal and up to three times in upconversion emissions of Er^{3+}ions was observed. M.R. Dousti et al. [7] have studied an efficient upconversion emission in Nd^{3+} -Ag doped tellurite glasses for enhanced upconversion emissions of Nd^{3+} under the excitation of 800 nm. Thus in addition of silver compound in telluride based glasses are investigated for new results due to low toxicity, low signal attenuation, absorption and impurities of optical glasses for biomedical and Fiber communications applications. The present work we have examined the influence of heat-treated samples of metallic NPs and erbium-doped tungsten tellurite glasses for the Raman and photoluminescence measurement. This study wills helpful Raman signal amplification for optical devices.

2. Experimental Method

The glass samples were prepared with high purity (99.9%) chemical powders according to compositions in mol% of $70TeO_2$-$15Li_2O$-$14WO_3$- $1Er_2O_3$-$x$$AgNO_3$ (x= 0, 0.5,1.0 in mol %). About 4-6 gm batches of starting materials were melted at 950^{0}C- 1050^{0}C in alumina crucible about 30-45 min. The melts were cast to form glasses by pouring it on stainless steel plate heated at 150^{0}C. These glass samples with and without Ag were annealed at 300^{0}C for 3h. All samples again submitted for heat treatment at 350^{0}C continuously for long duration to thermally reduce Ag^+ ions into Ag^0 which is needed to nucleate metallic nanoparticles and two were selected with Ag (x=0.5) for 18h and 30h for optical measurement. For the SPR absorption band of the glass was recorded by UV/VIS/NIR spectrometer (GBC Scientific equipment) in the range 400-900nm. Differential scanning calorimeter (DSC) was conducted with a Perkin-Elmer DSC 4000 instrument, at a heating rate 10^{0}C min^{-1}, between 150^{0}C and 450^{0}C .The Transmission Electron Microscope (TEM) operating at 200kV was employed to confirm the presence of the silver NPs. The Raman spectrum of the glass sample was measured in the range 200-1100 cm^{-1} by WI Tech Alpha 300R confocal Raman microscope with Ar^{3+} ions laser wavelength 514.5nm at power 15mW was used. The photoluminescence upconversion of glasses was obtained in range from 400nm to 700 nmby using a model Ava Spec 2048 Avantes, Fibre Optics Spectrometer excited by CW 808nm diode laser.

3. Result and discussions

Figure 1 shows absorption spectra in room temperature of without Er^{3+} ions doped TWLOAG (with Ag 0.5 mol%) glass heated *18 hrs* sample and different wavelength of the absorption bands of this sample clearly indicates the presence of silver. Thus surface plasmon bands (SPR) of the silver denoted the presence metallic AgNPs may be in form of spherical and non-spherical [8]. It is noted that of SPR band then SPR band will move to longer wavelength as if refractive index of the glass material is higher, [9]. Also altering heat treating duration and AgNPs concentration of glasses modifies the optical properties.

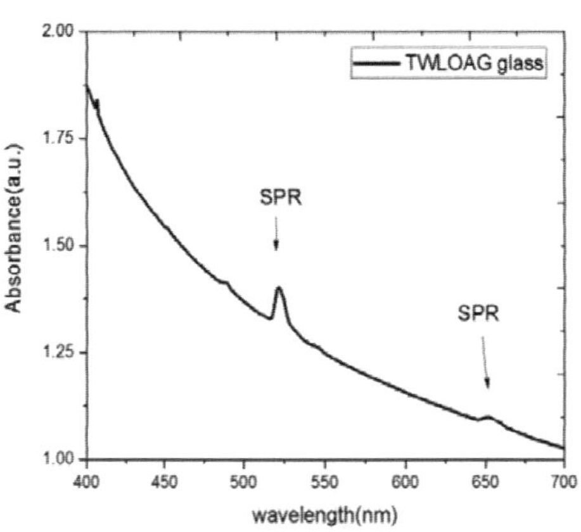

Figure 1. Surface Plasmon Resonance absorbance peaks TWLOAG (*18hrs*)

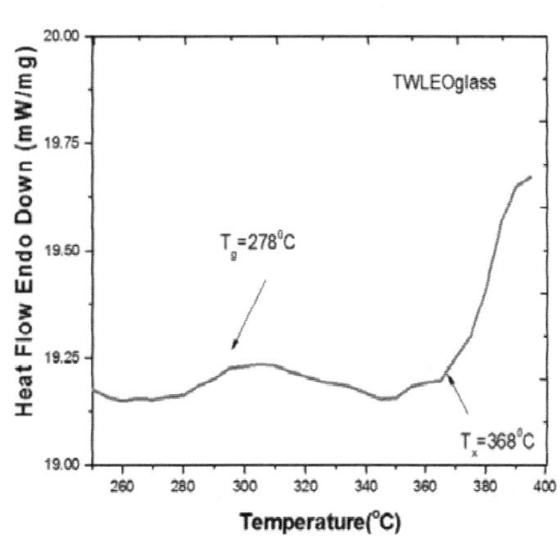

Figure 2.DSC curves of TeO_2-Li_2O-WO_3- Er_2O_3-$x$$AgNO_3$without heat treated glass

The pattern of DSC curve of TWLOAG glasses (annealed for 3hrs) heating at rate 10 °K/ min of as shown in figure 2. This sample curve suggest glass transition temperature T_g =330^{0}C and T_x glass exothermic peak at 380^{0}C.Thus small value ΔT fast nucleation and crystallization for AgNPs growing and therefore crystallization may starts at 370^{0}C. Hence, the glasses are heat-treated at 350^{0}C were the viscosity of the sample is enough to move, grow and aggregate for effective optical properties. In this glass Li_2O is taken between 5 to 10 mol% results in a decrease of T_g. [10,11] that lead to the cleavage of the networks formed by TeO_4 trigonal bipyramid units and the increase of non-bridging oxygen (NBO) atoms. Figure 3 showed the TEM image of the TWLEOAG *18h* glass sample with spherical and non- spherical silver NPs are gathered. The size of NPs varies between 35nm-40nm, with average value 38nm.

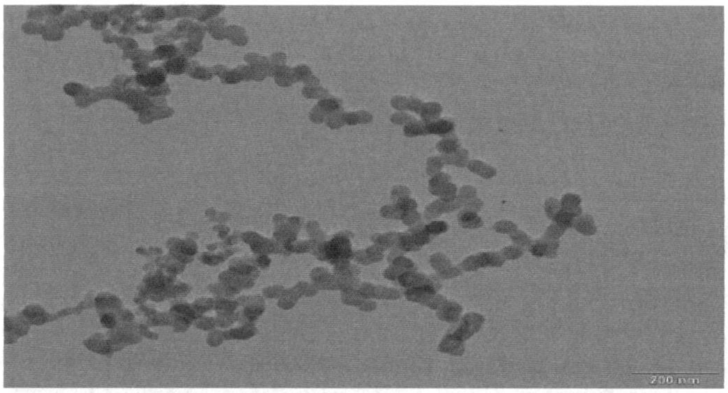

Figure 3. TEM image of Ag NPs in TWLEOAG (0.5) *18h*glass.

Raman spectra of TWLEOAG (Ag 0.5mol%) samples for 18hrs and 30hrs heat-treated glasses gives the main Raman peaks are cantered at 281, 367, 761and 930 cm^{-1} belongs to the linkages in the bulk matrix as shown Figure 4. Due to the limitation of Raman measurement for all glass samples at low-frequency Boson band region (< 200 cm^{-1}) may not be observed. However this band peaks are nearly same for different glasses and does not normally depend on chemical compositions [12] but it may affirm the presence of glass structure. Raman peaks above the wave number 200 cm^{-1}, increase

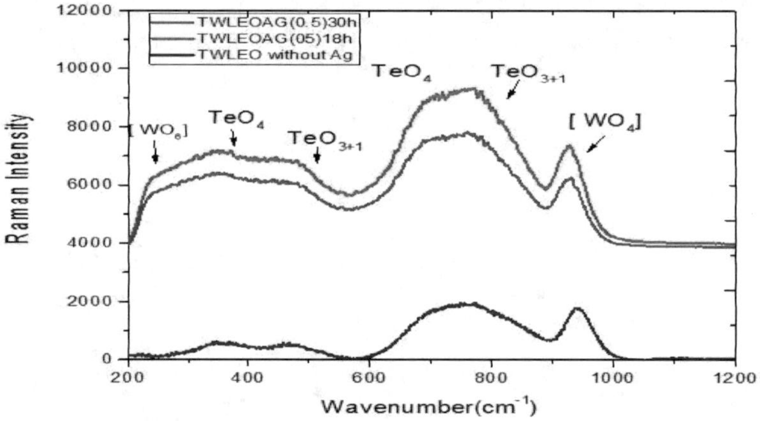

Figure 4. Raman spectra of *18h* and *30h*heat-treated TWLEOAG glasses

broad shoulder at 340cm^{-1} is occurs on account of symmetric and flexible vibration of Te-O-Te chain. Further band peak at780cm^{-1} is attributed to anti symmetric vibration of Te-O bond in TeO_3 trigonal pyramid (tp) units due to stretching vibrations of TeO_3 and /or TeO_{3+1} [13]. This non- bridge between oxygen and tellurium can be broken easily by rare-earth cations for weak binding energy. This band agrees well with the reported spectrum of the nearly same composition tellurite tungsten glass in ref. paper [14, 15]. The highest peak at about 930cm^{-1} is observed corresponds to the stretching vibrations of W=O and W-O- bonds associated with WO_4 and WO_6 groups [16].Compare to other oxide glass such as silicate, phosphate and borate glass [17], maximum phonon energy of this tellurite glass (930cm^{-1}) is smaller.

We have observed the Raman intensity enhancement of embedding silver NPs in TWLEOAG glasses heated in different duration. For selected sample TWLEOAG(0.5) 18 hrs give maximum fivefold enhancement compare to TWLEO (without AgNPs) glass due to local electromagnetic field which is surface-enhanced Raman scattering effect [18]. However this sample gives 20% more enhance Raman peaks intensity as compare to TWLEOAG(0.5) 30 hrs sample. In our measurement TWLEOAG(0.5) glass sample Raman broad peaks cantered at 210cm^{-1} is shifted to 213cm^{-1}and these are assigned to W-O stretching and O-W-O linkage bending vibration [16] and thereby numbers of NBO sites are proportionately increased. But band intensity around 352–355 cm^{-1}and 450–452 cm^{-1} showed the characteristics of the corner – linked octahedral in WO_3 as in ref. [19] for niobium oxide. Hence with the addition of Ag NPs in glass indicating the conversion of the TeO_4 (tbp) units into TeO_3 (tp) units lead to intensity of all Raman bands are first increased and decreased, and then maximum intense [20]. Thus in our Raman measurement, selected only 0.5mol% Ag NPs doped glass samples with different heat treated duration results can be compare to ref. [21] reported Raman Intensity for 1mol% AgNPs tellurite-based glass .

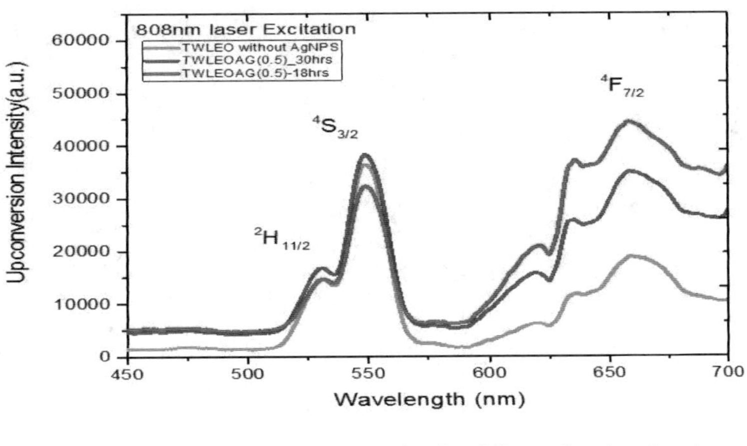

Figure 5. Emission spectra of TWLEOAG samples for the different heating duration under 808nm excitation

The red and green photoluminescence upconversion are observed under 808nm laser excitation of silver doped TWLEOAG glasses annealed for 18h and 30h and without silver doped glass for their comparison as shown in figure5. On increasing the annealing duration of 0.5mol% AgNPs doped TWLEOAG glass emits green at 550nm transition $^4S_{3/2}\rightarrow{}^4I_{15/2}$ intensity increase due to transverse SPR and on the other hand TWLEOAG(0.5) *18h* glass emits enhance red upconversion$^4F_{9/2}\rightarrow{}^4I_{15/2}$ intensity due to longitudinal SPR as reported in ref [22]. However the maximum red upcoversion enhancement (2 fold)is measured, which is due to local field enhancement or and energy transfer (ET) from Ag NPs to Er^{3+} ions but for broad red band emission ET process is more probable. Also more likely scattering or re-absorbance of excitation light lower the green emission at 550 nm intensity which is around plasmon band. Thus tune ability of green to red intensity enhancement can be maintained with increases annealing duration TWLEOAG glasses.

Figure 6 shows energy transfer from AgNPs to Er^{3+} ions, during ground state absorption 808nm photon the first excited at to the $^4F_{7/2}$ level of the Er^{3+} ion is and then quickly decays non-radiatively to the $^2H_{11/2}$ and $^4S_{3/2}$ levels [23-24] from which the upconversion transition of green emission takes place. After then fraction of Er^{3+} ions in the $^4S_{3/2}$ states would decay non-radiatively into the slightly lower $^4F_{9/2}$ energy level where the red upconversion transition originates. One additional path for the red emission comes from the energy transfer upconversion of Er^{3+} ion from $^4I_{13/2}$ level which is populated by non-radiative decay from the $^4I_{11/2}$ level. There are other processes that compete with upconversion processes such as radiative and non-radiative decays from AgNPs ions to $^4I_{11/2}$ level of Er^{3+} ions, and cross-relaxation of Er^{3+} ion pairs. But cross-relaxation of Er^{3+} ions is a major quenching mechanism at high Er^{3+}concentrations [25,26]. Main result of this work sample heat treated by *18hrs* presents enhancement in red upconversion photoluminescence as well as Raman signal. Furthermore, in presence of AgNP with RE ions doped TWLEOAG glasses show intensity variation in Raman and upconversion emission intensity denote no evidence of the formation of clusters.

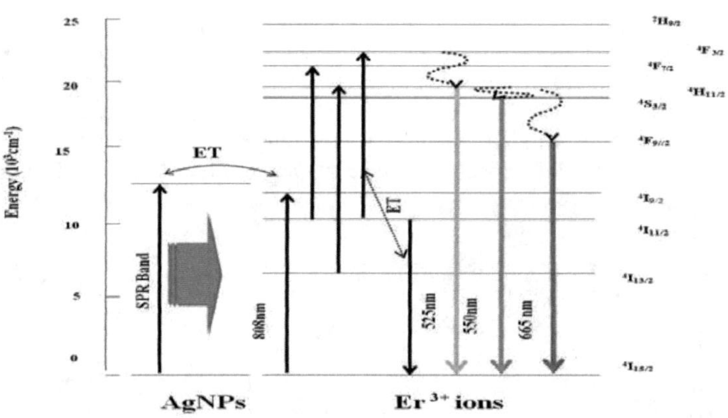

Figure 6. Energy-level diagram of AgNPs and Er^{3+}ions TWLOAG glass

4. Conclusions

The influence of silver nanoparticles on the Raman structure and photoluminescence of heat treated Ag-Er^{3+}codopedtellurite glasses were investigated. Absorption and TEM measurement confirm the Ag nanoparticles exist in the glass matrix. However, not much large temperature difference of DSC curves indicating that favour for Ag Nps nucleation growth and crystallization. The Raman spectra indicate TeO_4, TeO_3, TeO_{3+1}, WO_4, WO_6 structural units are small shifted and five-fold enhancement of the TWLEOAG(0.5)18hrs glass comparatively to without Ag doped glass has been observed. Also relative red upconversion intensity at 650nm due to SPR band enhances as compare to green upconversion intensity at 808nm excitation. In our work better result of TWLEOAG(0.5) *18hrs* (0.5% mol Ag)glass for Raman spectra and upconversion emission reveals the strong influence of Ag NPs on the different heat-treated glass will provide a potential application for optical devices.

Acknowledgments
The authors would like to thanks to RRCAT Indore Govt. of India, CIF Pondicherry University for providing the characterization facilities. The work is supported by the DST- SERB Project grants no EMR/2015/001017

IOP Conf. Series: Materials Science and Engineering **798** (2020) 012026 doi:10.1088/1757-899X/798/1/012026

References

[1] MallawanyA H E I 2002*Tellurite glasses Hand book*, CRC Press
[2] Weber M J1990 *J. Non-crystal. Solid* **123** 208
[3] Marcel P 2003 *Ann. Chim. Sci. Mater.* **28** 87
[4] Saddeek Y B and Latif L A E 2004 *Physica B* **348** 475
[5] Silva M A P, Dantelle G, Mortier M, Monteil A, Riberior S J L, Messaddeq Y, Briois V and Poulain M 2008 *J. Chem. Phys.* **128** 244516
[6] Amjad Raja J, Sahar M R, Dousti M R, Ghoshal S K and Jamaludin M N A 2013 *Optics Express* **21** 14282
[7] Dousti M R 2013 *J. Appl. Phys.* **114** 113105.
[8] Som T and Karmakar B 2011 *Solid State Sci.* **13** 887
[9] Som T and Karmakar B 2009 *Nano Res.* **2** 607
[10] Tastumisago M, Minami T, Kowada Y and Adachi H 1994 *Phys. Chem. Glasses* **35** 89
[11] Yousef E L S, Al-Salami A E and Hotzel M 2012 *Bull. Mater. Sci.* **35** 961
[12] Murugan G S and Ohishi Y 2005 *J. Non-Cryst. Solids* **351** 364
[13] Murugan G S and Ohishi Y 2004 *J. Non-Cryst. Solids* **341** 86
[14] Plotnichenko V G, Sokolov V O, Koltashev V V, Dianov E M, Grishin I A and Churbanov M F 2005 *Optics Letters* **30** 1156
[15] Lin H, Wang X, Lin L, Li C, Yang D and Tanabe S 2007 *Journal Phys. D: Appl. Phys* **40** 3567
[16] Bobovich Y O S and Yakhkid A K 1963 *J.Struct. Chem.***4** 851
[17] Berwal N, Dhankhar S, Sharma P, Kundu R S, Punia R, Kishore N 2017 *J. Mol. Struct. Soc.***1127**636
[18] Dousti M R, Sahar M R, Amjad R J, Ghoshal S K and Awang A 2013 *J. Lumin.* **143** 368
[19] Balachandran U and Eror N G 1982 *Journal Mater. Science Letters* **1** 374
[20] Maheshvaran K, Arun Kumar S, Sudarsan V, Natarajan V and Marimuthu K 2013 *J.Alloys Compd.* **561** 142
[21] Nurhafizah H, Rohani M S and Ghoshal S K 2017 *J. Non-Crystalline Solids* **455** 62
[22] Dousti M R, Sahar M R, Ghoshal S K, Amjad Raja J, Arifin R 2013 *J. Molecular Struct.* **1033** 79
[23] Jlassi I, Elhouichet H, Hraiech S and Ferid M 2012 *J. Lumin.* **132** 832
[24] Yeh D C, Sibley W A, Suscavage M and Drexhage M G 1987 *J. Appl. Phys.* **62** 266–275
[25] Liang L, Wu H, Hu H, Wu M and Su Q 2004 *J. Alloys Compd.* **368** 94–100
[26] Xiaoqiang Y, Summers C J and Park W 2012 *J. Appl. Phys.* **111** 3524

Role of polarization electric field on thermal conductivity of GaN/$In_{0.9}Ga_{0.1}N$/GaN superlattices

Subhranshu Sekhar Sahu* and Bijay Kumar Sahoo

Department of Physics, N.I.T, Raipur, Raipur-492010

*Corresponding author's e-mail address: ssahu.phd2017.phy@nitrr.ac.in

Abstract. Enhancement in thermoelectric efficiency of GaN/$In_{0.9}Ga_{0.1}N$/GaN superlattice (SL) needs higher electrical conductivity (σ) and Seebeck coefficient (*S*); but lower thermal conductivity (*k*). Both S and σ are improved due to the presence of polarization electric field (PEF) of this SL. In this work, the role of PEF on *k* of SL has been investigated and found that cross-plane and in-plane *k* are reduced due to PEF upto a certain temperature. Both cross-plane and in-plane thermal conductivities in the presence of PEF show cross over temperature termed as transition temperature(T_c). It is noted the T_c for cross-plane and in-plane thermal conductivity of GaN/$In_{0.9}Ga_{0.1}N$ superlattice (SL) are 510K and 530K, respectively which depends on In contents. Thus, the preferred value of S, σ and *k* of GaN/$In_{0.9}Ga_{0.1}N$ SL can be achieved as per requirement by changing In content; making it suitable for TE module for maximum power production at room temperature and above.

Keywords: Indium gallium nitride; Superlattice; Thermoelectric efficiency; Polarization electric field.

1. Introduction

Currently GaN/$In_xGa_{1-x}N$/GaN superlattice (SL) is an important structure for the manufacture of laser diodes (LD), light-emitting diodes (LED)and thermoelectric (TE) devices [1]. It is reported that the SL may be a good candidate for TE equipment due to high thermal stability and low thermal conductivity. Thermoelectric (TE) efficiency of a device is directly proportional to the coefficient of seebeck effect and electrical conductivity; but inversely proportional to thermal conductivity (k) of the material. A smaller reduction in k will generate significant improvement in TE efficiency. For high temperature TE device GaN/InGaN/GaN SLs are promising structures as it has good thermal stability and low thermal conductivity. Thus, reducing k further, device efficiency can be improved. Chen. Z. et al. reported that k of InGaAs/InGaAsP SL of thicknesses 10, 20, 30 and 40nm were 6.63, 2.09, 4.42 and 5.30 W/m-K at room temperature, respectively [2]. Scott T. Huxtable et al. reported that k of 3500nm thick $Si/Si_{0.9}Ge_{0.1}$ SL was found to be 9.6 W/Km [3]. Saha et al. reported that the thermal conductivity due to cross-plane of the $Ti_{0.7}W_{0.3}N/Al_{0.72}Sc_{0.28}N$ SL decreases from 4.5 to 1.7W/m-K as period thickness of SL decreases from 240nm to 4nm, respectively [4]. GaN/$In_xGa_{1-x}N$/GaN SL has a unique property called as polarization property. This polarization property produces electric field known as built in electric field. A. Sztein, et al. [5] reported that polarization field increases electrical conductivity and SL seebeck coefficient, although their work lacks the study of polarization effects on thermal conductivity. In this work, we have studied role of polarization field on k of this SLs.

2. Theoretical Development

GaN/InGaN/GaN SL has strain which generates piezoelectric polarization(pz). The pz polarization is calculated by $P^{pz} = [e_{31} \in_1 + e_{32} \in_2 + e_{33} \in_3]$ C/m^2. The constituent material has spontaneous (sp) polarization. The sp polarization is computed by $P^{sp} = [-0.042x - 0.034(1-x) + 0.037(1-x)x]$ C/m^2 [9]. Here x is *In* fraction, e_{ij} is piezoelectric coefficient and $\in_i$ is strain. Total polarization is $P = P^{sp} + P^{pz}$. The material parameters of $In_xGa_{1-x}N$ are taken from ref. [2]. In this SL, GaN acts as barrier whereas InGaN acts as well. The electric field due to total polarization in well E_w (or E_b in barrier) semiconductor layers are:

$$E_w = L_b(P_b - P_w)/(\varepsilon_b L_w + \varepsilon_w L_b) \text{ and } E_b = L_w(P_w - P_b)/(\varepsilon_b L_w + \varepsilon_w L_b).$$

Here L_w (L_b), P_w(P_b) and ε_w (ε_b) are respectively thickness of well (barrier) layer, polarization in well (barrier) layer and dielectric constant of well (barrier) material.

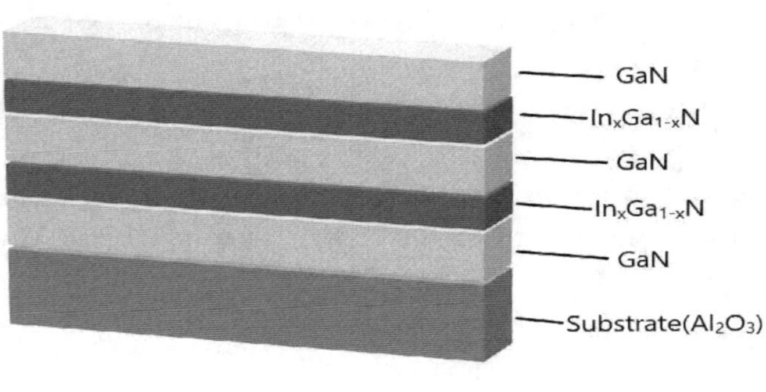

Figure 1. GaN/ $In_xGa_{1-x}N$/GaN SL

The field is of the order of ~ 10^9 V/cm. According to Heckmann principle Coupling of polarization electric (PE) field and strain modifies thermal properties of SL via elastic constant as constituents are piezoelectric materials. The elastic constant of $In_xGa_{1-x}N$ including polarization field has been found as, $C_{44,p}^{InGaN} = C_{44}^{InGaN} + [(e_{15}^2 + e_{31}^2 + e_{33}^2 + p_{sp}^2)/\varepsilon_0 \varepsilon]^{InGaN}$, where C_{44}^{InGaN} is elastic constant of $In_xGa_{1-x}N$ absence of polarization mechanism. Modified elastic constants tailor the phonon velocity, phonon scatterings in the presence of polarization field effect. Phonon velocity has been computed by $v = \sqrt{C_{44,p}^{InGaN}/\rho}$. The Debye frequency depends on phonon group velocity as $\omega_D = v[3N/4\pi V_0]^{1/3}$. N is atoms number in the volume V_0. and Debye temperature depends on Debye frequency as $\theta_D = h\omega_D / k_\beta$, where k_β is Boltzmann's constant and h is Planck's constant. It has been seen that polarization electric field not only enhances thermal parameters of $In_xGa_{1-x}N$ but also contributes their bowing constants.

In semiconductors, the heat energy is transported by acoustic phonons. The thermal conductivity is the sum of both phonon and electron thermal conductivity as $k = k_{ph} + k_e$, where k_{ph} is phonon thermal conductivity and k_e is electron thermal conductivity. Experiments reported that $k_e \cong 10^{-3} k_{ph}$. k_e can therefore be ignored and for a semiconductor $k \cong k_{ph}$. Callaway formula is used to determine k of each layer ($x = \hbar\omega / k_\beta T$)

$$k = \frac{{k_\beta}^4 T^3}{2\pi^2 \hbar^3 v} \int_0^{\theta_D/T} \frac{\tau_c x^4 e^x}{(e^x - 1)^2} dx. \tag{1}$$

Interfaces generate thermal boundary resistance (TBR) due to different materials. TBR can be computed by using the transmission coefficient of the materials expressed as [6]

$$\Gamma_{ij} = p\Gamma_{ij}^s + (1-p)\Gamma_{ij}^d \tag{2}$$

Where p is specularity parameter taking values from 0 to 1; Γ_{ij}^s is specular transmission coefficient and Γ_{ij}^d is diffuse transmission coefficient. Specular transmission coefficient and diffuse transmission coefficient calculated by
$\Gamma^s{}_{ij} = (4\rho_i v_i \rho_j v_j \cos\theta_i \cos\theta_j)(\rho_i v_i \cos\theta_i + \rho_j v_j \cos\theta_j)^{-2}$ and $\Gamma^d{}_{ij} = (C_j v_j)(C_i v_i + C_j v_j)^{-1}$
respectively;

where ρ_i density of the material , v_i phonon group velocity, θ_i angle of incidence, C_i specific heat of the material *i*. By estimating all above terms, it is easy to calculate TBR of the material i to j and it is computed by[7];

$$R_{ij} = [\frac{4\pi^2 \hbar^3 v_i^2}{k_b^4 \Gamma_{ij} T_i^3}][\int_0^x x^4 e^x / (e^x - 1)^2]^{-1} \tag{3}$$

First, every surface thermal conductivity k_{ph} is computed in the same way as the bulk material's thermal conductivity. Then TBR is computed. SL's in-plane and cross-plane thermal conductivity for a period of two layers can be determined by using measured layer thermal conductivity and TBR [8];

$$k_{in-plane} = (L_1 k_1 + L_2 k_2)/(L_1 + L_2)$$ and

$$k_{cross-plane} = (L_1 + L_2)/\{(L_1/k_1) + (L_2/k_2) + (R_1 + R_2)\}$$

Where L_1 is thickness of first material and L_2 is thickness of second material, while k_1 and k_2 are the corresponding surface thermal conductivities. R_1 and R_2 are the TBRs from first layer and second layer, respectively.

3. Results and Discussion

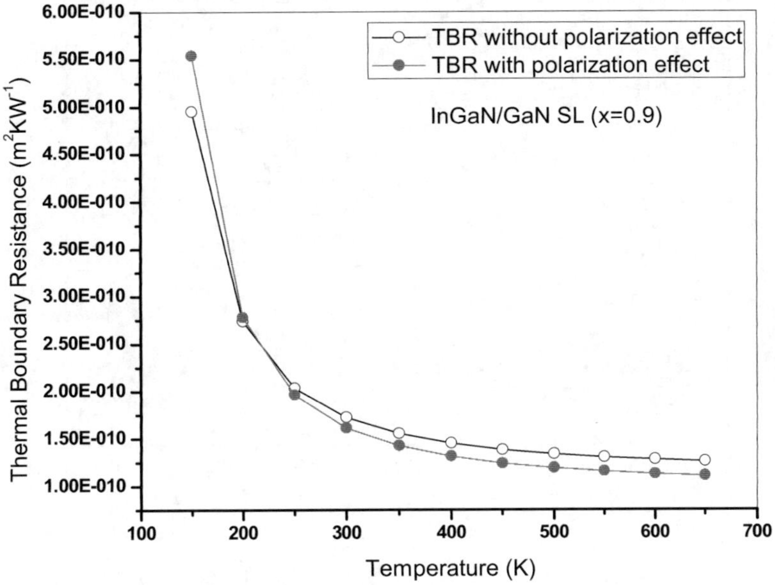

Figure 2. Thermal boundary resistance of GaN/ $In_xGa_{1-x}N$/GaNSL(x= 0.9).

The elastic constants, group velocity due to phonon, Debye temperature of GaN/ $In_xGa_{1-x}N$/GaN SL have been calculated including and excluding the effect of polarization field. It has been investigated that due to this interfacial polarization electric field all above parameters are significantly enhanced. Since all these parameters are used to estimate layer's thermal conductivity given by Callaway formula, thermal conductivity of each layer is modified. TBR in between the interfaces is also calculated through the equation 3. From fig.2 it is found that there is no significant variation between the TBRs induced by PEF and without PEF in low temperature limit. Below 200K temperature TBR without PEF is slightly smaller than TBR with PEF; however, above 200K temperature PEF reduces TBR further. Therefore, it is concluded in the high temperature limit TBR with PEF decreases than TBR without PEF.

IOP Conf. Series: Materials Science and Engineering **798** (2020) 012027 doi:10.1088/1757-899X/798/1/012027

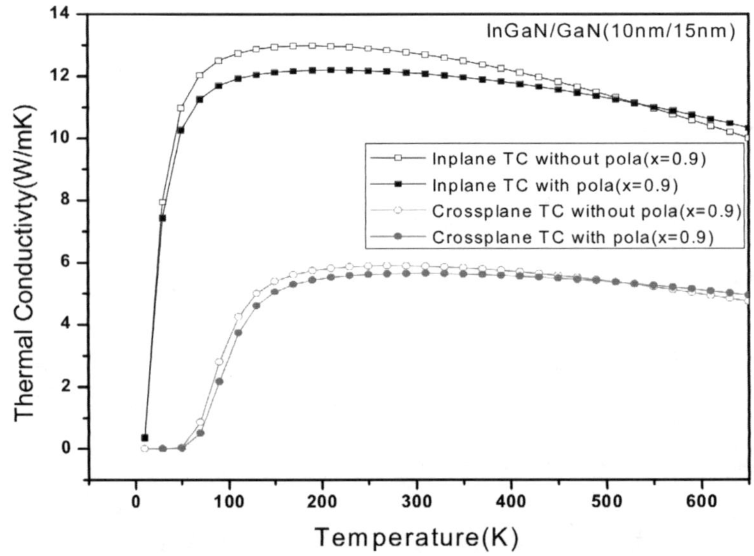

Figure 3. In plane and cross plane thermal conductivity of GaN/InxGa1_{-x}N/GaNSL(x= 0.9).

From figure 3, it can be seen, due to the influence of polarization electric field; in plane thermal conductivity of GaN/ $In_{0.9}Ga_{0.1}N$ SL decreases significantly up to the temperature 530K then above 530K in plane thermal conductivity increases due to PEF, whereas due to both the effect of interfacial polarization electric field and TBR, reduction of cross plane thermal conductivity occurs within the temperature range 60-510 K, when the In content of SL is maintained 0.9 unit and thickness of both well and barrier are maintained 10nm and 15nm respectively. Above 510K cross plane thermal conductivity with PEF exceeds than without PEF. In plane thermal conductivity with and without PEF show cross over temperature at 530K called as transition temperature(T_c) of in plane thermal conductivity, whereas cross plane thermal conductivity with and without PEF show cross over temperature at 510K called as transition temperature of cross plane thermal conductivity of GaN/ $In_{0.9}Ga_{0.1}N$ SL.

4. Conclusion

The interfacial polarization electric field effect on thermal properties of GaN/InGaN/GaN SL has been theoretically explored. The change in thermal conductivity (k) with temperature presence of polarization field and the effect of TBR predicts both in plane as well as cross plane thermal conductivity decrease with in temperature range 200K to 600K. This reduced thermal conductivity of GaN/InGaN/GaN SL significantly enhances the TE efficiency. Therefore, GaN/InGaN/GaN SLs are the best structures for the TE device preparations at room temperature and above.

Acknowledgments
With appreciation the author acknowledges the financial aid received by project No. EMR/2016/00101 9 to DST-SERB Govt. of India.

References

[1] Morkoc H 2013 *Nitride Semiconductor Devices* (New York: Wiley-VCH) p 25
[2] Chen Z, Yang J, and Chen Y 2006 Proceedings of the 1st IEEE International Conference on Nano/Micro Engineered and Molecular Systemsp 283
[3] Scott T, Alexis R, Chang-Lin T and Arun M 2002 *Appl. Phys. Lett.* **80**1737
[4] Saha B, Koh R, Comparan J and Sadasivam S 2016 *Phy. Rev.* B **93**045311
[5] Sztein A, Haberstroh J and Bowers J 2013 *J. Appl. Phys.***113** 183707
[6] Little W 1958*Can. J. Phys.* **37** 334
[7] Alvarez F, Alvarez-Quintana J and Jou D 2010 *J. Appl. Phys.***107** 084303
[8] Mei S and Knezevic I 2015 *J. Appl. Phys.***118** 175101
[9] Sahu S and Sahoo B 2019 *Thin solid films* **684** 59

IOP Conf. Series: Materials Science and Engineering **798** (2020) 012028 doi:10.1088/1757-899X/798/1/012028

Single Photon vs Two Photon Excitation of Silver Nanocluster

J. Tripathy,[1] Y-H Lu[2], Md. S. Islam,[2] R. M. Dickson[2]*

[1] School of Applied Sciences, KIIT Deemed University, Bhubaneswar, Odisha, India – 751013
[2] School of Chemistry & Biochemistry, Georgia Institute of Technology, Atlanta, GA USA – 30318

Email: *jagnya@gmail.com*

Abstract. Silver nanodots, nanoparticles and nanoclusters have been of immense interest to researchers for some time now. A lot has been published in this field from synthesis to optical properties of various types of silver nanoclusters. In this work we explore the impact of polarized excitation on modulatability and fluorescence enhancement of silver (Ag) nanocluster. Ag nanoclusters as previously reported are enhancible and modulatable, however, for pathological application purposes emission in NIR even IR is desirable, so that the fluorescence signal detection over the auto-fluorescence background signal in vivo mode becomes easier. In a preliminary work we have seen a rise in anisotropic values beyond the maximum steady state anisotropy of 0.4 when the sample is excited using two colored laser light. Though there are numerous multi-photon study results available in literature, this is first of its kind where we use two polarized excitations to sequentially probe the excitable dark state of the cluster and the emission is then analyzed to obtain a value of steady state anisotropy along with changes observed in signal enhancement owing to modulation of the secondary excitation source.Ag nanocluster samples encapsulated by 10- or 12-mer ss-DNA capable of producing red to NIR light upon excitation at shorter wavelengths were prepared in-house.

1. Introduction

Beginning towards the end of 20th century there had been a lot of scientific research on metal nano particles and nanoclusters. Among several different kinds of metal nanoclusters studied Silver being cheap is used more widely. However, toxicity of Silver had always been a concern. An alternative approach using nanoclusters encapsulated in DNA have since been used following the pioneering work by Dickson research group at Georgia Tech [1]. Nanoclusters particularly the encapsulated ones have been shownto have wide application owing to their largely reduced toxicity and enhanced photostability. Successive studies have demonstrated that these metal nanoclusters can have wide range of photoemissivity and may exhibit enhanced fluorescence and become modulatable [2,3]. A range of encapsulated silver nano-dots and nanoclusters has been reported with photoemission power ranging from Visible to IR. Reportedly, the stability and fluorescence enhancement achieved from nanoclusters designed to emit in NIR-IR range is weak compared to those emitting in visible. However, these long wavelength emitting nanoclusters are preferred over the other ones emitting in the short wavelength regime for pathological application and biological imaging owing to their non overlapping with the auto fluorescence from biological samples. In this preliminary study, we probe changes in inherent anisotropy property of Ag nanoclusters encapsulated in ss-DNA capable of emitting red light up on excitation with NIR laser. The change in anisotropy of the nanocluster induced via use of two laser excitation may be further exploited for biological imaging purposes. Anisotropy of the sample in this study has been determined using two color excitations, usually a combination of Visible (green) and NIR lasers. Though there have been several reports on two photon anisotropy, however, those studies employ two photons coming from same initial source whereas in our method sequential absorption of two photons from two separate excitation sources are exploited, and the effect of individual excitation polarization on the total anisotropy has been examined.

2. Experimental Method

DNA encapsulated nanosclusters (AgNC) were synthesized using ss-DNA obtained from Integrated DNA Technologies (IDT) following the previously described method by Dickson research group [4,5].

The UV-Visible absorption spectrawere acquired using a Varian Cary and a Shimadzu UV-2401 spectrophotometer. The fluorescence emission spectra were acquired using Jobin Yvon Horiba Fluoromax-3 and Photon Technology International QuantaMasterspectrofluorimeter.The sample concentration was then readjusted to give adequate fluorescence photon counts when excited with a single laser (primary) excitation of 532 nm (PicoQuant diode laser). Few microliters of the sample was placed on a coverslip and the fluorescence emissions due to a single polarized excitation of 532 nm(PicoQuant diode laser operated at 10 kHzrepetition rate) were acquired using anAvalanche Photodiode Detector (APD).

Dual laser excitation spectroscopy was carried out using a 532 nm pulsed laser as the primary excitation and a 803 nm (NIR) continuous wave (CW) laser as the secondary excitation source. 532 nm shining on the sample though a 60X water objective prepares an initially excited state population that rapidly decays to lower energy states. As reported in several papers by Dickson research group, a part of the initial excitedpopulation decays to an intermediate dark state which in itself is not a fluorescing state but a second longer wavelength (secondary) excitation can put the population back in the fluorescing state and thereby enhancing the fluorescence yield from the sample.The CW NIR laser was used to excite the dark state population and the resulting increase in fluorescence was measured using a Perkin-Elmer APD and analyzed using a photon counting module (SPC-630, Becker Hickl and National Instruments board). At the beginning the plane of polarization of the two lasers were made parallel to each other. Co-illumination by the 532 nm and the NIR laser also resulted in an extended fluorescence lifetime estimated in microseconds. The fluorescence anisotropy transients were obtained using the acquired polarized fluorescence emission andthe maximum steady state anisotropy values were determined for primary only (Single photon anisotropy) and dual laser excitation (two photon anisotropy).

3. Results and Discussion

The fluorescence emissionfrom the AgNCwas split into two components using a PolarizingBeam Splitter prior to collection through two identical APDs. Theperpendicular and parallel polarized components were collected simultaneously. The fluorescence photon counts thus recorded were then combined to look at the signal enhancement upon co-illumination with two lasers,and the effect of excitation polarization on signal enhancement was also analyzed. The total fluorescence emission from the AgNC upon additional excitation with a secondary NIR laser was found to be higher than that obtained with a single primary only excitation of 532 nm. Fluorescence signal enhancement was found to be maximum when polarization of the two lasers are identical (parallel to each other). The anisotropy decay profile was obtained under different conditions using the polarized emissions along with the calculated G-factor for the set-up. Figure 1 shows example of anisotropy decays obtained for Ag-DNA cluster in water for one photon and two photon excitations. The maximum steady state anisotropy (r_0) was then obtained by fitting the experimental curve to an exponential decay model using MATLAB and Python programs. The r_0 value thus obtained for dual laser (two photon) excitation was found to be higher thanthat for a single laser (one photon) excitation. Considering the preliminary stage of this study drawing any conclusion on the variation of the r_0 or the correlation timescales will be far fetching. However, the relative polarizations of the two laser sources seemed to affect the resulting anisotropy decay which needs to be explored further. We noted a change in the maximum (r_0) and the decay time upon a change in relative polarization of excitation source. While the maximum expected value of r_0 for single laser excitation was 0.4 for the dual laser excitation the maximum anisotropy $r_0$was estimated to be ~ 0.57 owing to an additional $\cos^2\theta$ dependence for the distribution of fluorophores in the sample. Interestingly, our findings though much in early stages suggest that r_0 could indeed be close to 0.57. We have consistently observed r_0 values > 0.4. Similar results had been reported for two photon anisotropy where the two photons were essentially similar [6,7]. However, in this study two photons of fundamentally different property were used. The preliminary results indicate effect of orthogonality of the excitation polarization on overall anisotropy of the nanoclusters and an in-depth analysis would be carried out to better understand the impact of relative polarization on anisotropy.

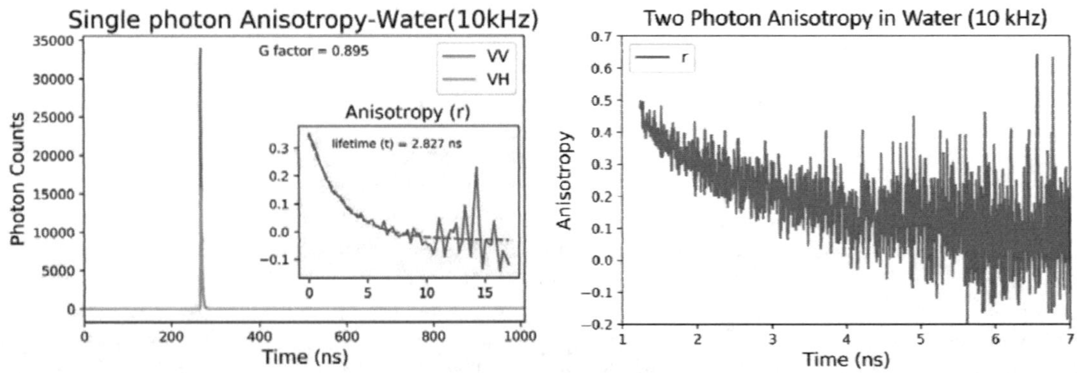

Figure 1 (Left) Polarized emission transients from AgNC encapsulated in DNA; VV representsparallellypolarized primary and secondary excitations, and VH represents primary excitation polarized perpendicular to secondary. The figure in the inset exhibit anisotropy decay as a function of time for primary only excitation of the cluster in water. (Right) Exhibits fluorescence anisotropy decay for the cluster in water upon dual laser excitation.

Though, the work is still much in its infancy we are consistently working at developing the experimental setup and, also streamlining the analysis process. We might be looking at signal deconvolution as a next step in order to subtract any photon contribution from other dark state induced effect following the second laser excitation in orderto better understand the current results.

4.Reference

[1] Jeffrey T. Petty,*,†, Jie Zheng,Nicholas V. Hud, and, and Robert M. Dickson* 2004 *Journal of the American Chemical Society* ***126*** **(16),***5207-5212*

[2]JeffreyT.Petty,*,† Chaoyang Fan,‡ SandraP.Story,† BidishaSengupta,† AshleeSt.JohnIyer,‡ Zachary Prudowsky,† and Robert M. Dickson*,‡ 2010*J. Phys. Chem. Lett.,**1**, 2524–2529*

[3]Vosch,T., Antoku,Y., Hsiang,J.-C., Richards,C.I., Gonzalez,J.I. and Dickson,R.M. 2007 *Proc. Natl. Acad. Sci. U.S.A.,* ***104***, 12616–12621.

[4]Hsiang JC, Fleischer BC, Dickson RM.2016 *J Phys Chem Lett.; 7:*2496–2501

[5]Richards CI, Choi S, Hsiang J-C, Antoku Y, Vosch T, Bongiorno A, Tzeng Y-L, Dickson RM. 2008*J Amer Chem Soc.;* ***130****:*5038–9.

[6] J. Fu et al. 2006 *Chemical Physics* ***321***,257–268

[7] AndreasVolkmer,* Vinod Subramaniam,† David J. S. Birch,* and Thomas M. Jovin† 2000 *Biophysical Journal 78(3)* 1589–1598

A Novel Method for the Removal of Uranium by Using Carboxyl Functionalized Graphene Oxide

Khilawan Patel[1], Rama Devi[1], Pradeep Kumar Dewangan[1], Vinayak[1] Alekha Kumar Sutar[2], Tungabidya Maharana*[1]

[1]Department of Chemistry, National Institute of Technology, Raipur, 492010, India

[2]Ravenshaw University, Cuttack, Odisha, 753003, India

*Corresponding author's e-mail address: khilawanpatel1209@gmail.com, tmaharana.chy@nitrr.ac.in

Abstract. Graphene oxide (GO) and functionalized carboxylic graphene oxide (COOH-GO) were successfully synthesized by modified Hummer's technique. The prepared GO and COOH-GO was characterized successfully by UV-Visible Spectroscopy, Fourier transforms infrared spectroscopy (FTIR), X-ray diffraction, Raman spectroscopy, Scanning electron microscopy (SEM)& Zeta potential. The removal of U(VI) heavy metal comparative study was done by using Graphite, GO & COOH-GO and the removal were confirmed by using LED fluorimeter. The effect of pH of medium, contact time, adsorbent dose, initial concentration of U(VI) were examined for the removal of U(VI). The extent of U(VI) removal has been found to be in the order of COOH-GO >GO> graphite. The U(VI) removal maximum efficiency was observed ~96% observed at pH 4.5. The higher removal efficiency is attributed to the higher negative surface charge of COOH-GO (zeta potential:-39.9 mV) in comparison to zeta potential of GO (-28.9 mV) &graphite (-21.6 mV).

Keywords: Removal of Uranium, Sorption study, Graphene Oxide, LED Fluorimeter.

1. Introduction

There are many different methods to remove heavy toxic metals from polluted water like ion-exchange, precipitation, electroplating, membrane separation, coagulation, disinfection (oxidation/ozonation), biological degradation, electro-kinetics, sorption, solution extraction and adsorption [1-3].Among all these techniques, the most efficient and widely used technique is sorption technique due to its low cost, versatility & simplicity [4-6]. All type of sorbent has been highly used for removal of toxic elements, such as metal oxides, modified silica, metal-organic frameworks, biomass, carbon-based materials, clay minerals and so on [7, 8]. Among these, carbon-based materials are better for the removal of U(VI) heavy metal in pollutant water due to its catenation property and high surface area [9]. Especially, the carbon-containing substances such as GO and its derivates have their excellent physio-chemical properties with high surface area 2630 m^2/g, ultra light weight, high mechanical strength, playing important role to remove heavy metal toxicity in polluted water [10-13].

GO was synthesized from graphite powder which has SP2 containing carbon. GO is one of the promising materials for the cleaning of environmental pollutants like heavy metals (Fe, Co, Ni, Cu, Pb, As, Cd, Hg & U, etc) as well as organic pollutants. It is arranged in a

hexagonal lattice with good electronic properties due to the л and л orbital's. It is the strongest material in the way of mechanical strength [14].
The application of GO and its derivative compounds are highly useful for the removal of various heavy metals. Further, GO make very stable dispersion in water and can be synthesized by wet or dry medium. The dry synthesis method consists of an oxidation reaction of graphene to atomic oxygen in an ultra-high vacuum [15-20]. However, in the wet synthetic method, the main source for the synthesis of GO is graphite due to its high natural abundances and low cost. In the wet synthesis approach, three main routes are possible. In the first route, graphite is converted to GO by mechanical method followed by oxidation while in second route exfoliation of graphite occurs in an aqueous medium followed by ultrasonic treatment [21, 22]. In the last route, both oxidation and exfoliation occur in the presence of highly acidic conditions as found in the Brodie-Staudenmaier-Hummers synthesis method. It can be seen that in all these paths, the structural properties of graphene oxide depends upon the degree of oxidation [23]. During the last century, the three major methods are proposed by Brodie, [24] Staudenmaier [25] and Hummers [26] respectively. In these basic synthetic methods, several modifications have been imposed by several researchers to improve the quality and yield of the product like Tour's group method, [27] free-water oxidation methods, like Sun method [28], Peng method [29], 4-Steps method [30], etc.
However, the furtherance of the removal study by using GO was carried out by using modified graphene oxide such as carboxylic group, alcoholic group, and epoxy modified GO. In these modified GO, different functional groups are highly decorated on the layer of graphene and if the targeted group concentration is lacking then the adsorption of heavy metal is negative. Zijie Li et al. have synthesized GO maximum sorption efficiency for U (VI) was 299 mg/g at pH 4 and Wang et al. have synthesized functionalized GO maximum sorption efficiency 398.4 $mg.g^{-1}$ at pH 6 [31, 32].
Nowadays, water pollution by heavy metals is more problems for living organisms. Thus, the removal of heavy metals in municipal wastewater and also in industrial areas like nuclear power plants, metallurgical industries, oil refinery industries, etc draws attention by several researchers. The heavy metal toxicity issue is truly globalized and has affected many areas. Therefore, in the current context, the approach has been made for the removal of heavy metals from water samples with high efficiency and by using low-cost material like GO. Thus, in the present investigation GO and COOH-GO were synthesized, characterized & used for removal of U(VI) in the water sample.

2. Experimental section

2.1. Reagents and materials used

Graphite fine powder (60 mesh, approx. size 250 μm) and $NaNO_3$ (99%) were purchased from LobaChemiePvt. Ltd. were used as received. $KMnO_4$ (>98.5%), Monochloroacetic acid (>99.0%), NaOH (>97%), H_2O_2 (30%), H_2SO_4 (98%) and HCl (35%) were purchased from Sigma Aldrich (Mumbai, India). Standard stock solutions of uranium 100μg/mL (1 lakh ppb)in 2-5% nitric acid (AccuStandard, USA) and all the chemicals and reagents were prepared with milliporedeionized (DI) water obtained from Merck Millipore water purifier. The pH was measured with a digital pHep® (Hanna instruments, USA). Uranium analysis was carried out by using LF 2a LED fluorimeter (Quantalase Inc. India).

2.2. Synthesis of GO

GO was synthesizedaccording to modified Hummers method by using graphite fine powder [33-35]. Graphite fine powder (4 gm) has been taken in a beaker (1000 ml) and was put on a magnetic stirrer for 1 hour at 300 rpm and then slowly added 480 ml of concentrated sulphuric acid followed by 0.5g of sodium nitrate under stirring. Then the solution mixture was cooled to 20°C by putting it in an ice bath. Further, 24g of $KMnO_4$ was added to the mixture solution under stirring and was stirred overnight at 35°C. Afterwards, 40 ml of H_2O_2 in 500 ml ice was added to the above prepared solution slowly to obtain a bright yellow colour precipitate. The precipitate was washed by using 0.5 M HCl to remove Mn^{2+}&further purified by washing repeatedly with millipore water under centrifugation at 4000 rpm until pH of washings becomes neutral. Further, the obtained black coloured GO was ultrasoniccated for 1 h and was dried overnight at 45°C.

2.3. Synthesis of COOH-GO

COOH-GO was synthesized by reduction of -OH groups of GO to -COOH group [36, 37]. In this method 0.05 g of prepared GO was taken in 50 ml millipore water and sonicated for 50 minutes. Further, 1.2 g of monochloroacetic acid and 1 g of sodium hydroxide was added to the GO solution and then sonicated for 4 hours. Then COOH-GO solution was neutralised with 0.5 M HCl solution and was purified by repeated washing with millipore water under centrifugation at 2000 rpm until the washings becomes neutral [35].Finally, the obtained COOH-GO was dried overnight at 45°C.

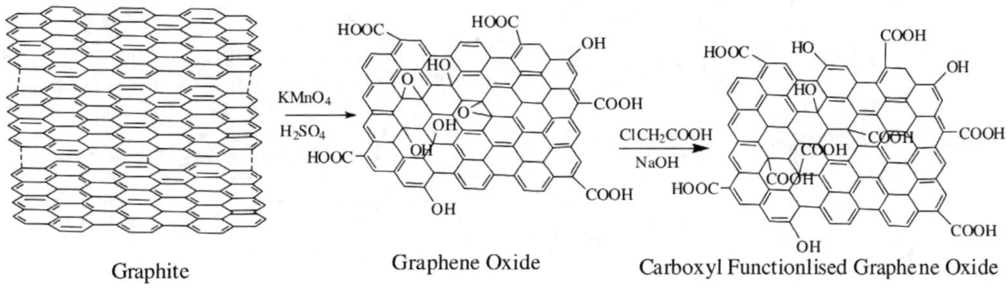

2.4. Characterization of graphite, GO and COOH-GO

Electronic spectra of Graphite, GO andCOOH-GO was recorded on a double beam UV-Visible spectrometer in the range from 200-900 nm (LABTRONICS MODE LT-2900). The different-different functional groups were confirmed by FTIR by using Cary 630 in the range from 400-4000 cm^{-1} (Agilent Technologies, UK). The morphological studies were carried out by SEM (SEM Zeiss EVO 18, USA). The X-ray diffractograms were recorded by PANalytical, XPERT-PRO by using CuKαradiation. Raman Spectra were recorded by Research India (RIAR-0403, Raman Spectrometer RIRAMAN532, India). The zeta potential was measured on a Zetasizer Nano (MALVERN, UK).

2.5. Batch sorption study for removal of U(VI)

Sorption study for U(VI) was studied in the batch experiment using a 500 ppb standard diluted uranium solution at the desired pH. LED fluorimeter was used for measurement of U(VI) in an aqueous medium. The sorption equilibrium capacity Q_e (mg/g) and percentage removal for U(VI) in aqueous medium were calculated by using equations, $Q_e = [(C_0 - C_e)\times V]/m$ & % of removal of U(VI) $= [(C_0 - C_e) \times 100]/C_0$respectively; where, C_e is thefinal concentration after sorption equilibration, $C_0$is the initial concentration of U (VI) ions in solution (μg/l), V is the volume of the uranium solution (ml) and m is amount of adsorbent (mg).

2.6. Selectivity of studies

The U(VI) removal was investigated in the prepared water sample solution by using sorbent materials COOH-GO, GO and graphite. All experiments were observed at room temperature.The U(VI) ion concentration was measured in filtrate of the sample water solution by LED Fluorimeter. The minimum detection limit of LED Fluorimeter is 0.2 ppb in the water sample.

3. Results and discussion

3.1. UV-Visible spectra of GO and COOH-GO

UV-Visible spectra were recorded for GO and COOH-GO (Figure 1) and the λ_{max}were obtained ~230 nm due to π-π* transition of >C=C< bond [36, 37]. One shoulder peak was observed ~290 nm due to the presence of n-π* transition of carbonyl (>C=O) functional group in GO. A slight shift in the absorption spectra of COOH-GO was due to increased oxygen domain carboxylic group concentration. The wavelength region 400 to 900 nm is not affected by the absorptions[38].

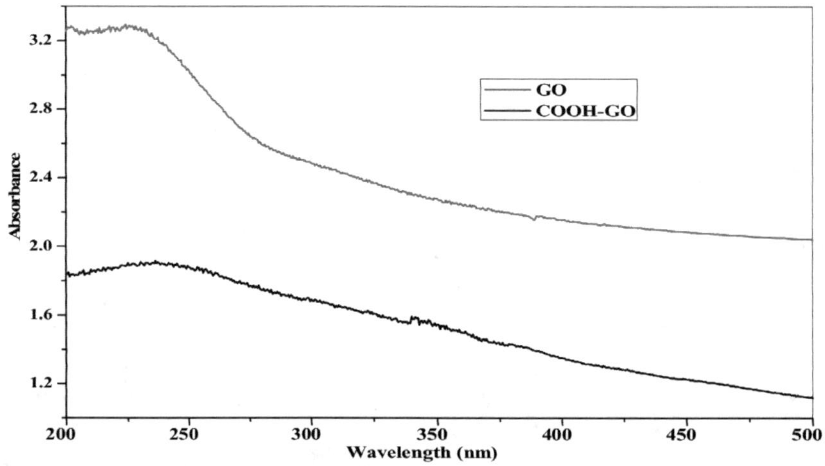

Figure 1. UV-Visible Spectra GO and COOH-GO

3.2. FTIR of GO and COOH-GO

The FT-IR spectra were reported in figure 2 for GO and COOH-GO. The FT-IR spectrum for GO contains peaks for hydroxyl (-OH), epoxide (-C-O-C-) and ketonic (>C=O) functional group corresponding to vibrational stretching frequencies around 3420 cm^{-1}, 1245 cm^{-1} and 1719 cm^{-1}, respectively. The two doublet peak 1719.2cm^{-1} and 1589.5 cm^{-1} observed due to keto-enol tautomerism in carboxyl/carbonyl functional groups, as previously reported by Pendolino et al. and Bagri et al. [39, 40] However, in the FT-IR spectrum for COOH-GO signals arises due to the above mentioned functional groups but with a slight shift as the hydroxyl (-OH) groups reduced to carboxylic carbonyl group increased.

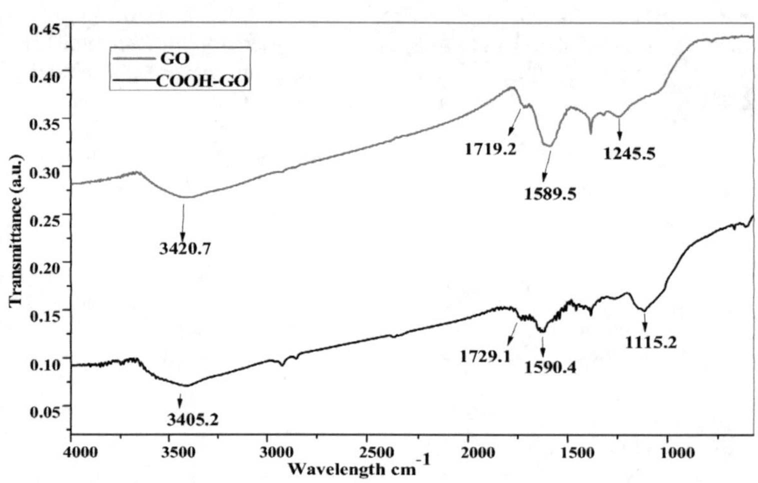

Figure 2. FT-IR Spectra of GO and COOH-GO

3.3. XRD study of GO and COOH-GO

X-ray diffractograms of the GO synthesized above has shown a single broad peak at $2\theta = 9.12°$ (Figure 3(A) and 3 (B)). The inter-planar distance (d) of the synthesized GO is 0.969 nm which is comparable to that reported by Jalili et al. [41]Further, the particle size of GO has been calculated as 7.72 nm by using Scherrer equation (Eq.1). It has been calculated that the synthesized GO contains 8 layers.

$$\mathbf{Dp = (0.94 \times \lambda) / (\beta Cos\theta)} \quad \ldots(1)$$

Where, Dp = Average size of crystallite, λ = X-Ray wavelength, θ = Bragg diffraction angle, β (Full Width Half Maximum) = Line broadening in radians.

The XRD pattern for COOH-functionalized GO has shown no sharp peak and thus it can be concluded that the inclusion of functional groups has introduced disorderness and amorphity leading to a more amorphous FGO.

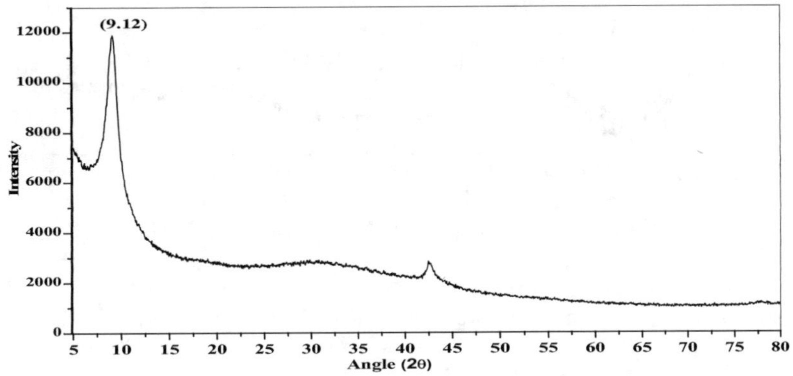

Figure 3. (A) X-Ray diffraction peak of graphene oxide

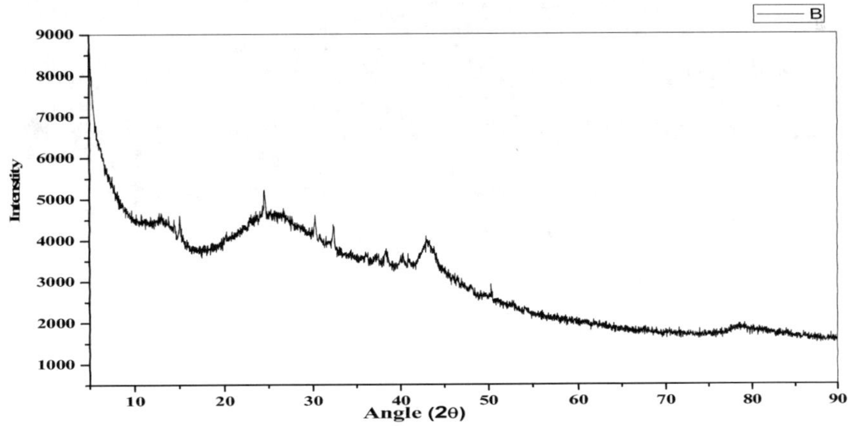

Figure 3.(B) X-Ray diffraction peak of COOH-GO

3.4. Raman spectra for GO and COOH-GO

Raman spectra of GO &COOH-GO havetwo peaksshown corresponding to G& D band. The G band (~1580 cm^{-1})is due to the ordered structureof sp^2 carbon and the D band (~1350 cm^{-1}) is due to the presence of sp^3 carbon atom which has arisen by the introduction of oxygen-containing functional groups(Figure 4). The intensity were observed for GO (I_D/I_G = 1.0232) & COOH-GO (I_D/I_G = 1.1435)[42, 43], respectively.

IOP Conf. Series: Materials Science and Engineering **798** (2020) 012029 doi:10.1088/1757-899X/798/1/012029

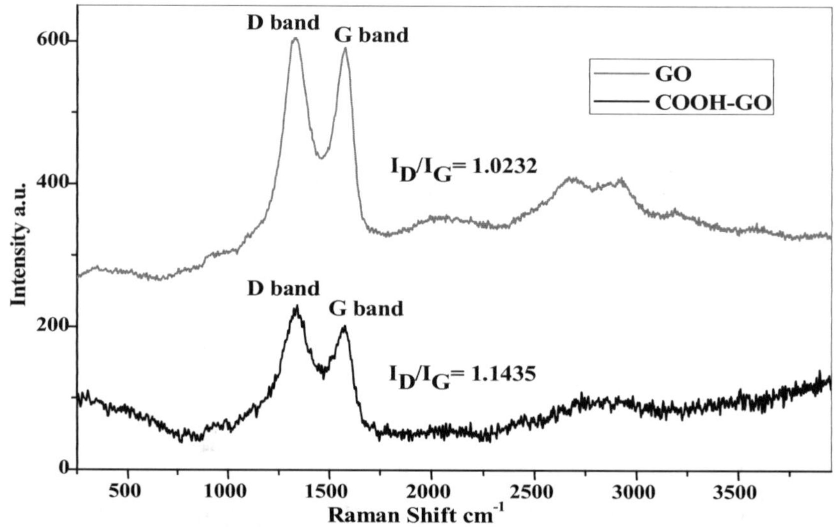

Figure 4. Raman Spectra of GO and COOH-GO

3.5. SEM Study of GO & COOH-GO

The surface morphology of nanomaterials was detectedby SEMfor graphite powder; GO and COOH-GO (Figure 5). In graphite the flake like structure was observed whereas in GO and COOH-GO have exfoliated porous like structures observed [44].

Figure 5. (A) SEM of graphite powder

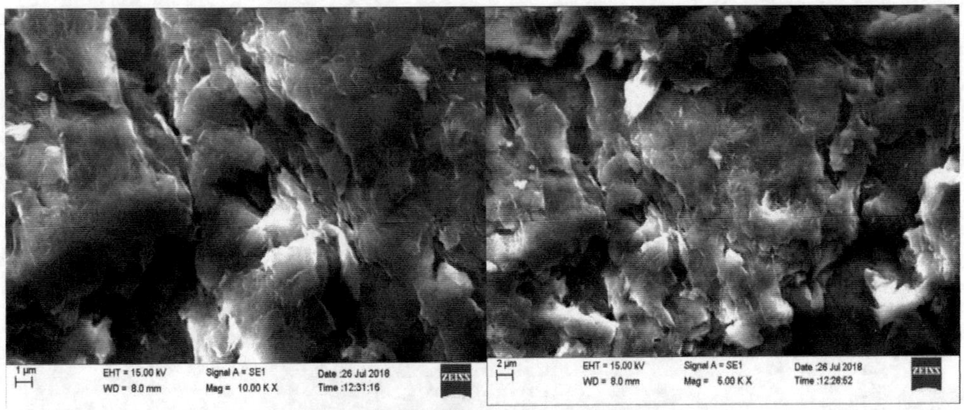

Figure 5.(B) SEM image of GO

Figure 5. (C) SEM image of COOH-GO

3.6. Zeta potential measurement of GO and COOH-GO

The colloidal dispersion of Graphite, GO and COOH-GO in DI water was prepared by 30 minutes sonication. The zeta potential for graphite (-21.6 mV), GO (-28.9 mV), COOH-GO(-39.9 mV), 20 ppb of U(VI) solution (-17.1), 100 ppb of U(VI) solution (+3.61) and 200 ppb of U(VI) solution (-7.54) were observed at pH 4.5. Graphite, GO and COOH-GO have shown more negative values corresponding to their negative surface charge. And thus, they can absorbUranium atoms owing to the fact that the surface charge of COOH-GO is more negative than GO and graphite and thus the removal of U(VI) was investigated practicallyin the order; COOH-GO > GO > Graphite [45].

3.7. Uranium batch sorption experiment

3.7.1. Effect pH for removal of U(VI)

Removal of heavy metal ions are strongly affected by pH of the solution. In current investigation, the effect of pH forU(VI) removal by COOH-GO, GO and graphite is provided in figure 6. The pH ranges have varied from 2 to 12 pH and pH was adjusted by using 0.01M HCl and 0.01M NaOH. The same kind of sorption trends were observed for sorbent substances such as graphite, GO and COOH-GO [42, 43]. Sorption (%) for U(VI) in pH range 2 to 12 was investigated and it is in the order: COOH-GO > GO > graphite. This investigation suggests that the occurrence of -COOH domain increase removal efficiency of heavy metal such as U(VI). The sorbent material developed electrostatic force of attraction on positively charged U(VI) [42].In contrast, at higher pH, the amount of U(VI) removal considerablydecreased for all three sorbents. The hydrolysed solution of Uranium containing

$(UO_2)_3(OH)_5^{3+}$, $UO_2(OH)^+$, $(UO_2)_2(OH)_2^{2+}$ and $(UO_2)_2(OH)_2$species results in sorption of U (VI) and thus its removal efficiency [46].

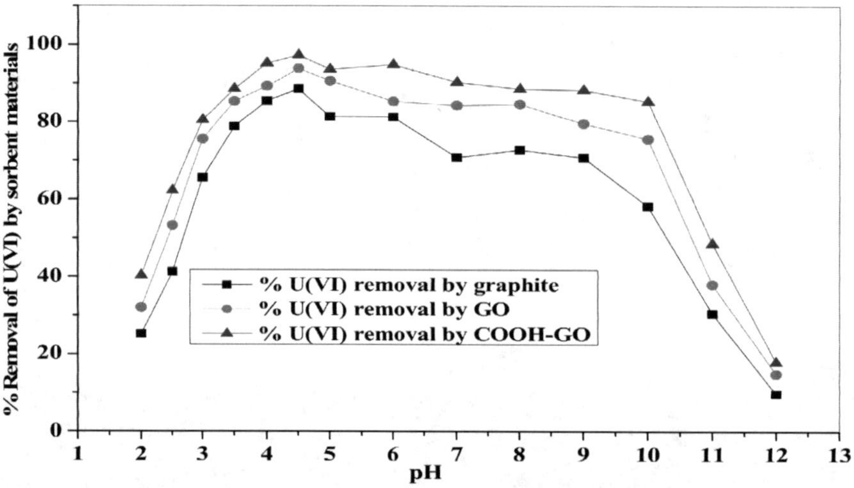

Figure 6. Effect of pH for removal of U(VI) (Experimental conditions U(VI) conc. = 100 ppb, Mass of sorbent = 10 mg, Volume of sample = 50 mL, pH range 2-12)

3.7.2. Effect of contact time for removal of U(VI)

The sorption activity was observed for time range from 10-140 minute (10, 20, 30, 40, 50, 60, 80, 100, 120 and 140 minutes) for 50 ml sample solution(Figure 7). The conc. of U(VI) solution was 100 ppb and adsorbent dose 10 mg at an optimum pH of 4.5. The percentage removal efficiency of U(VI) sorption increased steadily from 10 and 20 minutes and after that almost no change was observed. The highestremoval of U(VI) was found to be 96.2 % for COOH-GO, 93.3 % for GO and 87.6% for graphite powder. The maximum sorption capacity was achieved by COOH-GO as time gradient increases [42].

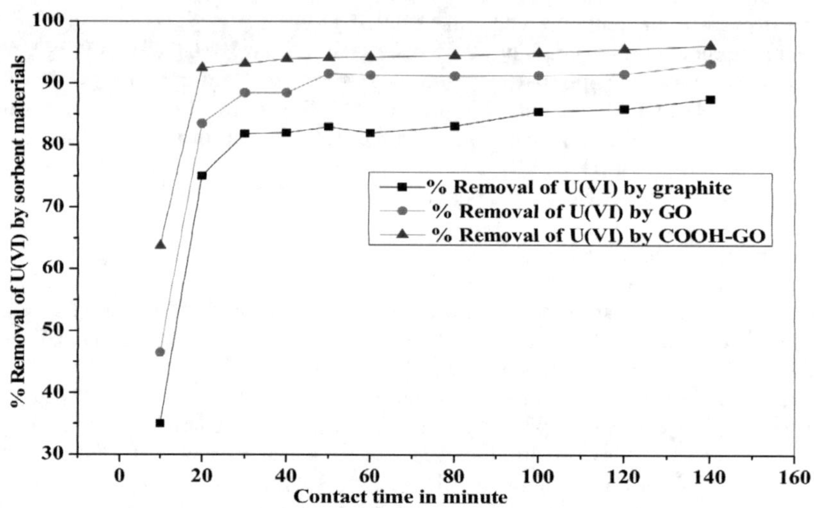

Figure 7. Effect of contact time for removal of U(VI)

(Experimental conditions U(VI) conc. = 100 ppb, Mass of sorbent = 10 mg, Volume of sample = 50 mL & pH = 4.5)

3.7.3. Effect of sorbent dosage for U(VI) removal

The sorption rate of U(VI) increased from 37% to 94 % with an increase in dosage of COOH-GO, GO and graphite from 5 to 100 mg.However, the sorption increases steadily when the dosage was varied from 5 to 10 mg and thereafter removal of Uranium slowly increases [Figure 8]. The maximum sorption of ~96 % was observed when the adsorbent dosage was 100 mg. The U(VI) sorption result indicates COOH-GO is much better than GO and graphite powder [47]Thus, 10 mg of sorbent was chosen for further experimentation. The sorption process was performed under a rotatory shaker without sonication treatment.

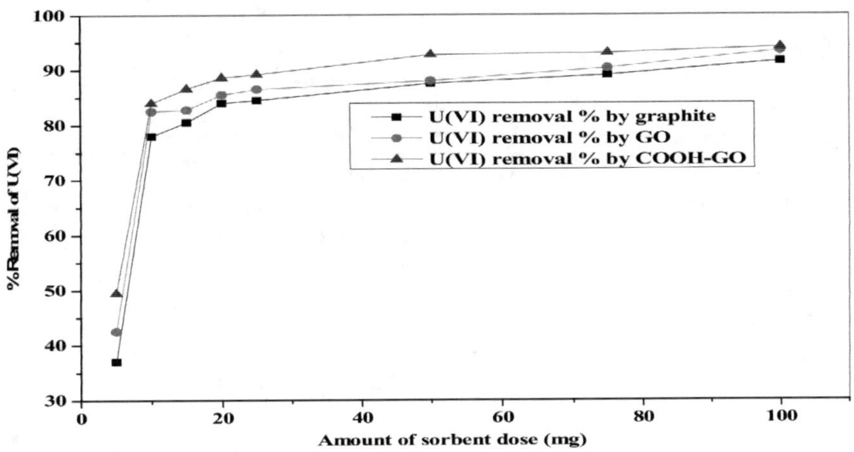

Figure 8. Effect of contact time for removal of U(VI) (Experimental conditions U(VI) Conc. = 100 ppb , Mass of sorbent = 10 mg, Volume of sample = 50 mL, pH 4.5 & Contact time = 20 minute)

3.7.4. Effect of initial U(VI) concentration on uranium sorption

The effect of initial concentration of U(VI) was also investigated by sorbent materials COOH-GO, GO and graphite by using 10-200 ppb of U(VI) solution (Figure 9) at pH = 4.5, adsorbent dosage = 10 mg and at room temperature. The COOH-GO, GO and graphite was added to U(VI) solution kept on rotator shaker and were shaken for 20 minute. The sorption rate of U(VI) increased gradually from 10-100 ppb and then increased slightly with an increase inU(VI) concentration from 100-200 ppb. The maximum sorption of 93.5 % was observed when the concentration of U(VI) was 200 ppb [42].

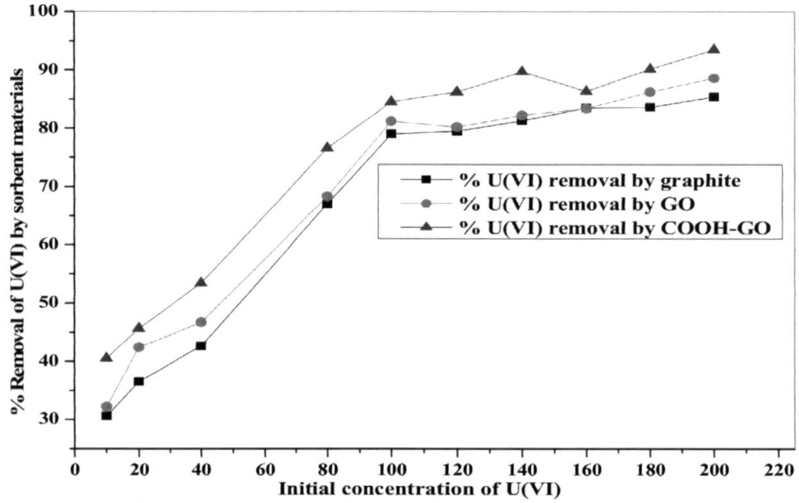

Figure 9. Effect of initial concentration for removal of U(VI) (Experimental conditions U(VI) concentration = 10-200 ppb, Mass of sorbent = 10 mg, Volume of sample = 50 mL, pH 4.5 & Contact time = 20 minute)

4. Conclusion

UV-Visible spectra, FT-IR, XRD, Raman and SEM analyses confirmed the formation of GO and COOH-GO. The removal efficiency of U(VI) is highly dependent on pH and contact time. The removal of uranium was found to be more for functionalised GOOH-GO than GO and graphite due to the presence of more oxygen containing domains in COOH-GO. In this study, COOH-GO have the maximum sorption capacity for U(VI) and was found to be 97.2% at pH 4.5 in comparison to GO and graphite. This fact was also verified from zeta potential study. The zeta potential study suggests the highly negative charge for COOH-GO due to carboxyl group on its surface which are responsible for the removal of U(VI).

Acknowledgements

The first author is thankful to Council of Scientific and Industrial Research (CSIR), File No. 09/1116(0008)/2018-EMR-I, Government of India for financial support in the form of Junior Research Fellowships. The authors are also thankful to National Institute of Technology, Raipur, India for providing research facilities.

References

[1] Lim J Y, Mubarak N M, Abdullah E C, Nizamuddin S, Khalid M and Inamuddin 2018 Recent trends in the synthesis of graphene and graphene oxide based nanomaterials for removal of heavy metals - A review *J. Ind. Eng. Chem.* **66** 29

[2] Nosier S A 2003 Removal of cadmium ions from industrial wastewater by cementation *Chem. Biochem. Eng. Q.* **17** 219

[3] Sani H A, Ahmad M B and Saleh T A 2016 Synthesis of zinc oxide/talc nanocomposite for enhanced lead adsorption from aqueous solutions. *RSC Adv.* **6** 108819

[4] Hosseini E, Zakertabrizi M, Habibnejad K A and Chang Z 2018 Mechanical and electromechanical properties of functionalized hexagonal boron nitride nanosheet: A density functional theory study *J. Chem. Phys.* **149** 114701

[5] Mustaqeem M, Saleh TA, ur Rehman A, Warsi MF, Mehmood A, Sharif A, Akther S 2018 Synthesis of Zn0. 8Co0. 1Ni0. 1Fe2O4 polyvinyl alcohol nanocomposites via ultrasound-assisted emulsion liquid phase. *Arab. J. Chem.*

[6] Dolatyari L, Yaftian MR and Rostamnia S 2016 Removal of uranium (VI) ions from aqueous solutions using Schiff base functionalized SBA-15 mesoporous silica materials *J. Environ. Manage.***169** 8

[7] Yu R, Shi Y, Yang D, Liu Y, Qu J and Yu ZZ 2017 Graphene oxide/chitosan aerogel microspheres with honeycomb-cobweb and radially oriented microchannel structures for broad-spectrum and rapid adsorption of water contaminants *ACS Appl. Mater. Inter.***9** 21809

[8] Zhang Z, Dong Z, Dai Y, Xiao S, Cao X, Liu Y, Guo W, Luo M and Le Z 2016 Amidoxime-functionalized hydrothermal carbon materials for uranium removal from aqueous solution. RSC Adv. **6** 102462

[9] Wang F, Li H, Liu Q, Li Z, Li R, Zhang H, Liu L, Emelchenko GA and Wang J 2016 A graphene oxide/amidoxime hydrogel for enhanced uranium capture *Sci. Rep.* ***6***19367

[10] Yu R, Shi Y, Yang D, Liu Y, Qu J and Yu ZZ 2017 Graphene oxide/chitosan aerogel microspheres with honeycomb-cobweb and radially oriented microchannel structures for broad-spectrum and rapid adsorption of water contaminants*ACS Appl. Mater. Inter.***9** 21809

[11] Zhao G, Wen T, Yang X, Yang S, Liao J, Hu J, Shao D and Wang X 2012 Preconcentration of U (VI) ions on few-layered graphene oxide nanosheets from aqueous solutions. *Dalton Trans.***41** 6182

[12] Zhao X, Zhang Q, Chen D and Lu P 2010 Enhanced mechanical properties of graphene-based poly (vinyl alcohol) composites. *Macromolecules***43** 2357

[13] Yang A, Zhu Y and Huang CP 2018 Facile preparation and adsorption performance of graphene oxide-manganese oxide composite for uranium *Sci. Rep.***13** 9058

[14] Cai Y, Wu C, Liu Z, Zhang L, Chen L, Wang J, Wang X, Yang S and Wang S 2017 Fabrication of a phosphorylated graphene oxide-chitosan composite for highly effective and selective capture of U (VI). *Environ. Sci. Nano.***4** 1876

[15] Sun Y, Yang S, Chen Y, Ding C, Cheng W and Wang X 2015 Adsorption and desorption of U (VI) on functionalized graphene oxides: a combined experimental and theoretical study. *Environ. Sci. Technol.***49** 4255

[16] Smith S C and Rodrigues D F 2015 Carbon-based nanomaterials for removal of chemical and biological contaminants from water: a review of mechanisms and applications *Carbon***91** 122

[17] Vinogradov N A, Schulte K, Ng M L, Mikkelsen A, Lundgren E, Martensson N and Preobrajenski A B 2011 Impact of atomic oxygen on the structure of graphene formed on Ir (111) and Pt (111) *J. Phy.Chem. C.* **115** 9568

[18] Yamamoto M, Einstein T L, Fuhrer M S and Cullen W G 2012 Charge inhomogeneity determines oxidative reactivity of graphene on substrates. *ACS Nano.***6** 8335

[19] Chung M G, Kim D H, Lee H M, Kim T, Choi J H, kyunSeo D, Yoo J B, Hong S H, Kang T J and Kim Y H 2012 Highly sensitive NO_2 gas sensor based on ozone treated graphene. *Sensors Actuat. B-Chem.***166** 172

[20] Huh S, Park J, Kim Y S, Kim K S, Hong B H and Nam J M 2011 UV/ozone-oxidized large-scale graphene platform with large chemical enhancement in surface-enhanced Raman scattering. *ACS Nano.* **5** 9799

[21] Stankovich S, Dikin D A, Piner R D, Kohlhaas K A, Kleinhammes A, Jia Y, Wu Y, Nguyen ST and Ruoff R S 2007 Synthesis of graphene-based nanosheets via chemical reduction of exfoliated graphite oxide *Carbon***45** 1558

[22] Bosch-Navarro C, Busolo F, Coronado E, Duan Y, Martí-Gastaldo C and Prima-Garcia H 2013 Influence of the covalent grafting of organic radicals to graphene on its magnetoresistance. *J. Mater. Chem. C.* **1** 4590

[23] Pendolino F, Armata N, Masullo T and Cuttitta A 2015 Temperature influence on the synthesis of pristine graphene oxide and graphite oxide. *Mater. Chem. Phys.* **164** 71

[24] Brodie B C 1859 On the atomic weight of graphite.*Philos. Trans. Royal Soc.***149** 249

[25] Staudenmaier L 1898 Verfahrenzurdarstellung der graphitsäure*Ber. Dtsch. Chem. Ges.* **31** 1481

[26] Hummers Jr W S and Offeman R E 1958 Preparation of graphitic oxide *J. Am. Chem. Soc.***80**

1339

[27] Marcano D C, Kosynkin D V, Berlin J M, Sinitskii A, Sun Z, Slesarev A, Alemany L B, Lu W and Tour J M 2010 Improved synthesis of graphene oxide. *ACS Nano.* **4** 4806

[28] Sun L and Fugetsu B 2013 Mass production of graphene oxide from expanded graphite *Mater. Lett.* **109** 207

[29] Peng L, Xu Z, Liu Z, Wei Y, Sun H, Li Z, Zhao X and Gao C 2015 An iron-based green approach to 1-h production of single-layer graphene oxide. Nature communications. **6** 5716

[30] Pendolino F, Armata N, Masullo T and Cuttitta A 2015 Temperature influence on the synthesis of pristine graphene oxide and graphite oxide *Mater. Chem. Phys.* **164** 71

[31] Li D, Müller M B, Gilje S, Kaner R B and Wallace G G 2008 Processable aqueous dispersions of graphene nanosheets. *Nat. Nanotechnol.* **3** 101

[32] Jasim D A, Lozano N and Kostarelos K 2016 Synthesis of few-layered, high-purity graphene oxide sheets from different graphite sources for biology *2 d Mater.* **3** 014006

[33] Yu S, Liu J, Zhu W, Hu Z T, Lim T T and Yan X 2015 Facile room-temperature synthesis of carboxylated graphene oxide-copper sulfide nanocomposite with high photodegradation and disinfection activities under solar light irradiation. *Sci. Rep.* **5** 16369

[34] Sun X, Liu Z, Welsher K, Robinson J T, Goodwin A, Zaric S and Dai H 2008 Nano-graphene oxide for cellular imaging and drug delivery *Nano Res.* **1** 203

[35] Mohamud H, Ivanov P, Russell BC, Regan PH and Ward N I 2018 Selective sorption of uranium from aqueous solution by graphene oxide-modified materials. *J. Radioanal. Nucl. Chem.* **316** 839

[36] Paredes J I, Villar-Rodil S, Martínez-Alonso A and Tascon J M 2008 Graphene oxide dispersions in organic solvents *Langmuir* **24** 10560

[37] Cao H, Wu X, Yin G and Warner J H 2012 Synthesis of adenine-modified reduced graphene oxide nanosheets. *Inorg. Chem.* **51** 2954

[38] Lai Q, Zhu S, Luo X, Zou M and Huang S 2012 Ultraviolet-visible spectroscopy of graphene oxides. *Aip Adv.* **2** 032146

[39] Pendolino F, Parisini E and Lo Russo S 2014 Time-dependent structure and solubilization kinetics of graphene oxide in methanol and water dispersions *J. Phys. Chem. C* **118** 28162

[40] Bagri A, Mattevi C, Acik M, Chabal Y J, Chhowalla M and Shenoy V B 2010 Structural evolution during the reduction of chemically derived graphene oxide *Nat. Chem.* **2** 581

[41] Jalili R, Aboutalebi S H, Esrafilzadeh D, Shepherd R L, Chen J, AminorroayaYamini S, Konstantinov K, Minett A I, Razal J M and Wallace G G 2013 Scalable one step wet spinning of graphene fibers and yarns from liquid crystalline dispersions of graphene oxide: towards multifunctional textiles. *Adv. Funct. Mater.* **23** 5345

[42] Sun Y, Yang S, Chen Y, Ding C, Cheng W and Wang X 2015 Adsorption and desorption of U (VI) on functionalized graphene oxides: a combined experimental and theoretical study *Environ. Sci. Technol.* **49** 4255

[43] Ding Z, Ma H and Chen Y 2014 Interaction of graphene oxide with human serum albumin and its mechanism. *RSC Adv.* **4** 55290

[44] Xie Y, Helvenston E M, Shuller-Nickles L C and Powell B A 2016 Surface complexation modeling of Eu (III) and U (VI) interactions with graphene oxide. *Environ. Sci. Technol.* **50** 1821

[45] Kudin K N, Ozbas B, Schniepp H C, Prud'Homme R K, Aksay I A and Car R 2008 Raman spectra of graphite oxide and functionalized graphene sheets. *Nano lett.* **8** 36

[46] Zhang Y, Ma H L, Zhang Q, Peng J, Li J, Zhai M and Yu Z Z. 2012 Facile synthesis of well-dispersed graphene by γ-ray induced reduction of graphene oxide. *J. Mat.Chem.* **22** 13064

[47] Yao W, Wang J, Wang P, Wang X, Yu S, Zou Y, Hou J, Hayat T, Alsaedi A and Wang X 2017 Synergistic coagulation of GO and secondary adsorption of heavy metal ions on Ca/Al layered double hydroxides. *Environ. Pollut.* **229** 827

Fabrication of L-Cysteine functionalized Graphene quantum dots as a fluorescent nanosensing probe for determination of mercury

Pradeep Kumar Dewangan[1], Fahmida Khan[*1], Vinayak Sahu[1], Komal kashyap[1], Ch Chandra[1], Khilawan Patel[1]

[1]Department of Chemistry, National Institute of Technology, Raipur, CG, 492010

[*]Corresponding author Email: fkhan.chy@nitrr.ac.in, pradeepdew2013@gmail.com

Abstract. A highly sensitive and selective nanosensing probe for mercury ion is reported. Herein highly luminescent cysteine(cys) functionalized Graphene quantum dots(GQDs) were fabricated through facile one-step pyrolysis method by using ethylene glycol and cysteine. The coating of cysteine not only increases quantum yield to 41% but also enhances selective detection of ppb level of mercury ion in the waste water. The fluorescence intensity of Cys-GQDs was sequentially quenched by different concentrations of mercury ion via forming non-luminescent complexes of Cys-GQDs-Hg(II). HR-TEM study clearly shows a monodisperse Cys-GQDs of sizes 2-5 nm with a spherical shape. Raman spectra of Cys-GQDs display the D and G bands at 1350 cm^{-1} and 1580 cm^{-1} respectively. Optimum pH is 8 for mercury detection. The minimum limit for accurate detection of mercury is 0.64 ppb

Keyword: Fluorescence probe, Cysteine, Mercury ion, Graphene quantum dots, pyrolysis

1. Introduction

Mercury (Hg) is a highly poisonous, nonradioactive transition heavy metal that is discharged into the environment from natural and anthropogenic sources [1-2]. It is indestructible and exists in inorganic, metallic and organic forms in nature [3]. It is discharged to the atmosphere from volcanic and geothermal activities, mining coal,oil, fossil fuel extractionand gas, mercury-dependent artisanal and dental clinics, small scale gold mining sector, thermometer factories[4-5]. The permissible detection of limit in potable water is 1ppb according to the world health organization [6]. Mercury has a high tendency to accumulate in the human body by ingestion of contaminated water and contaminated seafood. The mercury ion may cause serious problems in many human organs even at low concentration e.g. kidney brain, nervous system, endocrine system, and other organs [4-6]. The well-known case of the Minamata disease was due to the accumulation of the most common form of methylmercury, in Japan and Iraq[7]. Therefore, the discovery of highly sensitive and selective methods of detection of Hg(II) has challenges in recent years.

Up to now, for the detection of mercuryion many traditional methods have been reported. The commonly used analytical methods for detection of Hg(II) is inductively coupled plasma mass spectroscopy[8], chromatography-mass spectroscopy[9-10], graphite furnace atomic absorption spectrometry[11], surface-enhanced Raman scattering[12], atomic fluorescence spectroscopy[13-14], localized surface Plasmon resonance spectroscopy[15], resonance light scattering spectrometry[16], and electrochemical methods[17]. However, in general, these methods have some limitations, e.g. most of them required expensive instrumentation or complicated sample preparation, the need time-consuming pretreatment, complicated detection processes sophisticated, shortcomings dramatically restricted their application in daily monitoring for their handling expert is required. Therefore it is still of great challenge to develop a high sensitivity, low cost, short analysis time, portable and simple sensing platform for rapid on-site Hg(II) analysis. Therefore tunable fluorescent, low bio-toxic GQDs can be required alternatives.

Graphene quantum dots(GQDs) are the recent most attracting small fluorescent nanoparticles of the carbon family. It shows unique optical magnetic and exciting electrical properties governed by the quantum confinement and edge effects[18-21]. They have received tremendous attention among researchers of various fields due to its various applications in optoelectronic, FET, Bioimaging devices and biolabeling. GQDs shows excellent characteristic including excellent luminescent properties, low toxicity, photostability against photobleaching and blinking, cheap in cost, high biocompatibility, chemical inertness, eco-friendly, and high solubility in the various solvent[22-25]. Therefore GQDs as a new style carbon-based nano-material extend the strong potential for the promising candidate to substitute traditional semiconductor quantum dots. Although GQDs have recently fired up the enhancing concern in the sensing of the metal ion. The poor quantum yield and nonspecificity of original GQDs have limited the detection sensitivity. Various heteroatom doped GQDs e.g. N, S, B, N-P, N-S [26-30]doped GQDs have been prepared and applied in the detection ofheavy metal. However, these sensing probes have not good LOD and low quantum yield. The methods are a needs further treatment for better result. For this purpose cysteine is the better amino acid contain -COOH, -NH_2, and –SH group. Such a functional group makes a better sensing probe for selective detection of Hg(II). This advised great potential of cysteine as a good functional groupsto enrich the analytical carried out of GQDs because extra S-group compared to another amino acid.

We have synthesized the cysteine (Cys) capped GQDs. A green and simple method by pyrolysis method of cysteine and glycol was developed to synthesize Cys-GQDs with high quantum yield by pyrolysis method. Synthesized GQDs has been successfully applied as a highly selective and sensitivesensing probe for Hg(II) in the environmental sample. The -NH_2, -COOH, -SH group improves the fluorescence quantum yield. While S-atom enhances the selectivity of Hg(II) because S based ligand strongly binds with Hg(II) according to the hard-soft acid-base concept. So S- atom form a stable complex with Hg(II). Obtained GQDs selectively applied for detection of Hg(II) in the environmental sample with detection limit 0.64 ppb.

2. Experimental Sections

2.1. Materials and reagents

Cysteine, Histidine, glycine, Histidine, arginine, Lysine, L-tryptophan was purchased from HiMidia Laboratories Pvt. Ltd. Glutamate, L-glutathione, methionine were procured from Alfa Aesar. Ethylene glycol, acetic acid, acetone, disodium hydrogen phosphate, Hydrochloric acid, Sodium hydroxide, standard solution of metal ions (As(III), As(V), Mg(II), K(I), Ca(II), Na(I), Pb(II), Co(III), Zn(II), Cd(II), Hg(II), Ni(II), Cu(II), Fe(III)) were purchased from Merck limited Mumbai. By dilution of standard stock solutions of Hg(II), different concentrations were prepared. All aqueous solution was obtainedwith Millipore water.

2.2. Instrumentation

UV-Visible spectra were measured by carrying Eclipse spectrophotometer and the spectra were recorded from 200 nm to 800 nm. The optical property including fluorescence spectra measurement wasrecorded on an F97 fluorescence spectrophotometer. The fluorescence quantum yield and time-resolved fluorescence decay were calculated on an FL980 spectrofluorometer (Edinburgh, UK) with a 370 nm LED as the excitation source. Fourier transformed infrared spectroscopy (FTIR) study was measured with the Thermofisher DRS-FTIR instrument (10 mg catalyst, KBr wafer technique). The morphology and size of the Cysteine capped GQDs were recorded by using JEOL 2012 high-resolution transmission electron microscopy (HR-TEM) at a stimulating voltage 200 kV. Raman spectra recorded by Research India RIAR-0403 Spectrometer RIRAMAN532.

2.3. Synthesis of L-cysteine capped GQDs:

L-Cysteine capped graphene quantum dots (cys-GQDs) were prepared using ethylene glycol as a precursor in one stepby a pyrolysis method. In a 20ml round-bottom flask 0.3g L-Cysteine was taken with 10ml ethylene glycol up to 170°C on a hot plate with stirring. The color of the liquid changed from colorless to orange and finally reddish-brown. Then was cooled to room temperature and centrifuged by a 10000 rpm/minute centrifuge to remove insoluble matter. The supernatant was diluted with the same volume of Millipore water

.

3. Result and discussion

3.1. UV-visible study of cys-GQDs

The excitation properties of cys-GQDs were measured by UV-visible spectroscopy. UV-visible absorption spectra of cys-GQDs show a sharp peak at 370nm (Figure1). The sharp UV-vis absorption peak at 370 nm is indicated n-π^* type transitionof C=O, C=N [31]. In UV-Chamber aqueous solution cys-GQDs irradiated at 365 nm exhibit bright blue fluorescence.

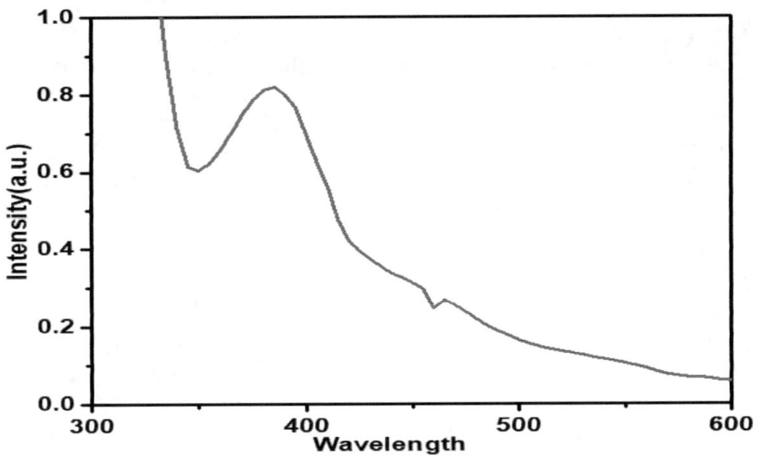

Figure 1. UV-visible spectra of cys-GQDs

3.2. HR-TEM Studyofcys-GQDs

To study of surface morphology and size GQDs HR-TEM is one of the most important techniques. HR-TEM image of cys-GQDs indicates that the size of most of the particle of quantum dots is in range 3 to 6 nm. Figure 2 (a) exhibit particles of cys-GQDs are monodisperse and spherical. Figure 2 (b) indicates that particles of cys-GQDs aggregated after the addition of Hg(II) ion. Aggregation of particles proves that interaction between Hg(II) ion and cys-GQDs occur.

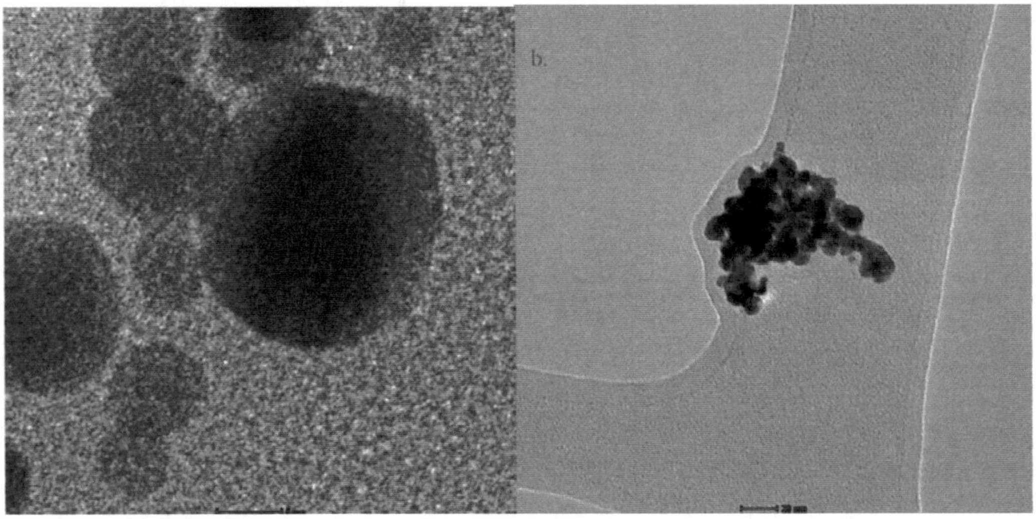

Figure 2. HR-TEM images of (a) size distribution of particle of cys-GQDs (b) cys-GQDs with the addition of Hg(II) ion.

3.3. Raman study of cys-GQDs

D-band and G-band are characteristics property of the graphitic structure of GQDs. Raman spectra cys-GQDs show two peaks at 1350 cm^{-1} and 1580 cm^{-1} which represent D and G band respectively (Figure 3 (a)) [32]. The disordered (d) band at 1350 occurs due to sp^3 hybridized carbon atom defect, while the crystalline(G) band at 1580 represents the in-plane vibration of sp^2 hybridized carbon atom. G- band also represent the aromatic character of cys-GQDs. The intensities ratio of D-band and G-band (I_D/I_G) is found to be 0.85 for cys-GQDs, which means that cys-GQDs is highly graphitic material. TEM image and Raman spectra confirm that GQDs is successfully prepared by this method.

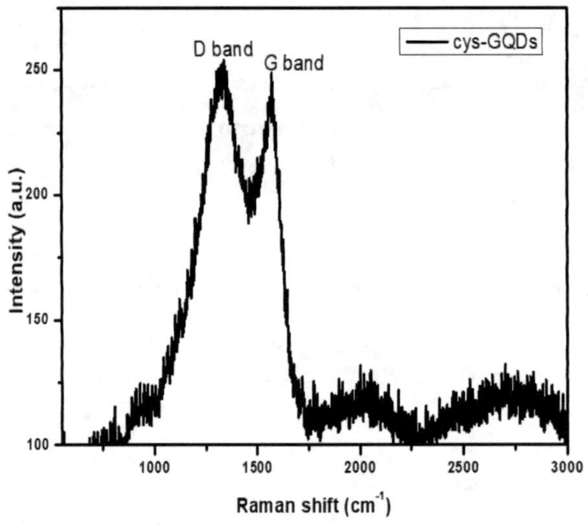

Figure 3. (a) Raman spectrum of csy-GQDs

Excitation spectra of GQDs

The emission fluorescence spectra were recorded to csy-GQDs by excitation range 310nm to 400 nm (Figure 3 (b)). 310 nm to 370 emission intensity of GQDs increases and wavelength shows a red shift after 370 nm fluorescence intensity decreases. Figure 3 (b) clearly indicated that maximum emission intensity occur when cys-GQDs was excited at 370 nm. Therefore excitation and an emission wavelength of cys-GQDs is 370 nm and 443 nm respectively.

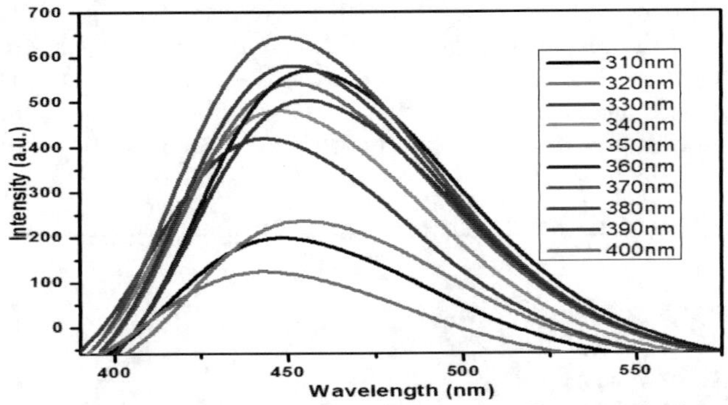

Figure 3. (b) Fluorescence emission spectra of cys-GQDs at different excitation wavelength ranging between 310 nm to 400 nm.

3.4. The mechanism of binding of Hg(II) ion withcys-GQDs:

After the characterizations, the sensitivity of cys-GQDs towards the detection of Hg(II) ion was examined by the change in photoluminescence intensitypresence and absence of Hg(II) ion. Fluorescence intensity gradually decreases by addition to various concentrations of Hg(II) ion in the aqueous solution of cys-GQDs. Change in fluorescence intensity after adding Hg(II) ion is occurs due to both dynamic and static quenching[33]. After the Collision between Hg(II) ion and cys-GQDs results in energy transfer as well as quenching, this process is known as dynamic quenching. Collision force to excited fluorescence material back to ground state. While static quenching Hg(II) ion interact with GQDs and form non-radiative co-ordinate complex(Hg-cysGQDs). GQDs transfer electron density on to Hg(II) ion results from fluorescent material to non-fluorescent co-ordinate complex.The fictionalization of GQDs by cysteine increases the chance of complex formation due to the addition of –COOH, -NH_2, -OH and –SH functional group. Sulpher based ligand is highly specific for bond formation with Hg(II) ion due to the well known soft base-soft acid interaction.

The quenching of fluorescence intensity of cys-GQDs after the addition of different concentrations Hg(II) ion ranging from 5 ppb to 60 ppb. The quenching of fluorescence intensity of cys-GQDs increases upon increasing Hg(II) ion concentration and 80% of the pure GQDs fluorescence intensity quenched at 60 ppb. A linear relationship between the intensity of cys-GQDs and Hg(II) ion well fitted in the stern-Volmer equation. The calibration curve in ranging 5ppb - 60ppb has an excellent linear relationship with a correlation co-efficient (R^2) of 0.995. quenching in fluorescence intensity calculated by using a stern-Volmer equation. The stern-volmer equation is followed as-

$$Io/I = 1 + k_{sv}[Q] = Kq\tau_0$$

Here, steady-state photoluminescence intensities are Iand I_o in the presence and absence of quencher, respectively. Quencher concentration is [Q] i.e. in this study mercury ion, Ksv represents quencher constant, fluorescence lifetime is τ_0; the quenching rate constant is K_q.K_{sv} value was calculated to be 6.50 and binding constant value K_b = 9.50 × 108 also be calculated by plotting log I_o–I/I and log[Q] the value.

IOP Conf. Series: Materials Science and Engineering **798** (2020) 012030 doi:10.1088/1757-899X/798/1/012030

3.5. The selectivity of the cys-GQDs for Hg(II) ion

The selectivity and specificity is the most important factor for fluorescence sensing purpose. In this study, different interferingmetal ions added to cys-GQDs for selectivity evaluation. The fluorescence intensity of cys-GQDs examined in the presence of As(III), As(V), Mg(II),K(I), Ca(II), Na(I),Pb(II), Co(III), Zn(II), Cd(II), Hg(II), Ni(II), Cu(II), Fe(III) ions (Figure 4). The concentration of each metal ions is 60 ppb. The interfering graph indicates that the fluorescence intensity of cys-GQDs is significantly quenched by only in case of presence Hg(II) ion, other metal ions have slight quenching. Here no metal ion interferes with the Hg(II) ion. Therefore cys-GQDs are successfully applied for sensing of Hg(II) ion. In the case of Hg(II) ion effectively quenching occur because of the special chemistry of cys-GQDs. Cys-GQDs have S-atom, which have a strong affinity to binding with Hg(II) ion. Hg(II) ion is soft acids and s- atom is soft base, so according toPearson's theory hard-soft acid-base S atom form co-ordination complex with Hg(II) ion is very rapidly. Therefore in the case of Hg(II) ion almost 80%quenching occurs.

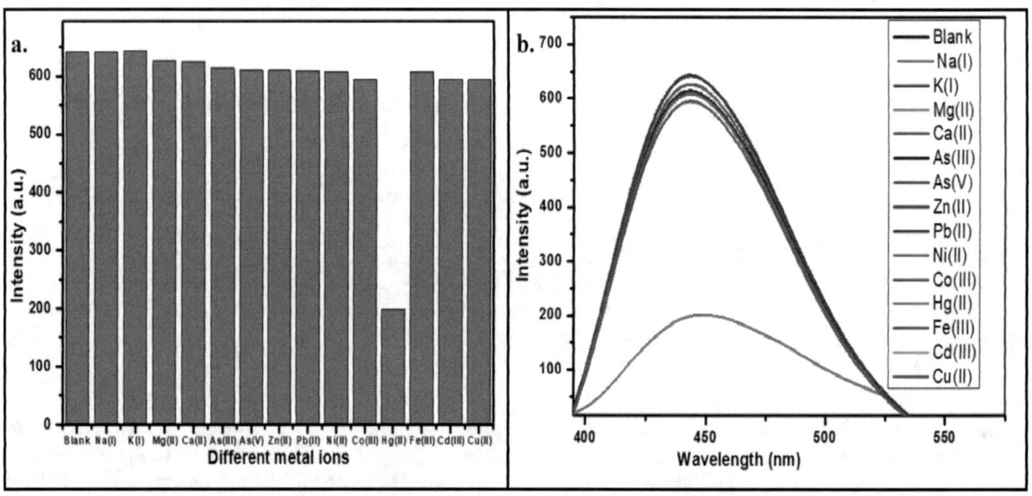

Figure 4. The selectivity sensing response of cys-GQDs in the absence and presence of different interfering metal ions.

3.6. Optimization of cys-GQDs as a fluorescence sensor for detection of zinc ion:

For the detection of analyte suitable pH and reaction time is an important factor.pH and reaction time influence the interaction between Hg(II) ion and cys-GQDs, therefore an optimum pH and time are required. For this purpose, optimization is important for better results.

Reaction time implies that interaction between bind site of cys-GQDs and Hg(II) ion required some time for the maximum quenching of fluorescence intensity. Fluorescence intensity of cys-GQDs analyzed at a different time (0 to 8 min). fig. implies that 0 to 3 min fluorescence intensity gradually decreases, then 3 min to 8 intensity is almost constant. Maximum quenching lies at 3 min that means all Hg(II) ions are bind with the active site of cys-GQDs. So optimum reaction time is 3 min, so this time is selected for the further investigation of Hg(II) ion.

In the order of detection of Hg(II) ion, the pH effect on the fluorescence intensity of cys-GQDs was investigated at different pH (2 to 12). The different pH of the buffer solution was prepared from dilute NaOH and H_3PO_4 solution. Figure 5 indicates the fluorescent intensity of cys-GQDs at a different pH. The fluorescence intensity gradually decreases from pH 2 to 8, after pH 8 fluorescence intensity increases. Maximum quenching occurs at a slightly alkaline medium. Fluorescence intensity of cys-

GQDs can be change due to protonation and deprotonation of donor atom (-COOH, NH_2, -SH)present in the surface of GQDs. At acidic medium maximum donor atom protonated, therefore less Hg(II) ion interact GQDs while alkaline medium electron density increases. At alkaline medium more donor atom available on the surface GQDs, which is reacted with Hg(II), therefore maximum quenching takes place at pH 8. So pH 8 is chosen for further detection of Hg(II) ion.
Fluorescence intensity cys-GQDs were investigated at different concentrations of Hg(II) ions by using a fluorescence spectrophotometer. Figure 5 indicated that fluorescence intensity regularly decreases with the increasing of concentration of Hg(II) ions. In this work, the concentration of Hg(II) ions is sensing from 10 ppb to 50 ppb. At 50 ppb approximately 80 % quenching occurs. Quenching of fluorescence intensity increases with the concentration of Hg(II) ion. Regularly quenching occur due to at high concentration of Hg(II) ions is interact with the maximum number GQDs and convert it in non-fluorescent material. Therefore fluorescence intensity of GQDs is decreasing.

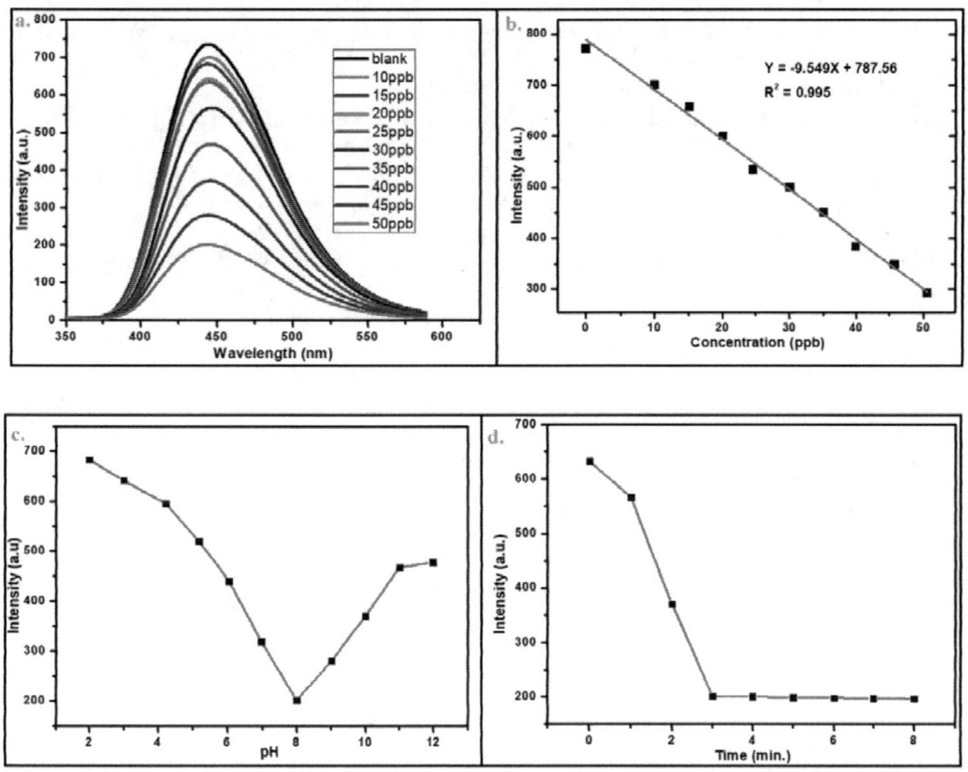

Figure 5. (a) Emission fluorescence spectra of Cys-GQDs with different concentrations of Hg(II)ion (ranging from 10ppb to 50ppb). (b) Calibration curve of fluorescence quenching of cys-GQDs in the presence of a different concentration of Hg(II)ion (from 10 ppb to 50ppb). (c) the response of pH curve on the PL intensity in an aqueous solution of cys-GQDs (ranging from 2 to 12) (d) Time-dependent fluorescence emission spectra of cys-GQDs at pH 8.

4. Conclusion

In this study, a highly selective and sensitive cys-GQDs base nanosensor developed for rapid and robust detection of ppb level of Hg(II) ion in an environmental sample. The preparation of GQDs is a very easy and one step pyrolysis method. The size of the GQDs is 2-6 nm. functionalization of GQDs surface by using cysteine amino acid enhances the quantum yield up to 41%. Thio functional group enhances the selectivity and sensitivity of cys-GQDs towards Hg(II). It is found that the detection mechanism for this method is based on both dynamic and static quenching. The detection of the limit of Hg(II) ion is 0.64 ppb. Result optioned in this sensing method shows that cys-GQDs provides a selectively and sensitivity with reliable, cheap, excellent repeatability for detection Hg(II) ion.

Acknowledgment
Authors gratefully acknowledge the Director of National Institute of Technology Raipur, for providing laboratory facilities andfinancial support from Council of Scientific and Industrial Research (CSIR) in the form of SRF with Grant No. 09/1116(0001)2016-EMR-I, Pusa, New Delhi.

Reference

[1] Ali M, Hery S, and Putri S A 2018. In EDP Sciences **73**-06002.
[2] Grandjean P, Satoh H, Murata K and Eto K 2010. Environmental health perspectives**118-**1137.
[3] Vianna A D S, Matos E P D, Jesus I M D, Asmus C I R F and Câmara V D M 2019 Cadernos de saude publica**35**-00091618**.**
[4] Guzzi G and La Porta C A 2008 Toxicology**244**-1.
[5] Riaz A, Khan S, Muhammad S and Shah M T 2019Mine Water and the Environment**1**-8.
[6] Ashraf M W and Mian A 2019 Bulletin of the Chemical Society of Ethiopia **33**-573
[7] Xu H, Xia Y K, Li C J, Zhang J Y, Liu Y, Yi W, Qin Z Y, Chen L, Shi Z F, Quan K and Yang Z X 2019Laboratory Investigation**99**-588.
[8] Zangmo T and Siripinyanond A 2019Analytica chimica acta**1085**-29.
[9] Mishra S, Tripathi R M, Bhalke S, Shukla V K and Puranik V D 2005Analytica Chimica Acta**551**-192.
[10] Zhang D, Yang S, Cheng H, Wang Y and Liu J 2019 Talanta**199**-620.
[11] Sun Z, Du J and Jing C 2016J. of Enviro. Sciences**39**-134.
[12] Leopold K, Foulkes M and Worsfold P J 2009Analytical chemistry**81**-3421.
[13] Sanchez-Rodas D, Corns W T, Chen B and Stockwell P B 2010 J. of Ana. Atomic Spectrometry**25**-933.
[14] Huang H, Qu C, Liu X, Huang S, Xu Z, Zhu Y and Chu P K 2011 Chem. Communications **47**-6897.
[15] Fan Y, Long Y F and Li Y F 2009 Ana. chimica acta **653**-207.
[16] Hasanjani H R A and Zarei K 2019 Biosensors and Bioelectronics **128**-1.
[17] Feng J, Guo Q, Liu H, Chen D, Tian Z, Xia F, Ma S, Yu L and Dong 2019Carbon**155**-491.
[18] Parlak M, Kayaaslan E, Ugus D, Bilen B, Erkturk H and Yazici T 2019 I. Soc. for Optics and Photonics**10929**-109290.
[19] Mombru D, Romero M, Faccio R and Mombru Á W 2019Physica E: Low-dimensional Systems and Nanostructures **113**-130.
[20] Pathan S, Jalal M, Prasad S and Bose S 2019J. of Materials Chemistry A **7**-8510.
[21] Kalita H, Harikrishnan V and Aslam M 2014 Int J Nanotechnol**11**-75.
[22] Wu Y, Wadia C, Ma W, Sadtler B and Alivisatos AP 2008 Nano Lett **8**-2551.
[23] Dewangan P K, Khan F, Shrivas K and Sahu V 2019 J. of Radioanalytical and Nuclear Chemistry **320**-757
[24] Alivisatos AP, Gu W and Larabell C 2005Annu. Rev. Biomed Eng. **7**-55.
[25] Yang Y, Xiao X, Xing X, Wang Z, Zou T, Wang Z, Zhao R and Wang Y 2019 Mat. R. Express**6**-095615.
[26] Gu S, Hsieh C T, Tsai Y Y, Ashraf Gandomi Y, Yeom S, Kih K D, Fu C C and Juang R S 2019 ACS Applied Nano Materials **2**-790.
[27] Du F, Sun L, Zen Q, Tan W, Cheng Z, Ruan G and Li J 2019Sensors and Actuators B: Chemical **288**-96.
[28] Chandra S, Laha D, Pramanik A, Ray Chowdhuri A, Karmakar P and Sahu S K 2016 Luminescence**31**-81.

[29] Sadhanala H K and Nanda K K The J. of Phy. Chemistry C **119**-13138.
[30] Han Y, Tang D, Yang Y, Li C, Kong W, Huang H, Liu Y and Kang Z Nanoscale7-5955.
[31] Fan M, Zhu C, Yang J and Sun D 2016Electrochimica Acta 216.
[32] Kamat P V 1993 Chemical Reviews **93**-267.

A review on the effect of processing techniques and functionalization of filler on mechanical properties of polymer nanocomposites

R. Verma[1], M.J. Rathod[1], R.K. Goyal[2]

[1]Department of Metallurgy and Materials Science, College of Engineering, Pune
[2]Department of Metallurgical and Material Engineering, MNIT, Jaipur

correspondence Email: vermar.meta@coep.ac.in, 020-25507635

Polymer nanocomposites with carbon fillers are widely used for electrical, mechanical and thermal applications due to their excellent intrinsic properties along with their light weight. Carbon fillers with higher aspect ratio have limitations of higher agglomeration and non-uniform distribution in polymer matrix due to high vander Waals forces acting between them and also due to lack of compatibility between polymer and filler. This results in poor properties of the fabricated composites, hence limiting their application. With proper selection of processing method and optimization of processing parameters, mechanical properties of the polymer nanocomposites can be improved. Functionalization of carbon fillercan enhance the compatibility of filler and the polymer matrix, which can further increase the mechanical properties like tensile strength and young's modulus. This can result in polymer nanocomposites which may have the result in substitution of metals in various applications. This paper reviews about effect of various processing techniques used for dispersion and fabrication of composites and functionalization of filler which can improvemechanical properties of fabricated nanocomposites.

Keywords: Polymer nanocomposites, functionalization, mechanical properties, fillers, matrix, mechanical properties

1. INTRODUCTION

Polymer nanocomposites have received much attention to enhance the properties of engineering polymers while retaining their ease of processing. Polymer nanocomposites need very low dispersant loadings as compared to conventional composites to enhance mechanical properties like modulus and strength [1,2]. Now- a- days, carbon fillers like carbon nanotubes, carbon fibers and carbon black are used due to their intrinsically high mechanical properties like high tensile strength and elastic modulus. Even though, carbon filler have high mechanical properties, there are various factors which affect the mechanical properties of carbon filler based polymer nanocomposites. Factors which affect the mechanical properties of polymer nanocomposites are a) Interfacial load transfer by Mechanical Interlocking, Chemical Bonding, Weak Vander waals Force, b) Dispersion of filler, c) Type and content of filler [3-5]. In order to form a nanocomposite material with the excellent mechanical properties, strong chemical bonding between the reinforcement and matrix is a necessary condition. There might be significant difference in mechanical behaviors between a long fiber-reinforced composite and nanofiber reinforced composite, even both have strong interfacial bonding [6].Xu *et al.* [7] studied the effect of duration of sonication on ultimate tensile strength of Graphitic nano fiber (GCF)/epoxy nanocomposites and they found that for lower sonication energy, nanocomposites show higher young's modulus and tensile strength as compared to higher sonication energy. This may be due to lesser damage to CNF for lower sonication energy. They also stated that for higher sonication energy, lower duration of sonication will show better mechanical properties.Salom*et al.* [8] studied the effect of functionalization of graphene nanoplatelets (GNP) by amine group in epoxy/GNP nanocomposites. Tensile strength of nanocomposites having functionalized GNP was 5 MPa larger than that of nanocomposites having un-functionalized GNP. Thus we can see that as the bonding between filler and matrix improve, mechanical properties of polymer nanocomposites increases. Bonding between filler and the matrix is affected by the choice of processing techniques and byimproving the compatibility between them. One of the ways to improve compatibility between filler and the matrix is functionalization of the filler. Therefore, this paper reviews about effect of functionalization and processing methods on mechanical properties of polymer nanocomposites.

2. Critical Length of fiber

Even with strong interfacial bonding, there might be significant difference in mechanical behavior between a long fiber-reinforced composite and nanofiber reinforced composite [9].Long fiber-reinforced composite can effectively stop the propagation of a major crack whereas short nano fibers may not have this kind of effect. Material failure is governed by a main crack nucleating from defects inside the matrix or at the interface between the matrix and the reinforcement. In other words, tensile strength of polymer nanocomposite depends on the fiber length rather than on amount of fibers [10]. Failure of the nanocomposite depends on the number of fiber ends, as under tensile loading condition, the crack initiate at the fiber end and passes through the fiber-polymer interface to the polymer matrix, therefore more no of fibers will result in increased number of stress concentrating points in the form of fiber ends [11,12]. For lower concentration of smaller sized filler there will be insufficient hindrances for crack propagation. While the crack will be mainly constrained by strong long fibers so the strength and toughness values along fiber direction in these composites are much larger than the values of the pure matrix. Zheng *et al.* [13] stated that the length of fiber must be longer than a certain length known as a load transfer length. This critical length L_c is mainly determined by the fiber tensile strength S_f, the fiber diameter d, and the interfacial shear strength τ_i between the fiber and the matrix.

$$Lc = \frac{d \times S_f}{\tau_i} \qquad \text{........... 1}$$

For strong interfacial bonding between the fiber and the matrix, the critical length should be smaller in order to prevent rupture of the fiber. This can be achieved by either functionalization of filler or by use of proper processing technique.

3. Functionalization of filler

CNTs have excellent tensile strength in axial direction, whereas in radial direction they are rather soft. First TEM observation of radial elasticity displayed that Vander Waal's forces can deform two adjacent nanotubes [14]. This radial direction elasticity of CNT is important for CNT nanocomposites for mechanical application where the mixed tubes are subjected to large deformation in the transverse direction when the load is applied to a composite structure [15].Though incorporation of carbon filler in polymer matrices increases their mechanical properties but very low amount of filler is desirable as increase in the filler content has drawbacks of increased cost of composite and difficulty of fabrication [16,17]. Decrease in mechanical properties of composite at higher filler content is due to agglomeration of CNT and CNF owing to their higher aspect ratio and vander Waals forces.To overcome this problem functionalization of CNT is done to enhance the bonding between filler and matrix [18,19]. Du *et al.*[20] reviewed that the for un functionalized CNT, composites having aligned CNTs show higher tensile strength and elastic modulus as compared to randomly oriented CNTs. Also these properties improve with increasing content of aligned CNTs. On the contrary, for randomly oriented CNTs, properties increases up to critical filler loading and then it start decreasing. Mechanical properties of nanocomposites having randomly oriented CNTs after critical loading may even reduce to values lower than pure polymer. This probably is due incomplete coverage of CNT by polymer matrix due to increase in viscosity of polymer matrix and large surface area of CNT. This makes uniform distribution and load transfer more difficult. Functionalization of CNT is done to enhance de-agglomeration and bonding between filler and matrix [21,22]. Various authors have observed different effects of functionalization on mechanical properties and it can be summarized that the mechanical properties of composites having treated MWCNT depends on the temperature, time and composition of functional group. There are different ways of functionalization of CNTs like covalent and non-covalent functionalization. Covalent functionalization is done by attaching desired groups on to the side walls of CNT permanently. Few of the functional groups are antharcene,3-aminopropyltriethoxisilane (3APTES), aminophenyl ($C_6H_4NH_2$), nitrophenyl ($C_6H_4NO_2$), poly (g-azidopropyl-L-glutamate)s (PAPLGs). Advantage of using these functional groups is that uniform dispersion of filler in the matrix can be achieved, however the limitation is that it introduces defects in CNTs. Non-Covalent functionalization is based on Vander- Waal's forces. Advantage of non-covalent functionalization is that theydo not damage or modify CNTs. Common examples of this type of functionalization are use of surfactants and non-covalent protein interactions [21-23]. Detail review of interfacial characteristics of CNT based polymer nanocomposites has been done by Chen *et al.* [24]

Madni*et al.* [25] discussed the effect of functionalization on pristine MWCNTs. PristineMWCNTs was in the form of agglomerates, but addition of surfactants like SOCT and DTAB resulted in deagglomeration of MWCNTs, best dispersion was observed where mixed surfactants were used and for lower concentration. This may be attributed to stable dispersion owing to synergistic behavior in a mixture of anionic and cationic surfactants [25]. Alshammari*et al.* [26] prepared PET/MWCNT composite by melt mixing method with as received MWCNT and functionally treated MWCNT, they found that the tensile modulus increased from 1073.3 MPa for pure PET to 1302.7 MPa and 1343.2 MPa for 2 wt% of as received MWCNT and treated MWCNT composites respectively. Tensile modulus of nanocomposite increased with addition of MWCNT as a result of high aspect ratio and high modulus of MWCNT. Higher tensile modulus of PET/treated MWCNT as compared to PET/as received MWCNT is may be due to shortening of MWCNT length causing lesser agglomeration and better dispersion of treated MWCNTand separated MWCNTs after treatment. But at the same time the tensile strength and % elongation decreased with the addition of MWCNT in PET matrix, this may be due to start of failure at MWCNT agglomeration.Similarly, Sadek*et al.* [27] studied the effect of surfactants on PVA/MWCNT composites and found that for low surfactant percentages (0.01 wt%), tensile strength, elongation at break and Young's modulus increased to 94.85 MPa, 160% and 67.05 MPa as compared to 40 MPa, 50% and 30 MPa for pure PVA. This may be because surfactant may have broken the MWCNT bundles into individual tubes resulting in increased hydrophobic interaction of CNT and PVA and uniform distribution of CNT in PVA matrix.

4. Processing Method for fabrication of polymer nanocomposites

Duration of mixing of filler and matrix and the processing method affects the mechanical properties of polymer nanocomposites. Compression molding has the ability to orient fiber in 0° as compared to extrusion method. It has been observed that fibers aligned at 0° showed maximum mechanical properties [28]. Also, compression molding has the ability to keep the filler in the matrix and have the laminate structure of composite which enhances the mechanical properties of composites. Higher shear stresses involved in extrusion causes the breakage of fiber which reduces the mechanical properties of composites [29]. Enqvist*et al.*[30] suggested that with increase in milling time, distribution of the filler improves and thus mechanical properties improve. The tensile strength of composites mixed for 1 h with 1 wt% MWCNTs is lower than that of composites mixed for 2 h with 0.5 wt% MWCNTs. This shows that, even lower concentration of filler may result in better mechanical properties with increase in milling time. In the same way, Zha*et al.* [31] studied the effect of sonication time on tensile strength and % elongation of PVDF/MWCNT nanocomposites. They found that with increase in the duration of ultrasonication, tensile strength increases while elongation at break decreases, showing enhanced mechanical strength of the composite but with reduced toughness. Tensile strength of nanocomposites was improved by 23% as compared to pure PVDF for sonication duration of 30 min. For same duration of sonication, elongation at break reduced by 50%. Increase in sonication time leads to uniform dispersion and interfacial adhesion between MWCNT and the matrix.

Zheng *et al.*[32] studied the morphology and tensile properties of PMMA/ MWCNT nanocomposites fabricated by anti-solvent precipitation (ASP) method where the solution of PMMA and MWCNT was precipitated out in methanol solvent and after drying it was hot compacted. Functionalization of MWCNT was done to find out its effect on tensile properties of nanocomposites. It was observed that the tensile strength of nanocomposites containing as received 0.5 wt% MWCNT and 1 wt% functionalized MWCNT resulted in similar (~14%) increase in the tensile strength. This according to authors is due to good dispersion of filler in the matrix in case as received MWCNT and decrease in the size of MWCNT due to functionalization. Mondal *et al.* [33] compared mechanical properties of Chlorinated Polyethylene (CPE)/CNF nanocomposite fabricated by two different methods namely solution method and solution cum melt mixing method. They observed that the tensile strength of the nanocomposites were 16.6 MPa and 11.75 MPa for 1 wt% MWCNT fabricated by solution cum melt mixing method and melt mixing method respectively as compared to 7.4 MPa for pure CPE. As sonication was involved in solution cum melt mixing method, surface roughness of MWCNT increased resulting in increased contact points between MWCNT and CPE polymer matrix which in turn resulted in better interfacial bonding and mechanical interlocking. Table 1 compares the mechanical properties of various polymer nanocomposites prepared by different processing techniques for same filler loading i.e. for 1 wt%.

Table 1: Comparison of mechanical properties of polymer nanocomposites prepared by different processing methods [6,25-26,32-35]

System	*Filler content*	*Method of processing*	*Tensile Strength[MPa]*	*Elongation at Break [%]*	*Young's Modulus [MPa]*
PET/MWCNT	1 wt%	Melt extrusion	30.0 ±3.8	2.6 ± 0.3	1252.6± 31.3
PET/treated MWCNT	1 wt%	Melt extrusion	25.6±6.0	2.7±0.8	1222.5±105
CPE/MWCNT	1 wt%	Melt mixing	11.7	-8.5	
CPE/MWCNT	1 wt%	Solution cum melt mixing	16.6	-3.2	
PVA/MWCNT	1 wt%	Solution mixing	~90	~160	~60
PC/MWCNT	1 wt%	Melt mixing	~ 74	~65 %	~1425
PVDF/CNF	1 wt%	Melt mixing	47.13	12.02	1284
EMA/SCF	1 wt%	Melt mixing	9.1	1203	1.9
UHMWPE/CNT	1 wt%	Ball milling	39.4	686	21.7
PP/MWCNT	1 wt%	Melt mixing	34.5		283

5. Conclusion

From literature review, it was observed that the mechanical properties depend on factors like interfacial load transfer between filler and matrix, type and content of filler and agglomeration of filler.Mechanical properties of nanocomposites can be improved by avoiding the agglomeration of filler and maintaining the proper length of filler material by employing proper processing methods.All papers suggested that elastic modulus increases with the increase in the filler content however tensile strength, toughness and strain at break decreases with increase in filler content, this may be due to agglomeration of fillers and poor interfacial bonding between filler and the matrix.

6. Reference

[1] J. Njuguna, K. Pielichowski, Polymer nanocomposites for aerospace applications: properties, 2003 *Adv. Engg. Mat.***5(11)**769-778.
[2] G. Zamfirova, V. Gaydarov, F. Faraguna, E. Vidović, A. Jukić, Influence of carbon nanotubes functionalization on the mechanical properties of polymethacrylate nanocomposites, 2016*Colloids and Surfaces A: Physicochem. Eng. Aspects***510** 169–175

[3] V.D. Punetha, S. Rana, H.J. Yoo,A.Chaurasia, J.T. Mcleskey, M.S.Ramasamy, N.G. Sahoo, J.W. Cho, Functionalization of carbon nanomaterials for advanced polymer nanocomposites: A comparison study between CNT and graphene, 2017*Prog. Poly. Sci.***67**1-47

[4] V. Rangari, Polymer Nanocomposite Materials for Structural Applications, 2011*Adv. Nanocompo.- Syn., Charac. Indus. App*61-84

[5] M. Modesti, A. Lorenzetti, D. Bon, S. Besco, Effect of processing conditions on morphology and mechanical properties of compatibilized polypropylene nanocomposites, 2005*Polymer***46**10237- 45

[6] Y. Zare, K.Y. Rhee, D.Hui, Influences of nanoparticles aggregation/ agglomeration on the interfacial/interphase and tensile properties of nanocomposites, 2017 *Composites B***122** 41-46

[7] L.R. Xu, V. Bhamidipati, W.H. Zhong, J. Li, C.M.Lukehart, Mechanical Property Characterization of Polymeric Nanocomposite reinforced by graphitic nanofibers with reactive linkers, 2004*J. Compo. Mat.***38 (18)** 1563 –68

[8] C. Salom, M.G. Prolongo, A. Tiribo, A.J.M.Martinez, I.A.Carcer, S.G. Prolongo, Mechanical Properties and Adhesive behaviour of epoxy-graphene nanocomposites, 2018, *Inter.J. Adhesion and Adhesives,***84**119-125

[9] S.Y. Fu, B. Lauke, Effects of fiber length and fiber orientation distributions on the tensile strength of short-fiber-reinforced polymers, 1996 *Compo. Sci. Tech.* **56**1179-90

[10] N. Sato, T. Kurauchi, S. Sato ,O. Kamigaito, _Micro failure behavior of randomly dispersed short fibre reinforced thermoplastic composites obtained by direct SEM observation,1991*J.Mat.Sci.***26** 3891-98

[11] Vijendra Singh, *Physical Metallurgy*, 2005 Standard Publishers Distributors 1st Ed. 452-466

[12]C. Zheng, N. Hossieny, C. Zhang, B. Wang, S.M. Walsh, Morphology and tensile properties of PMMA carbon nanotubes nanocomposites and nanocomposites foam, 2013 *Compo. Sci. Tech.* **82** 29-37

[13]I. Palaci, S. Fedrigo, H. Brune, C Klinke , M Chen, E. Riedo, Radial elasticity of multiwalled carbon nanotubes 2005*Phy.Rev.Lett.***94(17)** 175502 1-4

[14]K Saeed, Irahim, Carbon nanotubes-properties and applications: a review, 2013 *Carbon Letters***14(3)** 131-144

[15]W. Chanklin, J. Laowongkatr, L.P.F. Chibante, Electrical property validation of percolation modelling in different polymer structures of carbon based nanocomposites, 2018 *Mat. Today Commu.***17** 153-160

[16] J. H. Du, J. Bai, H.M. Cheng, The present status and key problems of carbon nanotube based polymer composites, 2007*Exp. Poly. Lett.***1 (5)** 253-273

[17]L. Wang, W. Peng, M. Sarafbidabad, Y. Zare, Explanation of main tunnelling mechanism in electrical conductivity of polymer/carbon nanotubes nanocomposites by interface percolation, 2019 *Poly. Bull.* 1-15

[18]V.D. Punetha, S. Rana, H.J. Yoo,A.Chaurasia, J.T. Mcleskey, M.S.Ramasamy, N.G. Sahoo, J.W. Cho, Functionalization of carbon nanomaterials for advanced polymer nanocomposites: A comparison study between CNT and graphene, 2019*Prog. Poly. Sci.***67** 1-47

[19]A. Graziano, S. Jaffer, M.Sain, Review on modification strategies of polyethylene/polypropylene immiscible thermoplastic polymer blends for enhancing their mechanical behavior, 2018 *J.Elast. Plast.* 1-46

[20]J. H. Du, J. Bai, H.M. Cheng, The present status and key problems of carbon nanotube based polymer composites, 2007 *Exp. Poly. Lett.***1(5)** 253-273

[21]S. Mallakpour, S. Soltanian, Surface functionalization of carbon nanotubes: fabrication and applications, 2016 *RSC adv.***6** 109916-35

[22]N.G. Sahoo, H.K.F. Cheng, H. Bao, Y. Pan, L.Li, S.H. Chan, Covalent functionalization of carbon nanotubes for ultimate interfacial adhesion to liquid crystalline polymer, 2011 *Soft Matter***7** 9505-14

[23]J. Liu, Y. Ye, Y. Xue, X. Xie, Y.W. Mai, Recent advances in covalent functionalization of carbon nanomaterials with polymers: Strategies and Perspectives, 2017 *J. Poly. Sci. A: Poly. Chem.***55** 622-631

[24]J. Chen, B. Liu, X. Gao, D. Xu, A review of the interfacial characteristics of polymer nanocomposites containing carbon nanotubes,2018*RSC Adv.***8** 28048-85.

[25]I. Madni, C.Y. Hwang, S.D. Park, Y.H. Choa, H.T. Kim, Mixed surfactant system for stable suspension of multiwalled carbon nanotubes, 2010 *Colloids and Surfaces A: Physiochem. Engg. Asp.***358** 101-107

[26]B.A. Alshammari, A. Wilkinson, Impact of carbon nanotubes addition on electrical, thermal, morphological and tensile properties of poly (ethylene terephthalate), 2016 *App. Petrochem. Res.* **6** 257-267

[27]S. Mallakpour, S. Soltanian, Surface functionalization of carbon nanotubes: fabrication and applications, 2016 *RSC Adv.* **6** 109916-35

[28]R.J. Kutiger, M.K. Alam, D.P. Anderson, R.L. Jacobsen, Processing and characterization of aligned vapour grown carbon fiber reinforced polypropylene. 2002 *Composites Part A, App. Sci. Manu.* **33** 53-62

[29]J. Tiusanen, D. Vlasveld, J. Vuorinen, Review on the effects of injection moulding parameters on the electrical resistivity of carbon nanotube filled polymer parts, 2012*Compo.Sci.Tech.* **72** 1741-52

[30]E. Enqvist, D.Ramanenka, P. Marques, J. Gracio, N. Emami, The effect of ball milling time and rotational speed on ultra- high molecular weight polyethylene reinforced with multiwalled carbon nanotubes, 2016 *Poly. Compo.* 1128-36

[31]J.W. Zha, Y.Gao, D.L. Hang, Y. Wen, R.K.Y.Li, C-Y Shi, A.M. Dang, Flexible electrospun polyvinylidene fluoride nanofibrous composites with high electrical conductivity and good mechanical properties by employing ultrasonication induced dispersion of multiwalled carbon nanotubes,2016*Compo.Sci.Tech.* **128** 201-206

[32]C. Zheng, N. Hossieny, C. Zhang, B. Wang, S.M. Walsh, Morphology and tensile properties of PMMA carbon nanotubes nanocomposites and nanocomposites foam, 2013 *Compo. Sci.Tech.* **82** 29-37

[33]S. Mondal, L. Nayak, M. Rahaman, A. Aldalbahi, T.K. Chaki, D. Khastgir, N.C. Das, An effective strategy to enhance mechanical, electrical and electromagnetic shielding effectiveness of chlorinated polyethylene-carbon nanofiber nanocomposites. 2017 *Composites Part B* **109**155-169

[34]A.S. Babal, R. Gupta, B.P. Singh, V.N. Singh, S.R. Dhakate, R.B. Mathur, Mechanical and electrical properties of high performance MWCNT/polycarbonate composites prepared by an industrial viable twin screw extruder with back flow channel, 2014 *RSC Adv.* **4** 64649-58

[35]A.C. Yilmaz, M.S. Ozen, E. Sancak, R. Erdem, O. Erdem, N. Soin, Analyses of the mechanical, electrical, and electromagnetic shielding properties of thermoplastic composites doped with conductive nanofillers, 2018*J.Compo.Mat.* **52(11)**1-10

A First principle study of structural and electronic properties of ZnO and ZnS Buckyball structures

Sakshi Sharma[1], A K Shrivastav[1], Anjali Oudhia[2] and Mohan L Verma[3]

[1]Department of Physics, National Institute of Technology, Raipur, India

[2]Department of Physics, Govt. Nagarjun P.G. Science College, Raipur, India

[3] Department of Applied Physics, FET-SSGI Shri Shankaracharya Technical Campus, Junwani, Bhilai, India

*Corresponding author's e-mail address: ssharma.phd2018.phy@nitrr.ac.in

Abstract- The nanostructured ZnO and ZnS in Buckyball form have been optimized using first-principle density functional (DFT) approach. For an improved understanding of the systems, investigation of their structural and electronic properties are performed. Study of Density of states (DOS) and partial density of states (PDOS) are carried out. The system obtained a well near reported values of energy and bond length, done previously. DFT study was found to be very much compatible with the theories of solid-state physics. This study provides a good comparison between the properties of ZnO and ZnS buckyball structures, which is reported for the first time. The study may be useful for group II-VI elements based applications on various fields such as OLEDs, Photovoltaics, sensors, biomedical applications etc.

Keywords: *Nanostructure; Buckyball; First principle study; Density functional theory; DOS; PDOS.*

1. Introduction

Group II-VI based materials have been investigated for different research works from the past 100 years. These are one of the most prominent materials for application-based research. Due to its large band gap and high electron mobility, it has various optical and electronic applications [1, 2]. Bulk material exhibits a wide direct band gap and high exciton binding energy in case of ZnO and ZnS. While ZnO and ZnS quantum dots display quantum confinement effect due to which bulk properties can be custom-made, which generates its wide-ranging applications in different fields like solar cells, gassensors and biosensors, optoelectronic luminescent materials etc [3].

Fullerene has a wider range of applications in the area of bio-medicines. It has a unique structure, which makes it a desirable candidate for the formation of a conjugate with protein and DNA, which is helpful for the development of anti-cancer treatment. It could be used as a drug delivery tool for the treatment of a particular cell. After the discovery of carbon buckyball structure, there are lots of development in the field of fullerenes both in experimental ways as well as in theoretical ways by using material modelling [4]. Here we are using material modelling for the examination of ZnO and ZnS buckyball structures and their comparative study is performed for the first time. Valence electrons play a significant role in defining a material. Valence electrons define various structural, electronic, magnetic, optical and transport properties in condensed matter physics. For understanding these properties there are various approaches available, but quantum mechanics has proven to be the best formulation to describe the interacting system of electrons and nuclei. As we know the Schrodinger equation is an essential quantum mechanical equation that defines a system of electrons and nuclei in terms of wave function ψ. The DFT based study provides an interpretation for intrinsic properties of a semiconductor.

In this paper, we have presented briefly the approximations to solve the problem of many-body using density functional theory and computational implementation of DFT is done using SIESTA (Spanish Initiative for Electronic Simulations on Thousands of Atoms) code [5].

2. Methodology

Here we performed simulations by using a computational program called Spanish Initiative for Electronic Simulations with thousands of atoms (SIESTA). This is based on density functional theory [6, 7]. The whole system was optimized by following all essential steps like mesh-cutoff, k-point, lattice-optimization etc. For correct results, it is essential that all the atoms remain fully relaxed while performing the optimization. Moreover for exchange and correlation potential; a Generalized Gradient Approximation (GGA) of Perdew– Burke–Ernzerhof is used [8].As GGA is comparatively more efficient to predict the energy gap of semiconductor than the LDA, so here we used GGA for our study [9]. Core electrons are modelled with Troullier–Martins norm-conserving pseudopotentials [10] the valence electrons functions are expanded in double zeta polarized basis set [11, 12] of localized orbitals and the real space grid is set to be 400 Ry. Until the Hellmann–Feynman forces acting on all components of each atom are smaller than 0.01eVÅ−1, the structure remains relaxed. This yielded a fully optimized ZnO buckyball which contained 24 Zn and 24 O atoms. Similarly, ZnS buckyball structure which contained 24 Zn and 24 S atoms.

3. Results and discussion

3.1 Structural Analysis:

Here the structural analysis of both ZnO and ZnS buckyball structures are done. For which some of the input parameters are set like Mesh-cut off, K-grid, CG steps etc. Some of the parameters are discussed below.

k-grid were acquired by using Monkhorst pack scheme with a value of (5x5x5) arbitrary units, these are among all three axes.

Table 1. Input parameters of ZnO and ZnS buckyball structure.

System used	Mesh-cutoff (Ry)	Fermi energy (eV)	Cell volume ($A^{\circ 3}$)
ZnO	400	-5.6007	384.077
ZnS	400	-4.6340	570.988

IOP Conf. Series: Materials Science and Engineering **798** (2020) 012032 doi:10.1088/1757-899X/798/1/012032

a) b)

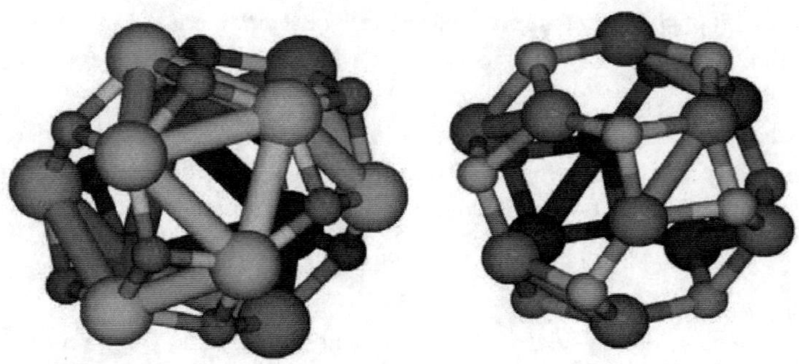

Figure 1. a) ZnO Bucky ball structure b) ZnS Bucky ball structure.

Figure1 (a) and (b) displays the buckyball structure of ZnO and ZnS by a ball and stick model. In Figure 1 (a) the yellow (bigger ball) and red (smaller ball) balls signify zinc and oxygen atoms respectively. While in Figure 1 (b) the brown (bigger ball) and yellow (smaller ball) balls signifies the Zinc and Sulphur atoms respectively. As reported in other papers the most symmetric buckyball structure among all comparing compounds, is the most stable one [13], here we obtained symmetric structures of both. The bond length and bond angles between the atoms are shown in the tables below.

Table 2. Bond lengths of ZnO and ZnS buckyball structures.

S.no.	Compound Name	Bond Name	Bond Length (□)
1.	Zinc Oxide	Zn-Zn	2.99
		Zn-O	2.02
		O-O	2.85
2.	Zinc Sulphide	Zn-Zn	3.20
		Zn-S	2.41
		S-S	3.50

Table 3. Bond Angles of ZnO and ZnS buckyball structures.

S.no.	Compound Name	Angle Name	Angle Value(degrees)
1.	Zinc Oxide	Zn-O-Zn	90.66
		O-Zn-O	86.63
		Zn-Zn-Zn	60.98
		O-O-O	Not present
2.	Zinc Sulphide	Zn-S-Zn	82.25
		S-Zn-S	93.58
		Zn-Zn-Zn	Not present
		S-S-S	Not present

IOP Conf. Series: Materials Science and Engineering **798** (2020) 012032 doi:10.1088/1757-899X/798/1/012032

As we know that electronegativity is nothing but a tendency to attract electrons shared in a covalent bond between atoms. Smaller is the atom, more is its electronegativity, this is because the distance between the electron and positively charged nuclei is less in them. Here we obtained that bond length of Zn-S is greater than that of Zn-O. This goes with the fact that Sulphur is less electronegative than oxygen. [14].Because of the difference between the bond length of ZnS and ZnO, the cell volume of ZnS (570.988 Å^3) is greater than that of ZnO (384.077Å^3). Which implies that there is a change in energy density due to increased cell volume. Also, it is observed that there is a difference in the Fermi level of the two compounds. ZnO (-5.6007 eV) is having higher fermi energy as compared to ZnS (-4.6340 eV).

3.2 Electronic properties Analysis:

Figure 2 (a) and (b) displays the calculated density of states (DOS) and Figure 3 (a) and (b) displays the projected density of states (PDOS) of ZnO and ZnS molecules. We can define the forbidden energy gap as the difference between the highest occupied molecular orbital (HOMO) and lowest unoccupied molecular orbital (LUMO) which is an essential mechanism to understand conductivity of material [15]. As the charge transport properties of the molecule predominantly depend on the forbidden energy gap, it is important to examine the variations in the molecular orbital energy level and forbidden energy gap. [16–18].

3.2.1. The density of states:

In Figure 2 (a) it could be observed that the forbidden energy gap of ZnO buckyball structure is around 0.4 eV. Whereas in Figure 2 (b), it could be observed that the forbidden energy gap of ZnS buckyball structure obtained is around 2.2 eV, both of these values are quite less than the reported band gaps in its bulk form.

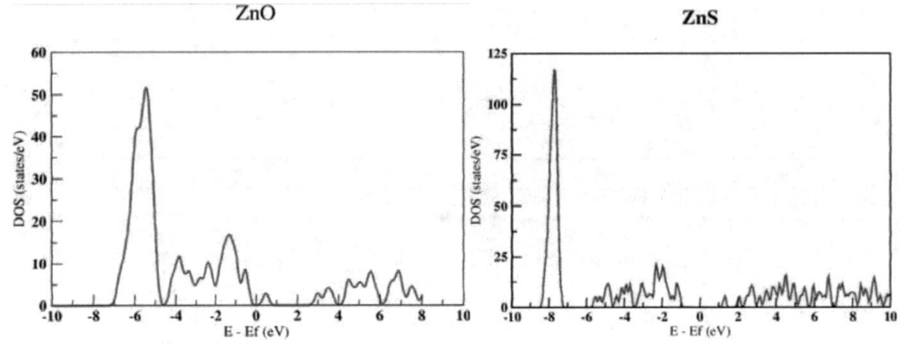

Figure 2. a) DOS plot of ZnObuckyball structure b) DOS plot of ZnS buckyball structure.

Defect levels are observed in the Valence band region which may contribute to the conductivity. Due to these defects, the band gap is reduced as compared to its bulk structure. This is the reason they are showing metallic behavior in case of ZnO. While in the case of ZnS it is still showing semiconductor properties.

3.2.2. Projected Density of states:

From the PDOS plots, it is obtained that in case of ZnO 3d orbital of Zn atom and 2p orbital of O atom shows maximum contribution in the valence band. 4s orbital in Zn and 2s orbital of O atom shows equal contribution in both valence band and conduction band. Similarly, in case of ZnS, 3d orbital of Zn atom and 3s orbital of S atom shows maximum contribution in the valence band. While 4s orbital of Zn atom and 3p orbital of S atom shows maximum contribution in conduction band [19].

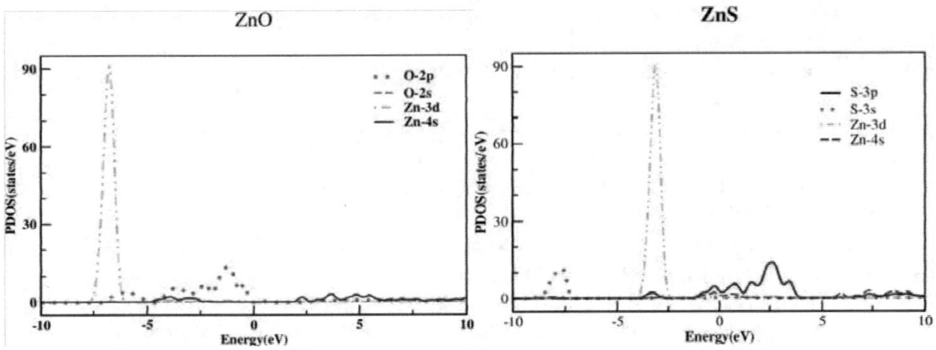

Figure 3. a) PDOS plot of ZnO structure b) PDOS plot of ZnS structure.

We optimized ZnO and ZnS structures which showed a density of states and partial density of states plots. The band gap of ZnO as evaluated here is 0.4 eV which is less than the reported bulk structure value of 3.37 eV, whereas for ZnS we got 2.2 eV which is also less than the reported band gap for its bulk structure value of 3.54 eV, but it gives way for material modelling to go for some different morphologies of ZnO and ZnS structures and improve its approach for practical use.

4. Conclusions

DFT was used for investigation of two new structures, each of ZnO and ZnS. The result obtained for the new buckyball structures of ZnO and ZnS shows that these two structures have reduced band gap as compared to its bulk structures. As reported in other papers the most symmetric buckyball is the most stable one, here we obtained symmetric structures of both. The molecular dimensions of both buckyballs are less than 100 nanometers. These are the very useful result as buckyball structures are having wide range applications in the fields of solar cells, sensors and biomedicine.

Acknowledgements

The authors appreciatively acknowledge the management of Shri Shankaracharya Technical Campus-SSGI for providing us computing facility in their research laboratory. We also acknowledge the kind support of the Department of Physics, N.I.T, Raipur.

References

[1] Lee J H, Ko K H and Park B O 2003 Electrical and optical properties of ZnO transparent conducting films by the sol gel method, *Journal of Crystal Growth* **247**119

[2] Chen J, Liu Y, Shao C, Mu R, Lu Y, Zhang J, Shen D and Fan X 2005 α-Fe_2O_3Nanotubes in Gas Sensor andLithium-Ion Battery Application,, *Adv. Mater.* **17** 582

[3] Zhang L, Yin L, Wang C, Lun N, Qi Y and Xiang D 2010 Origin of Visible Photoluminescence of ZnO Quantum Dots: Defect-Dependent and Size-Dependent, *J. Phys. Chem.* C **114** 9651

[4] Obayes H R, Al Obaidy A H, Alwan G H and Al Amiery A A 2015 Molecular simulation for novel carbon buckyball materials, *Cogent Chemistry* **1** 1026638

[5] Verma H, Rao B K, Verma M L and Chauhan J 2019 First principles study of breaking energy and mechanical strength of Kevlar-29, *Bull. Mater. Sci.* **42** 76

[6] Hohenberg P and Kohn W 1964 Inhomogeneous Electron Gas, *Phys. Rev.* **136** B864

[7] Sun C Q , Wang Y, Tay B K , Li S, Huang H , Zhang Y B 2002 Correlation between the Melting Point of a Nanosolid and the Cohesive Energy of a Surface Atom, *J. Phys.: Condens. Matter B* **106** 10701

[8] Perdew J P and Burke K 1996 Generalized Gradient Approximation Made Simple, *J. Phys. Rev. Lett.***77** 3865

[9] Jaffe J E , Snyder J A , Lin Z and Hess A C 2000 LDA and GG A calculations for high-pressure phase transitions in ZnO and MgO, *Phys. Rev.* B **62** 1660

[10] Trouillier N and Martins J L 1991 Efficient pseudopotentials for plane-wave calculations, *J. Phys. Rev.*B **43** 1993

[11] Junquera J, Paz O, Portal D S and Artacho E 2001 Numerical atomic orbitals for linear-scaling calculations, *J. Phys. Rev.* B **64** 235111

[12] Shen C, Wang J, Tang Z, Wang H, Lian H and Zhang J 2009 Physicochemical properties of poly(ethylene oxide)-based composite polymer electrolytes with a silane-modified mesoporous silica SBA-15, *Electrochim.* Acta **54** 3490

[13] Balakrishnan Akash, Rukkesh J, Bhavyal K, FloraRMary Nancy, Chamundeeshwari M, Palani S 2017 Bucky Balls: An Introduction, Discovery and Overview of Their Properties and Applications, *International Journal of Engineering Research and Reviews* **5** 37

[14] Lindquist B A and Dunning T H 2014 The nature of the SO bond of chlorinated sulfur -oxygen compounds, *Jr. Theoretical Chemistry Accounts* **133** 1443.

[15] Jothi M, Selvaraju K and Kumaradhas P 2013 Exploring the effect of External electric field on Octane dithiolate molecular Nanowire: A Quantum chemical and Charge density study, *J. Comput. Theor. Nanosci.* **10** 789

[16] Shafiee A, Salleh M M and Yahaya M 2011 Determination of HOMOand LUMOof [6, 6]-Phenyl C61-butyric Acid3-ethylthiophene Ester and Poly (3-octyl-thiophene-2, 5-diyl) through Voltametry Characterization, *Sains Malays.* **40** 173.

[17] Djurovich P I, Mayo E I, Forest S R and Thompson M E 2009 Measurement of the lowest unoccupied molecular orbital energies of molecular organic semiconductors, Org. *Elect.* **10** 515.

[18] Rangel T, Rignanese G M and Olevano V 2015 Can molecular projected density of states (PDOS) besystematically used in electronic conductance analysis, *Beilstein J. Nanotechnol.* **6** 1247.

[19] Verma Mohan L 2019 First Principles Approach to Study the Structural, Electronic and Transport Properties of Dimer Chitosan with Graphene Electrodes, *Journal of Electronic Materials* **48** 4007

IOP Conf. Series: Materials Science and Engineering **798** (2020) 012033 doi:10.1088/1757-899X/798/1/012033

Structural, Electrical and Magnetic Properties of Mechanothermally Synthesized Bi $(Co_{0.45}Ti_{0.45}Fe_{0.10})O_3$ Nanoceramic

Nripesh Kumar[1], Alok Shukla[1*] and Nitin Kumar[1]

[1]*Department of Physics, National Institute of Technology Mizoram, Aizawl 796012, India*

[*]Corresponding author: aloks.nitmz@gmail.com

Abstract. High-energy ball mill was exploited to synthesize $Bi(Co_{0.45}Ti_{0.45}Fe_{0.10})O_3$ nanoceramic, followed by solid-state cramicmethod. XRD (X-ray diffraction), SEM (scanning electron microscopy) and impedance spectroscopy methods were exploited to analyze elementary crystal structure, micrograph and comprehensive electrical characteristics of the as -synthesized nanoceramic. The impedance spectroscopy technique offered many exciting outcomes on the influenceof grain and grain boundary on the total resistiveproperties of the nanoceramic. The trend of frequency dependence of a.c. conductivity follows Jonscher's power law. The upgraded conductivity and dielectric property propose various applications in the embedded capacitors. PE loop tracer and VSM (Vibrating sample Magnetometer) were exploited to study the ferroelectric and magnetic characteristics of the synthesized nanoceramic. The B-site modified $BiFeO_3$ (with Co and Ti) exhibit ferroelectric and magnetic behavior at room temperature.

1. Introduction

Multiferroics demonstrate more than one ferroic order (i.e. ferroelectricity, ferromagnetism and ferroelasticity). Because of numerous applications in technological fields, multiferroic materials have gained a lot of attention in the area of solid-state or high temperature physics. Perovskites, containing the very general chemical formula like ABO_3, have the capability in providing the A-site or B-site or both A-site and B-site for substitution/doping with iso and non-isovalent elements. Because of such easy availability for doping on any or both sides, $BiFeO_3$ have gained the attention of many material scientists. [1–4]. The $6s^2$ lone pair in Bi^{3+} is accountable for ferroelectricity and Fe^{3+}-O-Fe^{3+} super-exchange formulation is accountable for G-type anti-ferromagnetic characteristic. Though, BFO has few boundaries like, large dielectric loss, low remanent polarization and high leakage current, besides, it is not very easy to synthesize the pure phase BFO material because of its small temperature range for phase steadiness. D-M interaction (i.e. Dzyaloshinskii –Moriya) is accountable for displaying weak ferromagnetism in the crystal symmetry of BFO, however the existence of a spin cycloid at 62 nm avoids net magnetization. BFO has rhombohedral crystal structure symmetry with the R3c space group and it has a T_C (i.e. Ferroelectric Transition Temperature) around 973 K and T_N (i.e. Antiferromagnetic Transition Temperature) around 643 K [5–9].

Ilmenite-type perovskite (Titanium based) compounds, commonly expressed as $MTiO_3$ (M= Co, Ni, Sr, Cd and Fe), have been considered as smart materials due to their multifunctional properties. For minimizing the leakage current issue, the substitution at the A, B or AB-sites of compound has been tried and accomplishedagreeable. Substitution effect in BFO with other perovskite materials of the Bismuth-site (i.e. A-site) and Iron-site (i.e. B-site) atoms such as La, Sm,Gd and Nd, Zr, Ti and Mn, for controlling the oxygen vacancies have also been tried and resulted many new outcomes. This current research work proposes that the partial co-substitution at the B-site (Co and Ti of selected percentages) of the (i.e. Fe^{3+} site)bismuth ferrite, by isovalent elements directtowards theenhancement of dielectric and magnetic properties. In the present work, polycrystalline sample of Co/Ti modified $BiFeO_3$ [i.e., $Bi(Co_{0.45}Ti_{0.45}Fe_{0.10})O_3$] (BCTFO) has been prepared by High Energy Ball Mill, followed by a solid-state ceramic route. The present report discuss the study on multifunctional property of as-prepared $Bi(Co_{0.45}Ti_{0.45}Fe_{0.10})O_3$ nanoceramic.

2. Experimental

$Bi(Co_{0.45}Ti_{0.45}Fe_{0.10})O_3$ceramic compound (lead free) was prepared by a High Energy Ball Mill, pursued by anSSR (solid-state reaction) methodology. Stoichiometric quantities of highly pure oxide powders, i.e.,cobalt oxide (99.5%), bismuth oxide (99.99%), iron oxide (99.5%) and titanium oxide (99.8%) were firstly stir together in a mortar & pestle manually in air mediaand later in methanol media. Further, it was followed by ball-millingwith zirconium oxide grinding jar for effective milling/grinding time of 30 h through high energy planetary ball mill (Retsch-PM 200, Germany). The ratio of mass to the balls was kept 1:10 at 400 rpm in the toluene medium. Subsequently, after 30 h of milling time, ball-milled powder was calcined comparatively at lower temperature (973 K) with already reported mixed oxide technique [10].Through XRDanalysis (Rigaku, Japan) exhibiting,CuKα radiation (wavelength, $\lambda = 1.5405$ Å) in a widespread range of Bragg's angle (20° to 80°) and at a very slow scan speed, single-phase of the powder material was formed. Surface morphology and preliminary elemental analysis was carried out by Field Emission Scanning Electron Microscope and EnergyDispersive X-ray Spectroscopy technique respectively, at room temperature.KBr hydraulic press was utilized to form pellet(diameter ≈12 mm and thickness ≈ 1.2mm) of material, followed by sinteringprocess at acontrolled temperature of 1023 K for 6 h to yield its denser form in the air medium. Using high quality silver paste, electrodes were made on sintered pellet for dielectric and electrical measurements, followed by heating at 200 K for 4 h to eliminate the PVA (binder) and humidity from the synthesized pellet. Both electrical and dielectric measurements were achieved over anextensivefrequency ranges from 1 kHz – 1 MHz and temperature rangesfrom 298K–573K,by operating a computer-controlled Phase Sensitive Multimeter (PSM-1735, Newton4th Ltd, UK). PE loop tracer and VSM (vibrating sample magnetometer) were deployed,to measure the ferroelectric and magnetic characteristic of synthesized nanoceramic at room temperature.

3. Results and Discussion

3.1. Structural Study

Figure 1 displays the plot of room-temperature XRD data for$Bi(Co_{0.45}Ti_{0.45}Fe_{0.10})O_3$ceramic,ball milled for 30 h of effective milling time. To attain the crystallographic information of the synthesized compound at room temperature, X-ray diffractometer was used. The basic crystalline nature and phase analysis of $Bi(Co_{0.45}Ti_{0.45}Fe_{0.10})O_3$compound was carried out by least-square refinement, applying the XRD data in 'POWDMULT' software[11]. Obtained parameters were, a=18.133Å, b=8.21 Å, c=12.73 Å and S.D.=0.010 with orthorhombic crystal symmetry.The value of parameters obtained are summarized in the Table 1. The crystallite size (D) of the as synthesized compound was calculated by using the Debye-Scherrer formula:

$$D = k\lambda/\beta cos\theta$$

where β (FWHM of a reflection peak) at the Bragg angle (2θ) and the wavelength (λ), k =0.89=Scherrer constant and D = Crystallite size. The value of average crystallite size was found to be 35.70 nm for the $Bi(Co_{0.45}Ti_{0.45}Fe_{0.10})O_3$. Furthermore, tolerance factor (T_F) value, for perovskite family (ABO_3) is depend on succeeding equation; [12]

$$T_F = \frac{(\langle R_A \rangle + R_O)}{\sqrt{2}(\langle R_B \rangle + R_O)}$$

where $\langle R_A \rangle$, $\langle R_B \rangle$&R_O represents the mean ionic radii of A(bismuth), B(cobalt, titanium, iron) and O (Oxygen) respectively. Calculated outcome of the tolerance factor is 0.83 for $Bi(Co_{0.45}Ti_{0.45}Fe_{0.10})O_3$that proposes much degree ofdeformation, for ideal perovskite compound the value of tolerance factor is one [10][13].

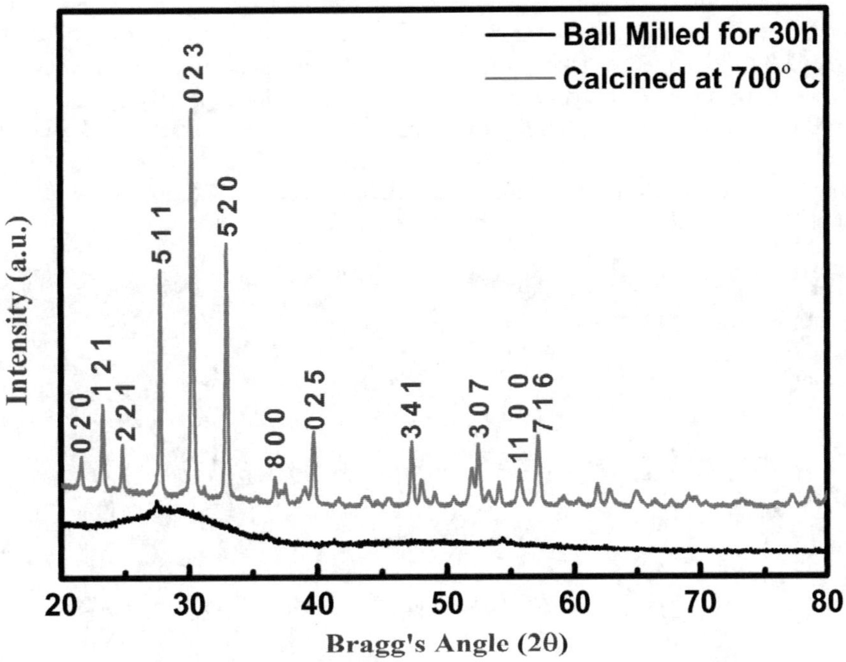

Figure 1. XRD plot of Synthesized nanoceramic compound

Table 1. XRD parameters

Peak No	2 Theta	I	I/I_0	I/I_0	D_{obs}	D_{cal}	h	k	l	Difference	Crystallite size (nm)
1	21.62	2154	17.16198	17	4.1069	4.1055	0	2	0	-.007	31.77
2	23.26	3689	29.39208	29	3.8209	3.8199	1	2	1	-.006	43.13
3	24.78	2495	19.87889	20	3.5898	3.5886	2	2	1	-.009	49.18
4	27.76	7714	61.46124	61	3.2109	3.2108	5	1	1	-.000	46.68
5	30.26	12551	100	100	2.9511	2.9515	0	2	3	.005	37.17
6	32.96	8504	67.75556	68	2.7152	2.7183	5	2	0	.038	36.26
7	39.74	2844	22.65955	23	2.2662	2.2670	8	0	0	.015	32.85
8	41.68	1360	10.83579	11	2.1651	2.1648	0	2	5	-.006	32.59

9	47.26	2555	20.35694	20	1.9217	1.9215	3	4	1	-.004	33.31
10	52.48	2472	19.69564	20	1.7421	1.7426	3	0	7	.014	35.75
11	55.7	1697	13.52083	14	1.6488	1.6487	11	0	0	-.002	24.75
12	57.14	2771	22.07792	22	1.6106	1.6103	7	1	6	-.013	24.93

3.2. Micro structural and elemental study

The SEM image of $Bi(Co_{0.45}Ti_{0.45}Fe_{0.10})O_3$nanoceramic isrepresented in Figure 2, whichdisplays that thesynthesized compound hasaverage grain size of≈ 250nm and were homogeneouslyspread over the surface. Further, the uniformly distributed shape, size and nature of the particleshowsthe presence of polycrystalline nature. Grains are regularly spread on the surface of the pellet and dimensions, spreading trend and cavities in the grains evidently advised higher density of pellet.Figure 2 alsodisplays EDX spectrum (elemental analysis) ofsynthesized nanoceramic.EDX outcomedisplays the occurrence of primarymaterialsand atomic rationearby to stoichiometric concentration of Bi, Co, Ti, Fe and O in the synthesized ceramic compound.

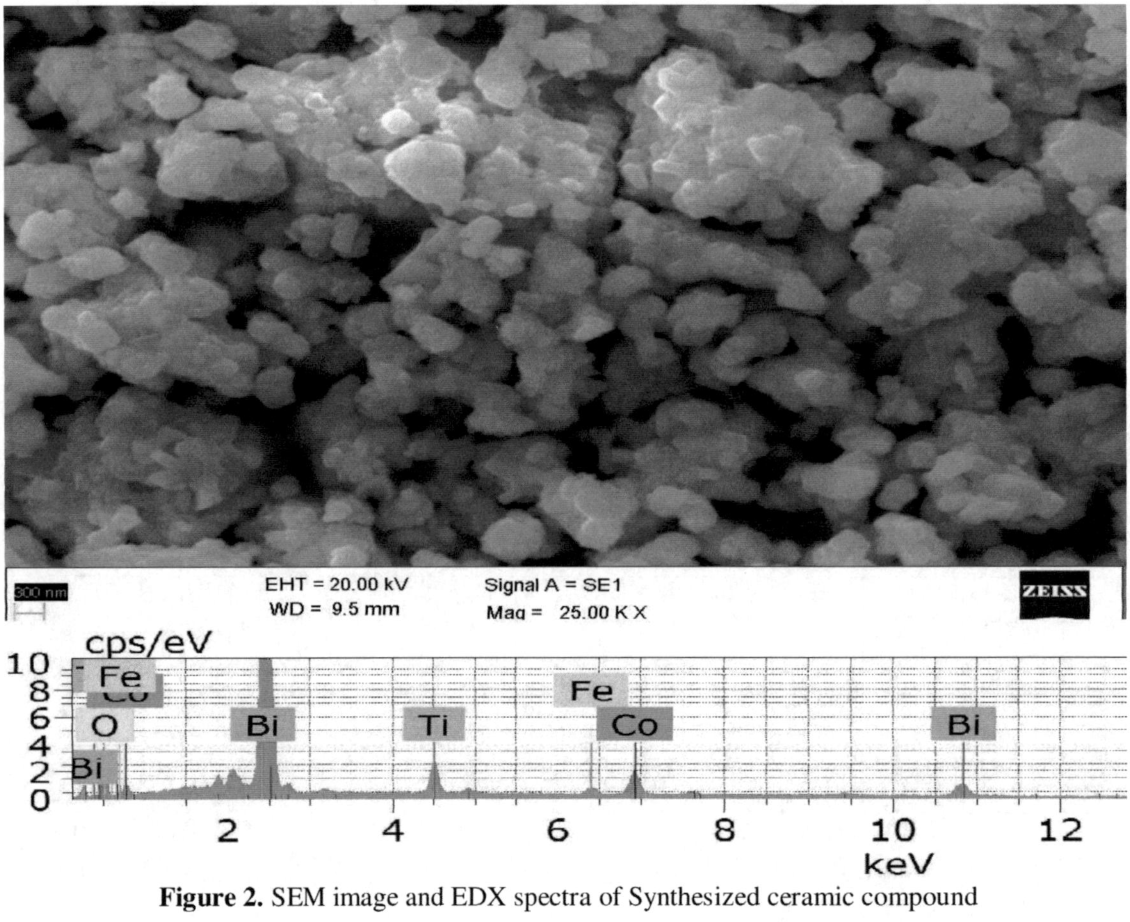

Figure 2. SEM image and EDX spectra of Synthesized ceramic compound

3.3. Temperature and Frequency Dependent Dielectric Study

Figure 3discloses the variationof dielectric parameterswith temperature (298–573 K) at selected frequencies of 1 kHz, 10 kHz, 50 kHz, 200 kHz, 500 kHz and 1 MHz for $Bi(Co_{0.45}Ti_{0.45}Fe_{0.10})O_3$.

Corresponding to the basiccharacteristic of dielectrics, existing analysis discloses the upsurge in dielectric constant dispersion spectrum concurrently with increasingnature at all designated temperature. The dielectric constant outcome decreases with an increment in the frequency. BCTFO [i.e. $Bi(Co_{0.45}Ti_{0.45}Fe_{0.10})O_3$]has high dielectric constant at low frequencies which decreases concurrently in the high-frequency range. The polarization (i.e. ionic, dipole and interfacial polarization)exist in the low-frequency region.An increment in the frequency is accountable for vanishing polarization.in the material. As anoutcome, the value of dielectric constant decreases gradually. Additionally, Figure 4showsdielectric parameter (i.e. tan δ and ε_r) with frequency dependence at selected temperatures. Outcome ofsuch plotsdecreases with increasing the frequency[14][15]. The synthesized compound does not demonstrate any phase transition in the experimental temperature range (298 K − 573 K). Hence, the examination of dielectric properties is the signature of perceiving the structural and ferroelectric phase transition. Also, for BCTFO compound, both dielectric parameters depending upon temperature and frequency are summarized in the Table 2 and Table 3 respectively.

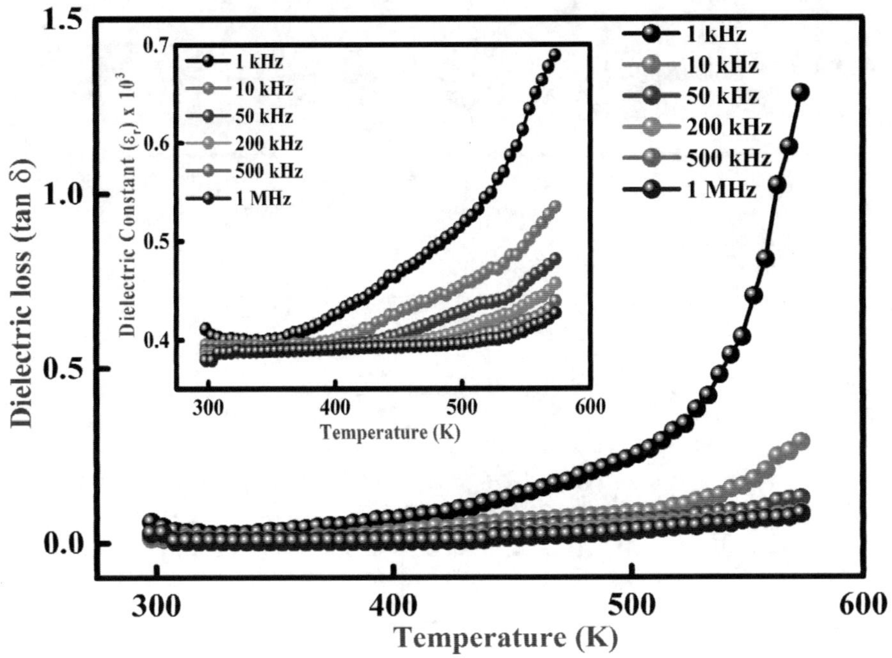

Figure 3. Temperature dependent dielectric plots

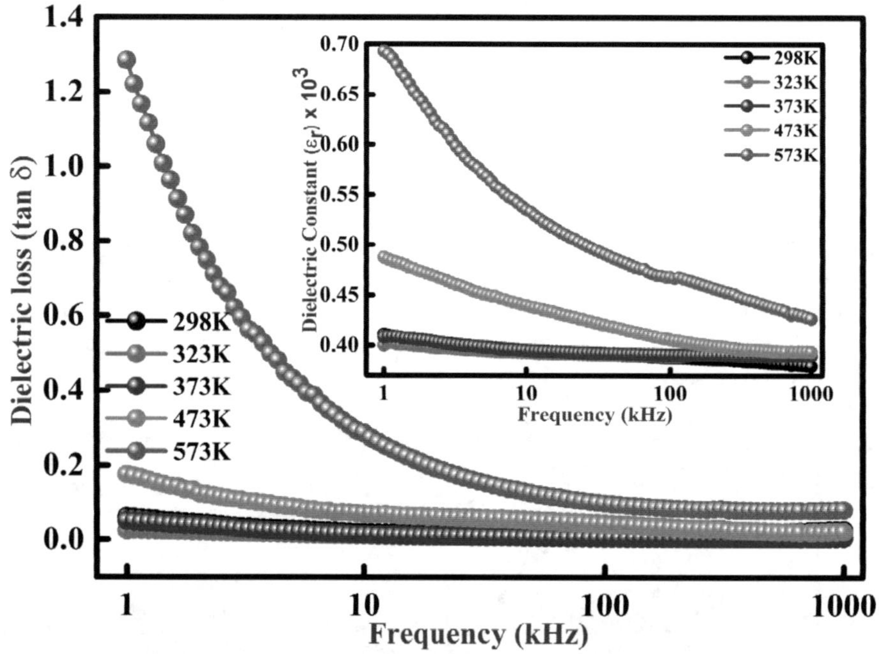

Figure 4. Frequency dependent dielectric plots

Table 2. Temperature dependent dielectric parameters

Milling Time	Temp. (K)	Dielectric Parameters	Frequency					
			1 kHz	10 kHz	50 kHz	200 kHz	500 kHz	1 MHz
30 h	298	ε_r (x10^3)	0.4109	0.3952	0.3899	0.3863	0.3829	0.3795
		tanδ	0.0607	0.0200	0.0134	0.0150	0.0200	0.0271
	573	ε_r (x10^3)	0.6888	0.5350	0.4815	0.4570	0.4389	0.4265
		tanδ	1.2843	0.2862	0.1260	0.0845	0.0789	0.0811

Table 3. Frequency dependent dielectric parameters

Milling Time	Frequency	Dielectric Parameters	Temperature				
			298K	323K	373K	473K	573K
30 h	1 kHz	ε_r (x10^3)	0.4109	0.4001	0.4082	0.4876	0.6930

		tanδ	0.0607	0.0241	0.0500	0.1756	1.2843
	1 MHz	**ε_r (x10^3)**	0.3795	0.3899	0.3899	0.3937	0.4265
		tanδ	0.0271	0.0056	0.0056	0.0223	0.0811

3.4. Electrical impedance spectroscopic study

With the intention of identifying the electrical and physical characteristics of prepared ceramic sample, electrical, impedance spectroscopic studiesare essentially needed. This unique method demonstrates to be advantageousfor establishing a co-relation among electrical and structural properties of multiferroics materials[16][17][18]. Figure 6 portrays the Nyquist plot (Z' v/s Z'', where Z' and Z'' are real and imaginary part respectively) for $Bi(Co_{0.45}Ti_{0.45}Fe_{0.10})O_3$ atselected temperatures.The Complex Impedance Spectroscopy technique (CIS) is a significant nondestructive method to investigate the electrical characteristics of synthesized compound over a widerange of temperature and frequency. Few impedance components (resistive and capacitive parameters) of the compound leads to a sequence of semicircles as soon as portrayed on a Nyquist formulism. Normally, the frequency dependence of the dielectric characteristics of a compound could defined as the subsequent terms:

Complex dielectric constant: $\varepsilon' = \varepsilon' - j\varepsilon''$ (1)

Complex Impedance: $Z^* = Z' - jZ'' = R_s - \frac{j}{\omega C_s}$ (2)

Phase sensitive multi-meter was exploited to collect thecomplex impedance data of the prepared material and further it was analyzed. For data collection i.e., output response (both real and imaginary components), an ac-signal was employedamongthe circular shaped andsilveredpellet.In conformity with thestudy, the semicircles are off-centre, and consequently, replicates a non-Debye type relaxation progressionappearedfor examinedcompound[19][20]. This characteristic property is considered as a result of the contribution of the possible liberationof space charge [21].

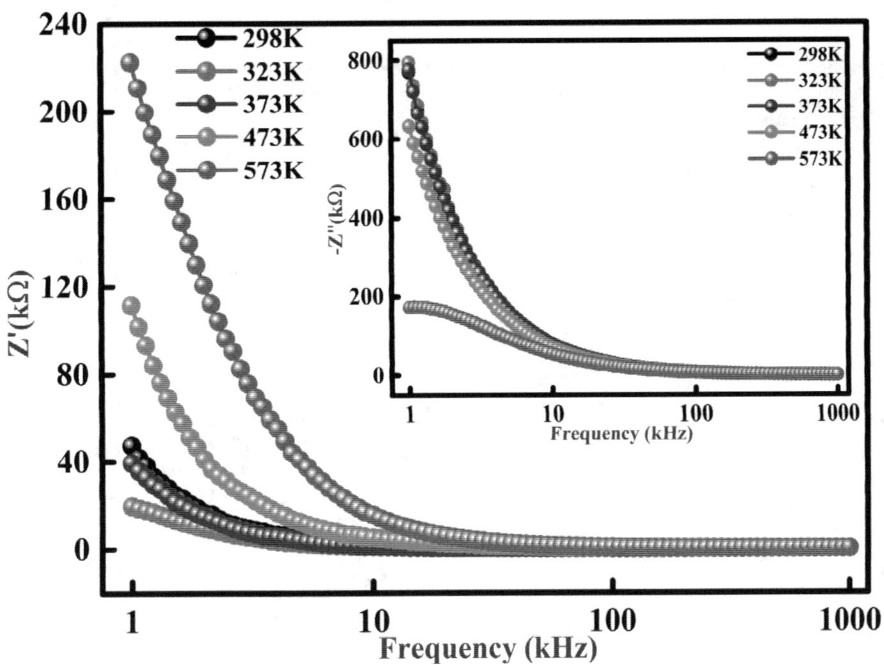

Figure 5. Frequency dependent dielectric plots

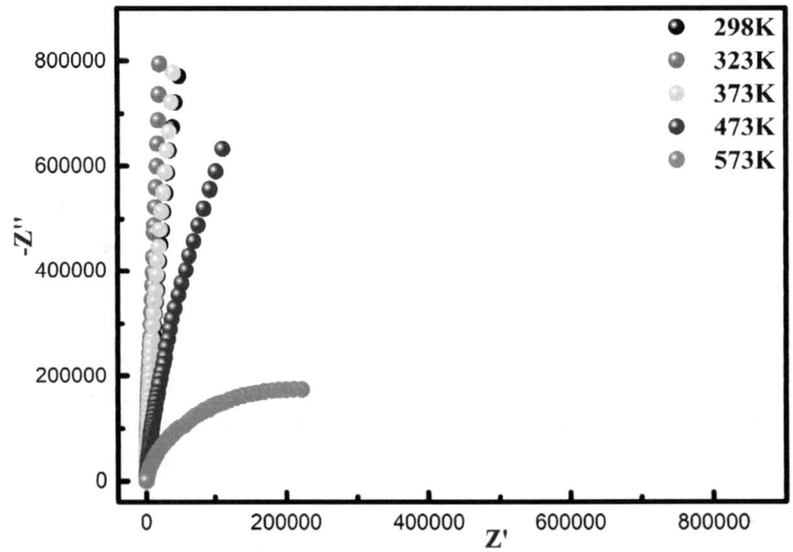

Figure 6. Nyquist plot of synthesized ceramic compound

3.5 Temperature dependent conductivity study

The outcome of ac-conductivity (σ_{ac}) was figured out exploring the dielectric data into an equation$\sigma_{ac}(\omega) = \omega\ \omega_o\ \varepsilon_r\ (\tan\delta)$ (where sign denotes their typical meaning) [22]. Figure 7leads the dissimilarity of ac-conductivity over an extensive range of frequencies (froml kHz to 1MHz) at certain temperature (298 K to 573 K) of the $Bi(Co_{0.45}Ti_{0.45}Fe_{0.10})O_3$[23].

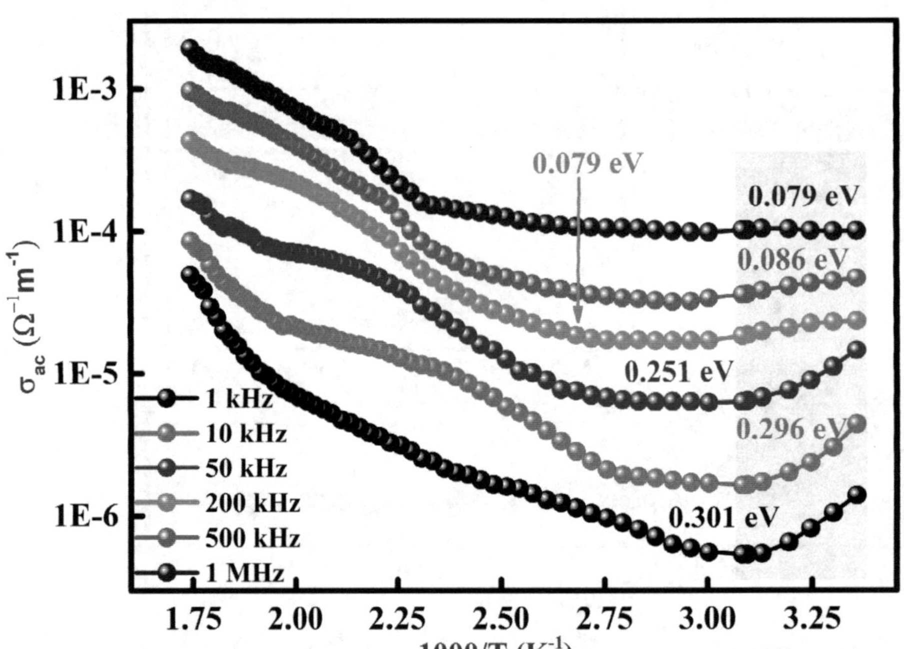

Figure 7. Temperature dependent σ_{ac} (ac-conductivity) plot of synthesized ceramic compound

The resultant of ac conductivity is nearly persistent at particular frequency at lower temperature. These activation energy outcome is the characteristic of oxygen vacancies diffusion [24]. The conductivity graphs can be fitted, using an empirical expression of $\sigma_{ac}(\omega) = \omega\, \sigma_\alpha + P\omega^\beta$, recognised as the universal power law of Jonscher [25]. Herein σ_α signifies the frequency-independent quantity that provides dc-conductivity (σ_{dc}), P is a thermally activated term and β is the frequency-dependent exponent and its value lies within ($0< \beta < 1$). The beginning of frequency dependence conductivity lies within the relaxation phenomena because of mobile charge carriers [26].

3.6 Ferroelectric and Magnetic properties

Figure 8 (inset)depicts P–E loop and M-H loop of BCTFO compound obtained atroom temperature.For P-E loop, applied electric field was 10 kV/cm.The existence of non-saturated polarization might be due to the heavy leakage current, occurringbecause of the inconsistentvalance of Fe(i.e.Fe^{3+} to Fe^{2+})[27][28][29]. In thiscurrent study, we could not performP–E loop experiment at higher electric field because of the conductive property of the synthesized compound. To examine the ferromagnetic characteristic (magnetic field v/s magnetization) of BCTFO sample, M-H loopdata wascollected with anoperatedmagnetic field of ±20 kOe at room temperature by exploiting VSM (Vibrating Sample Magnetometer).Moreover, for the examinedmaterial, the trend of M-H loop depicts linear dependence of magnetic field upon magnetization. However, the existence of high remnant magnetization and un-saturated hysteresis loops proposes the characteristic of high ferromagnetic tendency in BCTFO compound [14].

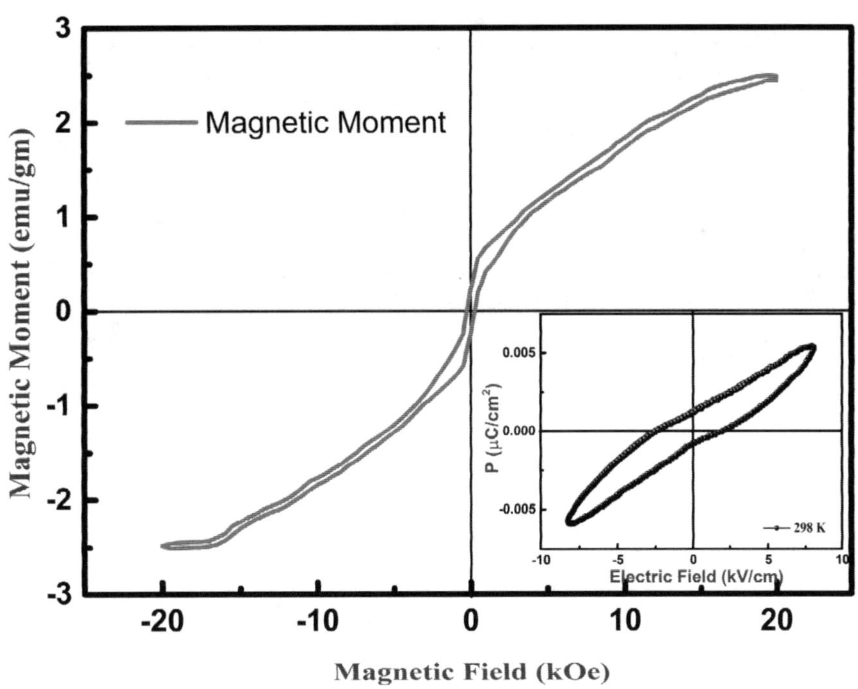

Figure 8. P-E loop and M-H loop of synthesized compound

4. Conclusion

In this work, High-energy ball mill was exploited to synthesize polycrystalline $Bi(Co_{0.45}Ti_{0.45}Fe_{0.10})O_3$ nanoceramic, followed by solid-state ceramic method. XRD technique confirmed the existance of single-phase orthorhombic crystal structure symmetryand EDX results were in well agreement with atomic concentration of the system. The crystallite size (average)was estimated to be 35.70 nm from the major peaks of BCTFO. Structural analysis shows that the structural change was observed with the substitution of Co/Ti. Impedance studies confirmed the presence of dielectric relaxation of non-Debye type. The ac-conductivity plotsconfirmed the Jonscher's universal power law at selected temperatures. Additionally,Nyquistoutcome has depicted the existence of semi-circular arc at higher temperature. Co-substitution of Co/Ti has been found useful, as it reduces the dielectric loss and increase the magnetic property which is useful for possible multifunctional applications of $BiFeO_3$.

Acknowledgement

Author Dr. Alok Shukla gratefully acknowledges the financial support received from SERB-DST, Government of India, New Delhi, in the form of Research Project no. EMR/2015/002420.

Conflict of interest

The authors declared that they have no conflict of interest.

References

[1] Suresh P, Babu P D and Srinath S 2016 Role of (La, Gd) co-doping on the enhanced dielectric and magnetic properties of BiFeO3ceramics *Ceram. Int.***42** 4176–84
[2] Behera C, Das P R and Choudhary R N P 2014 Structural and electrical properties of mechanothermally synthesized NiFe2O4 nanoceramics *J. Electron. Mater.***43** 3539–49
[3] Wu H, Lin Y B, Gong J J, Zhang F, Zeng M, Qin M H, Zhang Z, Ru Q, Liu Z W, Gao X S and Liu J M 2013 Significant enhancements of dielectric and magnetic properties in Bi(Fe1-xMgx)O3-x/2 induced by oxygen vacancies *J. Phys. D. Appl. Phys.***46**
[4] Sheibani S, Ataie A, Heshmati-Manesh S and Khayati G R 2007 Structural evolution in nano-crystalline Cu synthesized by high energy ball milling *Mater. Lett.***61** 3204–7
[5] Eerenstein W, Mathur N D and Scott J F 2006 Multiferroic and magnetoelectric materials *Nature***442** 759–65
[6] Spaldin N A and Fiebig M 2005 The renaissance of magnetoelectric multiferroics *Science (80-.).***309** 391–2
[7] Catalan G and Scott J F 2009 Physics and applications of bismuth ferrite *Adv. Mater.***21** 2463–85
[8] Ederer C and Spaldin N A 2005 Weak ferromagnetism and magnetoelectric coupling in bismuth ferrite *Phys. Rev. B - Condens. Matter Mater. Phys.***71** 1–4
[9] Pandey R, Pradhan L K and Kar M 2018 Structural, magnetic, and electrical properties of (1–x)Bi0.85La0.15FeO3-(x)CoFe2O4 multiferroic composites *J. Phys. Chem. Solids***115** 42–8
[10] Kumar N, Shukla A, Behera C and Choudhary R N P 2016 Structural, electrical and magnetic properties of Bi(Ni0.45Ti0.45Fe0.10)O3 *J. Alloys Compd.***688** 858–69
[11] Wu E 1989 POWD, an interactive program for powder diffraction data interpretation and indexing *J. Appl. Crystallogr.***22** 506–10
[12] Goldschmidt V M 1926 Die Gesetze der Krystallochemie *Naturwissenschaften***14** 477–85
[13] Shukla A, Kumar N, Behera C and Choudhary R N P 2016 Structural and electrical characteristics of (Co, Ti) modified BiFeO3 *J. Mater. Sci. Mater. Electron.***27** 7115–23
[14] Kumar N, Shukla A and Choudhary R N P 2018 Development of lead-free multifunctional materials Bi(Co0.45Ti0.45Fe0.10)O3 *Prog. Nat. Sci. Mater. Int.***28** 308–14

IOP Conf. Series: Materials Science and Engineering **798** (2020) 012033 doi:10.1088/1757-899X/798/1/012033

[15] Kumar N, Shukla A, Kumar N, Hajra S, Sahoo S and Choudhary R N P 2019 Structural, bulk permittivity and impedance spectra of electronic material: Bi(Fe0.5La0.5)O3 *J. Mater. Sci. Mater. Electron.* **30** 1919–26

[16] Gupta P, Padhee R, Mahapatra P K, Choudhary R N P and Das S 2018 Structural and electrical properties of Bi3TiVO9ferroelectric ceramics *J. Alloys Compd.* **731** 1171–80

[17] J. R. Macdonald 1987 *Impedance spectroscopy* (New York: John Wily & Sons)

[18] Liu J, Duan C G, Mei W N, Smith R W and Hardy J R 2005 Dielectric properties and Maxwell-Wagner relaxation of compounds ACu3Ti4O12(A=Ca,Bi2/3,Y2/3,La2/3) *J. Appl. Phys.* **98** 1–6

[19] Kumar N, Shukla A, Kumar N, Choudhary R N P and Kumar A 2018 Structural, electrical, and multiferroic characteristics of lead-free multiferroic: Bi(Co0.5Ti0.5)O3-BiFeO3 solid solution *RSC Adv.* **8** 36939–50

[20] Trivedi P, Katba S, Jethva S, Udeshi M, Vyas B, Vagadia M, Gautam S, Chae K H, Asokan K and Kuberkar D G 2015 Modifications in the electronic structure of Rare-Earth doped BiFeO3 multiferroic *Solid State Commun.* **222** 5–8

[21] Gupta P, Mahapatra P K and Choudhary R N P 2020 Structural, dielectric, impedance, and modulus spectroscopy of BaSnO3-Modified BiFeO3 *J. Phys. Chem. Solids* **137** 109217

[22] Pati B, Choudhary R N P and Das P R 2013 Phase transition and electrical properties of strontium orthovanadate *J. Alloys Compd.* **579** 218–26

[23] Senthilkumar B, Kalai Selvan R, Vinothbabu P, Perelshtein I and Gedanken A 2011 Structural, magnetic, electrical and electrochemical properties of NiFe2O4 synthesized by the molten salt technique *Mater. Chem. Phys.* **130** 285–92

[24] Sivakumar N, Narayanasamy A, Ponpandian N, Greneche J-M, Shinoda K, Jeyadevan B and Tohji K 2006 Effect of mechanical milling on the electrical and magnetic properties of nanostructured Ni 0.5 Zn 0.5 Fe 2 O 4 *J. Phys. D. Appl. Phys.* **39** 4688–94

[25] A.K. Jonscher 1996 *Universal Relaxation Law* ed Chelsea Dielectrics Press (London)

[26] Jonscher A K 1977 The 'universal' dielectric response *Nature* **267** 673–9

[27] Kumar N, Shukla A, Kumar N, Sahoo S, Hajra S and Choudhary R N P 2018 Structural, electrical and ferroelectric characteristics of Bi(Fe0.9La 0.1)O3 *Ceram. Int.* **44** 21330–7

[28] Zhang X, Sui Y, Wang X, Wang Y and Wang Z 2010 Effect of Eu substitution on the crystal structure and multiferroic properties of BiFeO3 *J. Alloys Compd.* **507** 157–61

[29] Lazenka V V., Lorenz M, Modarresi H, Brachwitz K, Schwinkendorf P, Böntgen T, Vanacken J, Ziese M, Grundmann M and Moshchalkov V V. 2013 Effect of rare-earth ion doping on the multiferroic properties of BiFeO3 thin films grown epitaxially on SrTiO3 (1 0 0) *J. Phys. D. Appl. Phys.* **46** 175006

Fabrication and Characterization of Polypyrrole Nanoparticles for Supercapacitor Electrode Materials

Rama Devi, Khilawan Patel, Vinayak Sahu, Komal Kashyap, Kavita Tapadia, Tungabidya Maharana*

Department of Chemistry, National Institute of Technology Raipur, 492010, India

*Corresponding author's e-mail address: tmaharana.chy@nitrr.ac.in, ramachandrakar29@gmail.com

Abstract: In the present work Polypyrrole nanoparticles are synthesized by oxidative polymerization method in the presence of aqueous nitric acid(HNO_3) as solventandferric chloride ($FeCl_3$) as oxidizing agent. During polymerization process the solvent and oxidizing agents played a crucial role in modifying the nanostructures of polypyrrole (PPy). Scanning electron microscopy (SEM) revealed theuniform globule morphology and narrow size distribution of produced PPy particles in acid medium. Combined effect of solvent and oxidant are advantageous for the production of PPy nanoparticles (PPyNPs) with a yield of 90%.Four point probe method and strong UV-Visible band confirmhigh electrical conductivity of PPy due to a large π-conjugation in the polymer structure. The nature and composition of PPyNPs were confirmed by X-ray diffraction (XRD) spectrometer and Fourier transform infrared spectroscopy (FTIR). Electrochemical properties were confirmed by cyclic voltammetry (CV) and galavanostatic charge discharge (GCD) spectrum.CV and GCD spectra illustrate the excellent electrochemical properties of PPyNPs which show higher specific capacitance of 398F g^{-1} at 100mAg^{-1} current density.It has wide applications in electronic and electrochromic devices, sensors, membrane separation,corrosion protection, chromatographic stationary phase, rechargeable batteries, drug delivery,supercapacitors,light-weight batteries and microwave shielding etc.

Keywords: Polypyrrole; Cyclic voltammetry; Galvanostaticcharge discharge; FTIR spectroscopy; Nanoparticles.

1. Introduction

In recent years, conducting polymers a specific class of polymers, viz. polyaniline, polypyrrole, polythiophene, are an attractive class of versatile materials. Among this conducting polymers, polypyrrole (PPy) is a promising polymer for many applications [1] such as supercapacitors[2], solar cells [3],biosensor [4], electrode batter [5], gas sensors [6],micro actuators[7], electrochromic devices [8], electromagnetic shielding devices [9], anticorrosion coatings [10], light emitting diodes (LED) [11], etc. because of their higher electrical conductivity, high charge densities, high doping-rate during charge-discharge process, color changes associated with their oxidation state, variable electro-magnetic parameters and lower cost(compared to noble metal oxides), facile synthesis through electrochemical and

chemical processing, good environmental stability and tunable electrical and optical properties,than many other conducting polymers [12-15].
Polypyrrole nanoparticles have been manufactured by electrochemical and chemical methods in the presence of numerous solvents such aswater, chloroform, methanol, ethanol, aqueous HNO_3,HCl, acetonitrile, tetra hydro furan, benzene and ethylene glycol with numerous oxidizing agents like ($FeCl_3$), hydrogen peroxide, ammonium persulfate (APS),ferric nitrate,etc.[16-18] . However, synthesis of polymers by electrochemical process is a complex process that affects the yield of PPy through numerous factors such as the concentration and existence of monomer-electrolyte, solvent, pH, temperature, electrode, cell conditions andapplied potential. Therefore, optimizing all these parameters in one experiment is difficult, but chemical polymerization method does not require any specialized instrument and is applicable for producing these polymers in large quantities. It is rather a fast process for high yield synthesis of polypyrrole [19-20]. Recently, nano sized conductive polymers such as nanorods, nanotubes, nanofibers and nanowiresare beingpaid attentiondue to their high surface area and small size, leading to superior application for optoelectronic nano-devices [21].
Thus, in the present investigation, a systematic study has been carried out toobserve the structure morphology, conductivity and electrochemical properties of polypyrrole thin films synthesized by using the chemical oxidative polymerization method.

2. Experimental

2.1. Materials

Pyrrole monomer, Sodium dodecyl sulfate (SDS), N-methyl-2-pyrrolidone (NMP) and poly tetrafluroethylene (PTFE)were purchased from Sigma Aldrich. The commercially available pyrrole was purified by a simple distillation process to obtain a colorless pure liquid. Ferric chloride ($FeCl_3$) purchased from Loba chemic, Mumbai, India and Nitric was purchased from and Merck limited, Mumbai, India were used as received.

2.2. Synthesis of PPy nanoparticles

A basic and widely used technique for the synthesis of polypyrrole nanoparticles is as follows: first, the mixture of pyrrole monomer (2.5mM) and SDS (0.412 g) was added to a 75mL HNO_3 (0.35 M) aqueous solution in an ice bath at 0°C. The oxidant solution was prepared by mixing $FeCl_3$ (5 mM) in 25 ml aqueous HNO_3 (0.35 M) and subsequently added drop wise to the above solution kept in the ice bath.The reaction mixture was placed in the ice bath maintained at 0°C for 24 h with continuous stirring. Eventually, the PPy nanoparticles (PPyNPs) produced were isolated by filtration and washed away with excess of de-ionized water. The reaction mixture was centrifuged four or five times after the reaction until the solution was colorless.

2.3. Fabrication of PPyNPs enrobed glassy carbon electrode

A GCE was polished to a mirror with 0.3 and 0.05 μm polishing paper and alumina pastes before modifica tion. The electrode was then ultrasonically rinsed in ethanol and sequentially deionized water. The electrode materials were prepared by mixing 85% of the PPyNPs, 10% of activated charcoal and 5% of poly tetrafluroethylene (PTFE) as binder in few drops of NMP then the obtain slurry (5μL) were lowered with a microsyringe on the surface of glassy carbon electrode (GCE) and allowed to dried at 70°C in the oven for 20 minutes. The electrode was placed closed firmly over a small bottle during the process, thus gradually evaporating water and producing more consistent films. The resulting electrodes have been cleaned with water and prepared for further tested at room temperature. The cyclic voltameter and galavanostatic charge discharge characterization of the electrode was carried out in the presence of 1.0 M sulfuric acid electrolyte.

2.4. Characterization

The structure and morphology of PPyNPs was analyzed by scanning electron microscope (ZEISS EVO-50 SEM machine) operating at an accelerating voltage of 20kV. Fourier transform infrared spectra of prepared sample were taken by an Agilent Technologies Cary 630 FTIR spectrophotometer. XRD data of PPy powders were obtained from X-ray diffractometer (PANalytical XPERT –PRO).The conductivity of the PPyNPs were measured by four-probe conductivity meter(Model: CCS-01 programmable current source and a DMV-001microvotameter supplied by SES Instrument Pvt. Ltd. India).

Electrochemical measurements were performed on MetrohmAutolab (MULTI AUTOLAB M204) using three electrode assemblies consisting of a platinum plate being used as counter electrodes, a standard calomel electrode (SCE) being used as a reference electrode and PPyNPs enrobed glassy carbon electrode (PPy-GCE) being used as a working electrode. The electrochemical output of the working electrode was evaluated by using CV and GCD tests performed between 0.0V to 0.8V in 1M H_2SO_4 electrolytic solution.

3.Result and Discussions

3.1. Morphological analysis

The scanning electron micrographs (Fig.1 (a) & (b)) of the PPyNPsshow their globular structure.It Can be clearly seen form the SEM micrographs that the spherical PPy nanoparticles are of <150 nm size and are almost uniform in structure and distribution [22].

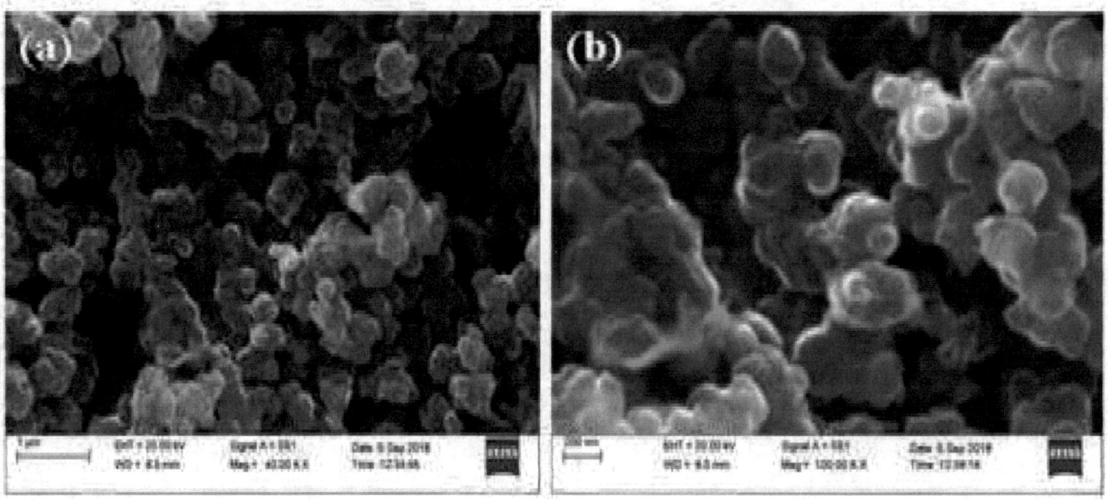

Figure 1. (a) & (b) SEM images of PPy nanoparticles at different magnification

3.2. FTIR analysis

The FTIR spectrum of PPy nanoparticles (Fig. 2) showed charge carries bipolaron bands at 928 and 1210 cm^{-1} indicating that the PPy formed is in the oxidized state [23]. The strong peaks between1500 to 1400cm^{-1} are assigned to fundamental vibration of pyrrole ring [24-25]. The peak 3445cm^{-1} is assigned to N-H stretching vibration from pyrrole [26].

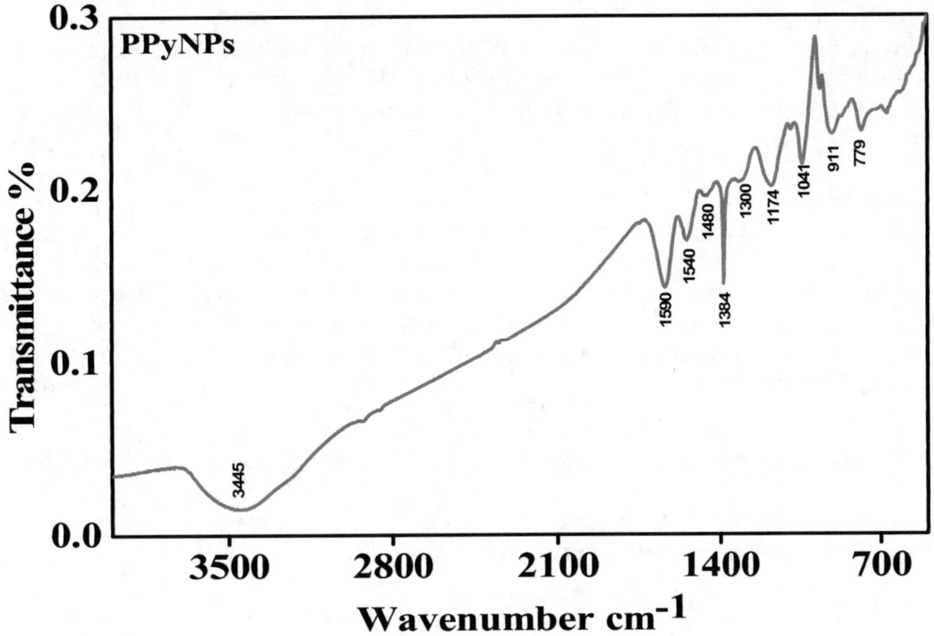

Figure 2. FTIR images of PPy nanoparticles.

3.3. XRD analysis

XRD diffractogram (Fig.3) of the PPy nanoparticles showed a broad peak centered at 2θvalue around 26 confirming amorphous nature of PPyNPs[25-26].

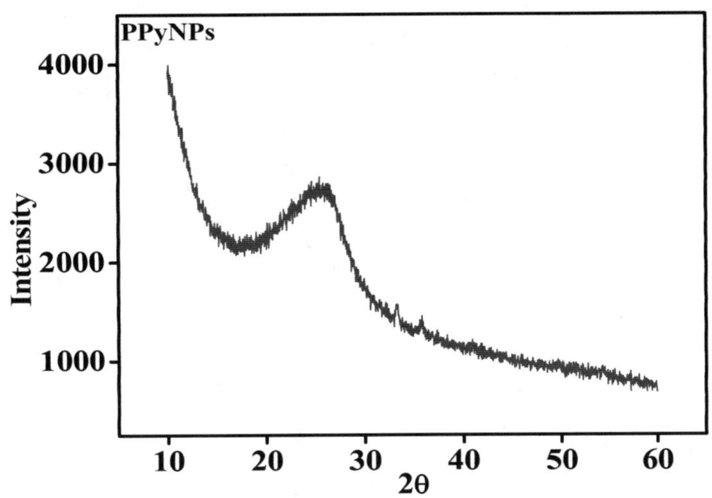

Figure 3. XRD spectra of PPy nanoparticles.

3.4. Conductivity studies

The conductivity of PPy nanoparticles as high as 39 S/cm has been obtained which is due to the formation of nano sized particles because of the presence of sodium dodecyl sulfate (SDS) which act as an anionic surfactant. In the present study, the conductivity of PPyNPs is much higher than that of other reported method [27].

3.5. Electrochemical studies

The electrochemical properties of the PPy nanoparticles were examined by cyclic voltammetry and galvanostatic charge discharge spectroscopy. The CV spectra of PPy-GC electrode at different scan rates are shown in Fig.4a. With the increase in the scan rates, current density was also increased. The shape of the curve is elliptical, which shows itsinvolvementof high current.Fig. 4(b) displays the GCD curves of the PPy enrobed GC electrode.The specific capacitance of PPy-GCelectrode was measured by galvanostatic charge discharge spectra using the equation (1)

$$Cs = \frac{I \times \Delta t}{m \times \Delta v} (1)$$

Where Δtrepresents discharge time, ΔV *is* discharge voltage, I represent current and *m* is mass of electrode material.

The specific capacitance of PPy-GC electrode at 100mAg^{-1} current density is about 398F g^{-1}which confirms the good electrochemical performance of PPy nanoparticles.

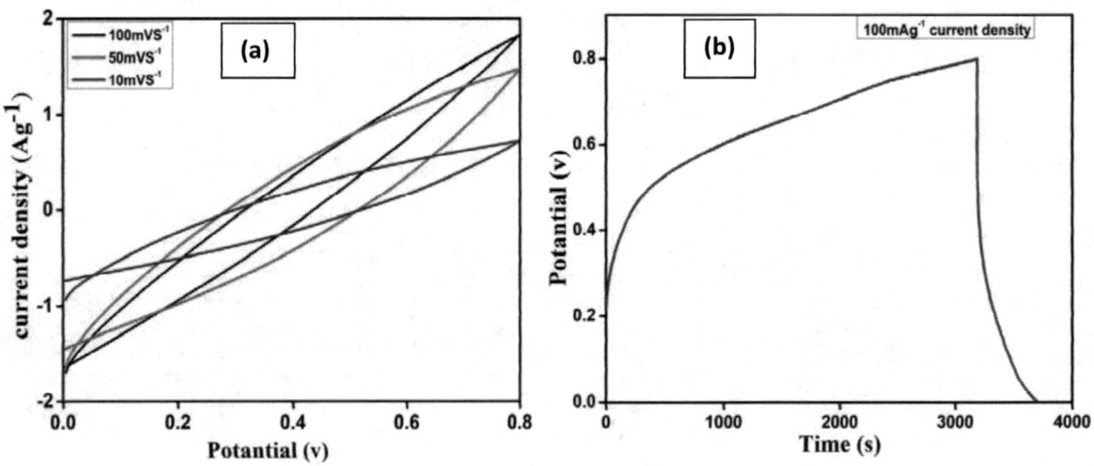

Figure 4. (a) Cyclic voltammetry (CV) spectra of PPy nanoparticles with different scan rate; (b) Galvanostatic charge discharge (GCD) spectra of PPy nanoparticles at 100mAg^{-1} current density in 1M H_2SO_4 electrolyte.

4. Conclusions

The polypyrrole was synthesized using oxidative chemical polymerizationmethod in the presence of $FeCl_3$as oxidizing agent. The PPyfilms had a uniform spherical morphology which was confirmed from SEM. The composition and amorphous nature of the PPy nanoparticles are confirmed by FTIR and XRD techniques. CV and GCD spectra illustrate the excellent electrochemical properties of PPy nanoparticles which show higher specific capacitance of 398F g^{-1} at 100mAg^{-1} current density. The proposed work has been applied successfully in electronic and electrochemical devices because of higher electrical conductivity of PPyNPs.

Acknowledgement
The authors are extremely thankful to National Institute of Technology, Raipur for providing MHRD scholarship and necessary laboratory facilities.

References

[1] DengJ,WangX, GuoJ,and Liu P,2014*Ind. Eng. Chem. Res*, **53**3513680-13689.
[2] Yang L, Shi M, Jiang J, Liu Y, Yan C, Liu H, and Guo Z, 2019Mater. Lett,**244** 27-30.
[3] Wang G, Dong W, Yan C, Hou S, and Zhang W, 2018Mater. Lett,**214** 158-161.
[4] Apetrei RM, Cârâc G, Bahrim G, and Camurlu P, 2019Int. J. Polym. Mater,**68**17 1058-1067.
[5] Liu Q, Luo Y, Chen W, Yan Y, Xue L, and Zhang W, 2018, Chem. Eng, **347**, 455-461.
[6] Liu B, Liu X, Yuan Z, Jiang Y, Su Y, Ma J, and Tai H,2019, Actuat. B-Chem ,**295** 86-92.
[7] Zondaka Z, Harjo M, Khorram MS, Rasti P, Tamm T, and Kiefer R, 2018, React. Funct. Polym, **131**, 414-419.
[8] Dulgerbaki C, Maslakci NN, Komur AI, and Oksuz, AU, 2018, Eur. Polym. J, **107**, 173-180.
[9] Sambyal P, Dhawan SK, Gairola P, Chauhan SS, and Gairola SP, 2018, Curr. Appl. Phys, **18**5 611-618.
[10] Hao L, Zhu K, Lv G, and Yu D, 2019, Prog. Org. Coat, **136**, 105251.
[11] KandulnaR, Choudhary RB, and Singh R, 2019, J. Inorg. Organomet, **29**3 730-744.
[12] Kim S, Jang LK, Park, HS, and Lee JY, 2016, Sci. Rep. 6 30475.
[13] Zhang L, Du W, Nautiyal A, Liu Z, and Zhang X, 2018,Science China- Materials, **61**3 303-352.
[14] Zang L, Liu Q, Yang, C, Chen J, Qiu J, and Song G, 2018,Polym. Bull, **75**8 3311-3323.
[15] Shen C, Sun Y, Yao W, and Lu Y, 2014, J. Polym, **55**12 2817-2824.
[16] Yen SJ, Chen EC, Chiang RK, and Wu TM, 2008,J. Polym. Sci. Pol. Phys, **46**13 1291-1300.
[17] Zang J, Li CM, Bao SJ, Cui X, Bao Q, and Sun CQ, 2008, **41**19 7053-7057.
[18] Rahaman M, Aldalbahi A, Almoiqli M, and Alzahly S, 2018, j. Polym, **10**6 632.
[19] Kłucińska K,Jaworska E, Gryczan P,Maksymiuk K, and Michalska A, 2015,**51**63 12645-12648.
[20] Effati E,Pourabbas B, and Zakerhamidi MS, 2019, RSC. Adv, **9**30 16977-16988.
[21] Ghosh S,Maiyalagan T, and Basu RN, 2016, **8**13 6921-6947.
[22] Baig U, Gondal MA, Ilyas AM, and Sanagi MM, 2017,J. mater. Sci. Technol, **33**6 547-557.
[23] Waghuley SA, Yenorkar SM,Yawale SS, and Yawale SP, 2008, Sens. Actuator. B- Chem, **128**2 366-373.
[24] Cherusseri J, and KarKK,2016,Chemical Physics, **18**12 8587-8597.
[25] SinguBS, and Yoon KR,2018,J. Alloy. Compd, **742** 610-618.
[26] Chougule MA, Pawar SG, Godse PR,Mulik RN, Sen S, and Patil VB, 2011, Soft Nanoscience Letters, **1**01 6.
[27] Santim RH, de Aquino HA, and Malmonge JA, 2016,Mater. Sci. Forum, **869** 391-395.

IOP Conf. Series: Materials Science and Engineering **798** (2020) 012035 doi:10.1088/1757-899X/798/1/012035

Quantum Hall Effect in Trilayer Graphene

J Brahma[1] and S Sahoo[2]
Department of Physics, National Institute of Technology Durgapur, Durgapur – 713209, West Bengal, India.
[1]E-mail: brahmajitu@gmail.com
[2]E-mail:sukadevsahoo@yahoo.com

Abstract: Graphene is a two-dimensional sheet of carbon atoms with properties that are superior to other materials. Due to these superior properties, a lot of research is being done with this material. One of those properties is the Quantum Hall Effect (QHE). Monolayer graphene shows very interesting behaviour in the presence of a high magneticfield (up to 30T) and at very low temperature (~0.2K). Its unconventional Landau Level (LL) spectrum of massless Diracfermions leads to a new type of quantum Hall effect, known as half-integer quantum Hall effect or anomalous QHE.We expect uniqueresults can also be seen in trilayer graphene (TLG).In this work, we study the integer quantum Hall effect in trilayer graphene theoretically.

1. Introduction

In recent years, graphene [1,2] has attracted many researchers due to its unique electronic properties. Thediscovery of quantum Hall effect in graphene [3,4] has opened up a large research field. In case of two-dimensional system electrons obey the Schrödinger equation and the Hall conductivity is quantized at integer multiples of e^2/h. However, in case of graphene it is found that the electrons behave as massless Dirac fermions and the Hall conductivity is quantized at integer multiple of e^2/h but with a shift of $1/2$ which comes from the Berry phase $\Phi_B = \pi$ due to the cyclotron motion of the massless Dirac particles [3-6]. This unconventional quantum Hall effect is termed as half-integer quantum Hall effect. Generally, the electronic structure of multilayer graphene depends on the number of layers. Due to the unusual electronic structure, trilayer graphene [7-13] has attracted the scientific community recently.The quantum Hall effect is also found in trilayer graphene. In this paper, we study how the Berry phase affects the Hall conductivity and calculate the Hall conductivity for trilayer graphene.

2. Tight binding model of trilayer graphene

A trilayer graphene is composed of three single layer graphene, each sublattice in a top layer is located directly above the same one in the bottom layer. The low energy Hamiltonian for a trilayer graphene around the ***K*** point can be written as [7]

$$\boldsymbol{H}_{\mathrm{p}}=\begin{pmatrix} 0 & \upsilon\pi^\dagger & t_\perp & 0 & 0 & 0 \\ \upsilon\pi & 0 & 0 & t_\perp & 0 & 0 \\ t_\perp & 0 & 0 & \upsilon\pi^\dagger & t_\perp & 0 \\ 0 & t_\perp & \upsilon\pi & 0 & 0 & t_\perp \\ 0 & 0 & t_\perp & 0 & 0 & \upsilon\pi^\dagger \\ 0 & 0 & 0 & t_\perp & \upsilon\pi & 0 \end{pmatrix}, \qquad (1)$$

where $t_\perp$ is the nearest interlayer neighbour π-orbital hopping amplitude. Then we can write the difference equations as

$$\varepsilon a_n = t_\perp(a_{n-1} + a_{n+1}) + \upsilon\pi^\dagger b_n, \qquad (2)$$

$$\varepsilon b_n = t_\perp(b_{n-1} + b_{n+1}) + \upsilon\pi a_n, \qquad (3)$$

where$(a_1, b_1, \cdots, a_N, b_N)$ are eigenvectors, $\varepsilon$is the eigenvalue and the boundary condition $a_0 = a_{N+1} = b_0 = b_{N+1} = 0$.

Let $c_n = a_n + b_n e^{-i\varphi}$ and $d_n = a_n - b_n e^{-i\varphi}$where $\varphi = \tan^{-1}(p_y/p_x)$, then

$(\varepsilon - \upsilon|p|)c_n = t_\perp(c_{n-1} + c_{n+1})$, (4)

$(\varepsilon + \upsilon|p|)d_n = t_\perp(d_{n-1} + d_{n+1})$, (5)

with the same boundary condition $c_0 = c_N + 1 = d_0 = d_{N+1} = 0$. The energy spectrum can be written as

$\varepsilon^{\pm}_{r.p} = \pm\upsilon|p| + 2\,t_\perp \cos\left(\frac{r\pi}{N+1}\right)$,(6)

where $r = 1, 2, \cdots, N$.

3. Berry phase intrilayer graphene

Berry phase is a geometrical phase in which the value of the phase depends upon the space and trajectory of the system and not on thesize andshape of the electron orbit.Due to graphene's unusual property, it shows a unique energy spectrum. In the spectrum we have observed that there is no gap between the valence and the conduction band at the ***K*** points in the Brillouin zone. This kind of spectrum is due to a nonzero value of the Berry phase.This nonzero phase occurs due to the trajectory of electrons in Brillouin zone where the system is degenerate [14].

In quantum Hall effect of graphene [3-6], the Hall conductivity $\sigma_{xy} = \pm(ge^2/h)n$ shows plateau at certain values of *n*. Here $g = 4$ is due to spin and valley degeneracy and *n* is Landau level (LL) index. For a nonzero Berry phase there is shift of the Landau levels which leads to the shift of $1/2$ in the Hall conductivity of graphene

$\sigma_{xy} = \pm\left(\frac{4e^2}{h}\right)\left(n + \frac{1}{2}\right)$. (7)

If anelectron orbits a closed contour in the momentum space, a Berry phase shift $\Phi_B = J\pi$is gained [15-17].For trilayer graphene, we have $J = 3$ and the Berry phase is $\Phi_B = 3\pi$. Thus the expression for Hall conductivity of trilayer graphene [8, 18]can be written as

$\sigma_{xy} = \pm\left(\frac{4e^2}{h}\right)\left(n + \frac{3}{2}\right)$,(8)

where $n = 0,1,2,3,\ldots$Here the $\pm$ value corresponds to electrons and holes respectively.

4. Results and discussion

The unique band structure of graphene leads to a nonzero Berry phase $\Phi_B = J\pi$. For monolayer we have $J = 1, \Phi_B = \pi$, for bilayer $J = 2, \Phi_B = 2\pi$ and for trilayer $J = 3, \Phi_B = 3\pi$. The Hall conductivity of graphene depends on the number of layers. We have calculated the Hall conductivity of trilayer graphene for different filling factors. Wehave found that the quantum Hall conductivity of trilayer graphene is quantized atfilling factors$\nu = \pm 6, \pm 10, \pm 14, \pm 18, \ldots$ which is similar to monolayer graphene but the plateaus at $\nu = \pm 2$ is missing. The Hall conductivity of trilayer graphene is represented in Table 1 and its variation with filling factor is shown in Figure 1. We hope these results will open the new routes for studying both integer quantum Hall effect and fractional quantum Hall effect in trilayer graphene in the near future.

Acknowledgement
SSahoo acknowledges Council of Scientific and Industrial Research (CSIR), Govt. of India for providing financial assistance through the sanction No. 03(1473)/19/EMR-II Dated 05/08/2019 for this work.

Table 1. Calculation of quantized Hall conductivity of trilayer graphene

n	$\boldsymbol{\nu} = \pm 4\left(\boldsymbol{n} + \frac{3}{2}\right)$	$\boldsymbol{\sigma_{xy}} = \boldsymbol{\nu}\left(\frac{\boldsymbol{e^2}}{\boldsymbol{h}}\right)(\boldsymbol{\Omega^{-1}})$
0	6	0.0002323
1	10	0.0003873
2	14	0.0005422
3	18	0.0006971
4	22	0.0008521
5	26	0.0010069
6	30	0.0011619

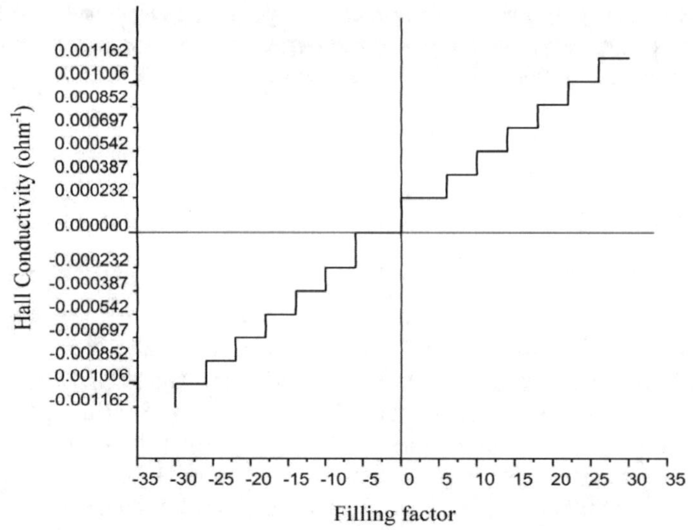

Figure 1. Variation of quantized Hall conductivity of trilayer graphene with different filling factors.

References

[1]Novoselov K S*et al*2004 *Science***306** 666
[2] Castro Neto A H*et al* 2009*Rev. Mod. Phys.* **81** 109
[3] Novoselov K S *et al*2005*Nature* **438** 197
[4] Zhang Y*et al* 2005*Nature* **438** 201
[5] Sahoo S 2011 *Indian J. Pure & Appl. Phys.***49** 367
[6] Cea P 2012 *Mod. Phys. Lett. B* **26** 1250084
[7] Min H andMacDonald A H 2008 *Progress of Theoretical Physics Supplement***176** 227
[8] Mohammadi Y 2019 arXiv: 1912.00823 [cond-mat.mes-hall]
[9] Mohammadi Y, Moradian R andTabar F S 2014 *Solid State Commun***193** 1
[10] Henriksen E A, Nandi D andEinstein J P 2012 *Phys. Rev. X***2** 011004
[11] Min H and MacDonald A H 2008*Phys. Rev. B.***77** 155416
[12] Kumar A*et al* 2011 *Phys. Rev. Lett.* **107** 126806
[13] Cobaleda C et al 2013 *J. Phys.: Conf. Ser.* **456** 012006
[14] Mikitik G P andSharlai Yu V2009 *Low Temp. Phys.***33** 439
[15] MikitikGP andSharlaiYu V 2008 *FizikaNizkikhTemperatur***34** (10) 1012
[16] Jhang S H*et al* 2011 *Phys. Rev. B.***84** 161408 (R)
[17] Novoselov K S*et al* 2006 *Nature Physics***2**177
[18] Yuan S, Roldan R andKatsnelsonM I 2011 *Phys. Rev. B.***84** 125455

Parametric uncertainty quantification in natural frequency of sandwich plates using polynomial neural network

P. K. Karsh[1]*, H. P. Raturi[2], R. R. Kumar[3], S. Dey[2]

[1]Department of Mechanical Engineering, Parul Institute of Engineering & Technology, Parul University, Vadodara, India
[2]Department of Mechanical Engineering, National Institute of Technology Silchar, India
[3]School of Engineering, Jawaharlal Nehru University, New Delhi

* Corresponding author's e-mail: pradeepkumar.karsh270071@paruluniversity.ac.in

Abstract. In this paper, uncertainty quantification in natural frequencies for laminated soft core sandwich plates is presented by employing finite element (FE) coupled polynomial neural network (PNN) approach. The computational efficiency and accuracy is achieved by using PNN as surrogate model. Latin hypercube sampling method is employed for training of data in PNN model. The stochastic first three natural frequencies of sandwich plates are studied for individual variation in input parameters. The stochasticity in individual input parameters are considered in order to assess their influence on global response of the structure. The algorithm discussed in this article is observed to be converging with the previously published literature (for deterministic case) and validated with full scale Monte Carlo simulation (MCS) i.e. original finite element approach (for stochastic case). The computational time and cost reduced significantly by employing the present surrogate based FE approach compared to that of conventional Monte Carlo simulation approach.

Keywords: Sandwich plate, Finite element, Monte Carlo simulation, Polynomial neural network, Uncertainty quantification.

1. Introduction

In present scenario, speed and efficiency are the most important parameter for any system (either machine or structure). It results in continuously increasing demand of better material. Scientists and researchers are continuously trying for a better and efficient material according to its applications such as aerospace, automobile, submarines and other task defined functions. Therefore, the material developed needs to be highly task specific and withstand in any service condition. That is why lightweight robust structures are highly preferred, predominantly in structural applications where high amount of vibration is suspected to produce due to dynamic loads. Sandwich plates are one of them which is widely used in design and construction of aerospace structures. A sandwich plate consists of face-sheets (both upper and lower) separated by core in between them [1]. Face-sheets consist of laminate of different orientation stacked one over another to get specific property. All the laminates and core are bonded together to serve the purpose. It also shows superior properties like weight sensitiveness, good resistance for environmental degradation, economic, ecofriendly. All these merits make sandwich plate suitable for enormous structural application such as aircrafts, submarine, auto-industries, civil construction, orthopedic devices [2-3]. Sandwich plates are more widely used in cost sensitive than that of weight sensitive applications with respect to monolithic plates. The cost effective sandwich plates requires sandwich core of low cost material which exhibits better weight sensitiveness as well. The development and automation in production processes make possible the production of low cost sandwich plates. In general, a sandwich plate is made up of facesheet at outer layers and the core inserted in between them wherein the facesheets are fabricated with stiff and strong materials but of relatively lower thickness than that of light and thicker core. It is not necessary for the core as well as facesheet material of a sandwich plate to be homogeneous. In a sandwich plate, a core is having less strength and more energy absorption capability while face sheet is having more strength and ductility. Recently, interest in the sandwich plate was observed to be considered on the core of the lattice structure. Stretching causes the deformation in members of a truss, presented in lattice structured core [4]. Due to its inherent material anisotropy and manufacturing inaccuracies, such structures are always subjected to spatial variabilities. In real-life applications, it is difficult to get the accurate design specifications of complex structures like sandwich plate with laminated facesheet because of the implicit anisotropy and complexity during manufacturing process leading to inexorable uncertainties. Due to material and geometric uncertainties, overall performance of sandwich plates significantly deviate from deterministic results. Here, it is aimed to address the effect of source-uncertainty on the free vibration behaviour of sandwich plates. So, it is required to understand the fundamental principles of stochastic mechanics. The variability in material and geometric properties affect the stiffness of the plate leading to deviation in the performance from deterministic to probabilistic [5]. Thus, it is imperative to consider such variations during design and analysis to ensure better safety, reliability and serviceability of the structures. The bodies having finite mass and elasticity is prone to vibration. The oscillatory motion of the system about an equilibrium point is named as vibration. Vibration can be categorised into linear and nonlinear, damped and undamped, forced and free vibration. When the mechanical system is subjected to external force, it is named as forced vibration, while in case of free vibration only initial displacement is given to the system. In free vibration, no external force is applied to the system after the initial disturbance. The resonance is the phenomenon in which the natural frequency of the system coincides with the forcing frequency leading to catastrophic failure of the structure. Hence, it is required to avoid the specified frequency range for the safe design of the mechanical system and to know that range it is necessary to determine the natural frequencies of the system. The accurate material and geometrical properties are very difficult to obtain due to the unavoidable source-uncertainties. Hence, it becomes mandatory to consider the variabilities in material and geometric properties for safe design of the structure [6-7]. Stochastic analysis in engineering fields attained global attentions for the researchers. In stochastic analysis, the main focus is drawn to determine the variability in responses due to variabilities in material properties, geometric properties, operational conditions and environment variations. Uncertainties or variabilities are associated with the every phase of the modelling and simulation. There are two types of uncertainty namely aleatory uncertainty and epistemic uncertainty. The aleatory uncertainty is random in nature and it cannot be reduced. The range or value of aleatory uncertainty is known, but the exact value is varying according to the time and unit. Epistemic uncertainties occur in the system due to lack of knowledge or information of the modelling [8]. Sandwich plates are widely used in design and construction of Aerospace craft. In such application these materials are subjected to wide environmental changes such as pressure, temperature, density, humidity. This inevitable change in surrounding affects the vibrational response of the structure. Therefore it is essential to include the actual operating condition in order to get the changes in vibration characteristic of sandwich plates. The vibration response of the Aerospace craft is usually carried out in atmospheric condition rather than actual unevenly varying condition for the cause of convenience. Thus, it is essential to consider the material and geometric uncertainty in order to accommodate above mentioned environmental changes as well as other inaccuracies occurring during design and fabrication of the sandwich plate. In most of the scientific literatures, researchers have concentrated on deterministic natural frequency analysis [9-13], however some researchers [14-23] have considered the stochastic behaviour in their study. No study has been carried out for uncertainty quantification of natural frequency for laminated sandwich plate by employing FE coupled PNN surrogate model approach.

2. Mathematical Formulation

The strain in structure occurs as per following strain-displacement relation [10]

$$\{\bar{\varepsilon}(\varpi)\} = \left[\frac{\partial U(\varpi)}{\partial x}\frac{\partial V(\varpi)}{\partial y}\frac{\partial W(\varpi)}{\partial z}\frac{\partial U(\varpi)}{\partial x}+\frac{\partial V(\varpi)}{\partial y}\frac{\partial U(\varpi)}{\partial z}+\frac{\partial W(\varpi)}{\partial x}\frac{\partial V(\varpi)}{\partial z}+\frac{\partial W(\varpi)}{\partial x}\right] \quad (1)$$

i.e. $\{\bar{\varepsilon}(\varpi)\}=[H(\varpi)]\{\varepsilon(\varpi)\}$. where, $[H]$ represents the unit step function. The displacement vector in generalized form may be written as

$$\{S(\varpi)\} = \sum_{i=1}^{n} N_i(\varpi)S_i(\varpi) \quad (2)$$

where, $\{s\} = \{u_0 v_0 w_0 \theta_x \theta_y u_u v_u w_u u_l v_l w_l\}^T$, and N represents the shape function. Equation (1) can be used to write strain vector as

$$\{\varepsilon(\varpi)\} = [A(\varpi)]\{S(\varpi)\} \quad (3)$$

where, [A] is strain-displacement matrix. As per Hamilton's principle, the dynamic equilibrium equation may be written as

$$[R(\varpi)]\{\bar{S}\} = \lambda^2[M(\varpi)]\{\bar{S}\} \quad (4)$$

where, $[R(\varpi)]$ is the stochastic natural frequency and $[M(\varpi)]$ is global mass matrix which can be expressed as

$$[M(\varpi)] = \sum_{k=1}^{n_u+n_l} \iiint \rho_k(\varpi)[N]^T[J]^T[N][J]dxdydz = \iint [N]^T[K(\varpi)][N]dxdy \quad (5)$$

where, $\rho_k(\varpi)$ represents the stochastic mass density. The stiffness matrix $[K(\varpi)]$ may be written as

$$[K(\varpi)] = \sum_{k_l}^{n_u+n_l} \rho_k(\varpi)[J]^T[J]dz \quad (6)$$

2.1 Polynomial neural network model

An input-output dataset in optimal PNN structure [24] can be expressed as

$$(X_i, Y_i) = (x_{1i}, x_{2i}, x_{3i}, \ldots x_{ni}, y_i) \quad (7)$$

where, i = 1,2,3 … n. Polynomial regression equation is determined for each pair of input variable xi and xj model output Y

$$Y = A + B_{xi} + C_{xj} + D_{x2i} + E_{x2j} + F_{xixj} \quad (8)$$

where, i, j = 1,2,3...n and co-efficient of polynomial equation is represented by A, B, C, D, E, F. The estimated output $\hat{y}_i$ is determined as

$$(x_1, x_2, x_3, \ldots x_n) = A_0 + \sum_{i=1}^{n} B_i x_i + \sum_{i=1}^{n}\sum_{j=1}^{n} C_{ij}x_i x_j + +\sum_{i=1}^{n}\sum_{j=1}^{n}\sum_{k=1}^{n} D_{ijk}x \quad (9)$$

where i, j, k = 1,2,3...n. X ($x_1, x_2, \ldots, x_n$) = input variable vector

P($A_0, B_i, C_{ij}, D_{ijk}, \ldots$) = vector of coefficients. The basic architecture of PNN is represented in figure 1 while the methodology of stochastic natural frequency analysis of sandwich plate by using surrogate model is presented in figure 2.

3. Results and Discussion

In this study, source uncertainty is quantified for natural frequencies analysis of simply supported skewed (30°) sandwich plate considering individual variation of input parameters. The dimensions of the sandwich plate under consideration are as: length = 100 mm, width = 100 mm and thickness = 10 mm. The core and face-sheet thickness is considered as 0.9 cm and 0.1 cm, respectively. Both upper and lower face-sheet is having 8 laminate of 90°/0°/90°/0°/90°/0°/90°/0° orientation angle. Stochasticity in skew angle is taken as Δ =5% whereas Δ =10% is taken for other parameters, unless otherwise mentioned. The material properties considered in this analysis are:

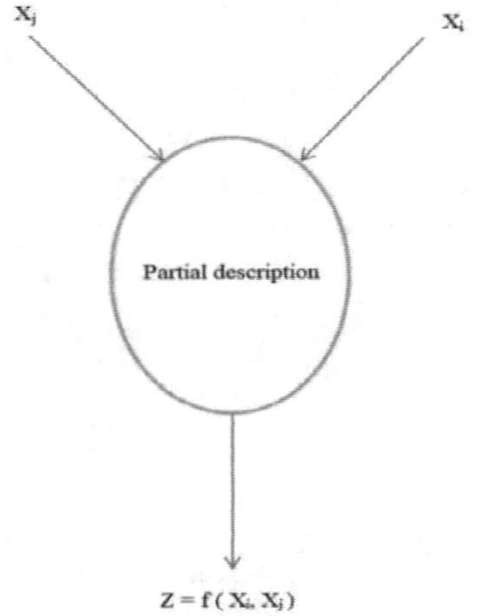

Figure 1. Taxonomy for architecture of PNN

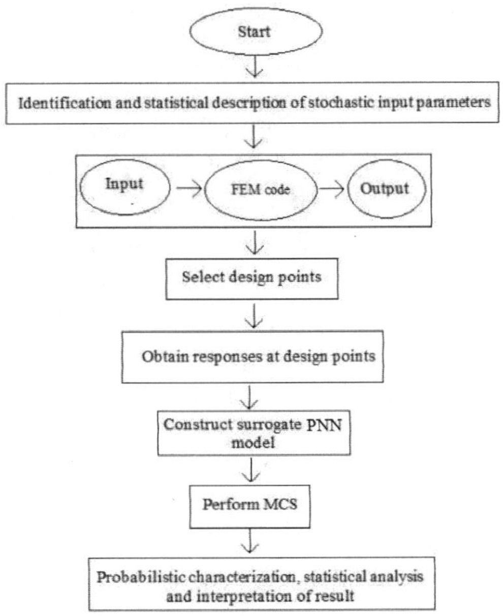

Figure 2. Flow diagram of stochastic free vibration analysis by employing PNN Model

Fig. 2

(a) For core:

E1 = E2 = E3 = 0.5GPa, G12 = G13 = 0.4GPa, G23 = 0.2GPa, $\upsilon_{12} = \upsilon_{13} = \upsilon_{23} = \upsilon_{32} = 0.27$, $\upsilon_{21} = \upsilon_{31} = 0.006$, ρ =1000 kg/m^3 and

(b) For face-sheet:

E1 = 38.6GPa, E2 = E3 = 8.27GPa, G12 = G13 = 4.14GPa, G23 = 1.656GPa, $\upsilon_{12} = \upsilon_{13} = \upsilon_{23} = \upsilon_{32} = 0.26$, $\upsilon_{21} = \upsilon_{31} = 0.006$, ρ =2600 kg/m^3.

3.1 Validation

The deterministic FE code is validated with the published literature [7,10] and it is observed to be in good agreement as presented in Table 1.

Table 1 Non- dimensional deterministic natural frequencies of a four-layered (90°/0°/90°/0°) anti-symmetric composite sandwich plate

Skew angle	Mode	Present analysis	Kulkarni and Kapuria [10]	Wang et al. [7]
30°	1	1.8889	1.9209	1.9410
	2	3.4827	3.5353	2.9063

Two types of validation are required in surrogate assisted probabilistic analysis. First one is FE code validation and second one is surrogate model validation. FE code is already validated and the deterministic result is presented in table 1, whereas PDF and scatter plot is portrayed in figure 3 for surrogate model validation. The least deviation of PDF plot for MCS and PNN validates the surrogate model. For further validation of the surrogate model, scatter plot is shown. Least deviation of scatter plot from diagonal line indicates the predictability and applicability of PNN surrogate model.

3.2 Numerical results

Uncertainty in first three natural frequencies of sandwich plate is quantified and presented in figure 4 to figure 7. The different degree of stochasticity in individual parameter is considered to get the global response of the structure due to local source uncertainties. In Figure 4, the natural frequency response of sandwich plate having variation in only ply-orientation angle with different degree of stochasticity is depicted. The mean natural frequencies for first and second mode are almost same with insignificant variation in response bound whereas uncommon behaviour is observed for third mode. In third natural frequency mean frequency as well as response bound increases with increase in stochasticity. Figure 5 shows the common behaviour of response bound for stochastic natural frequencies, considering variation in only skew angle. The response bound increases with increase in degree of stochasticity with common mean value.

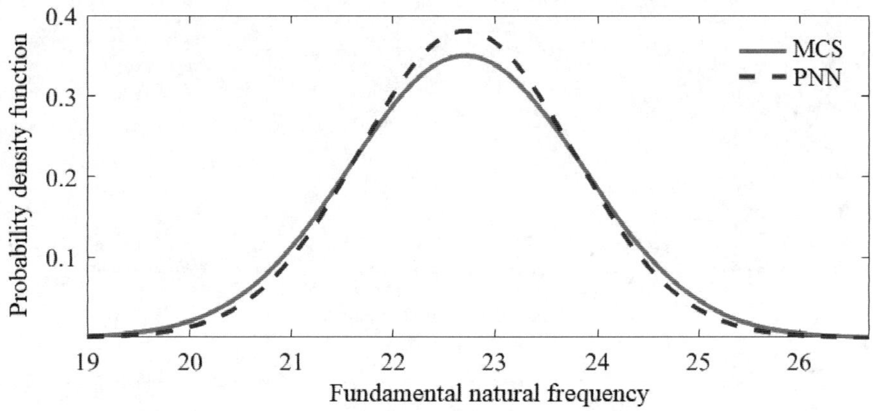

(a)

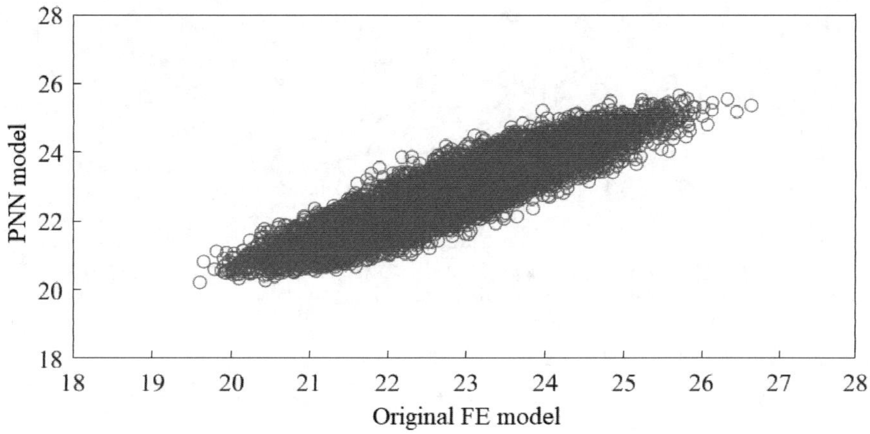

(b)

Figure 3. (a) Probability density function (PDF) and (b) Scatter plot for combined variation of all input parameters

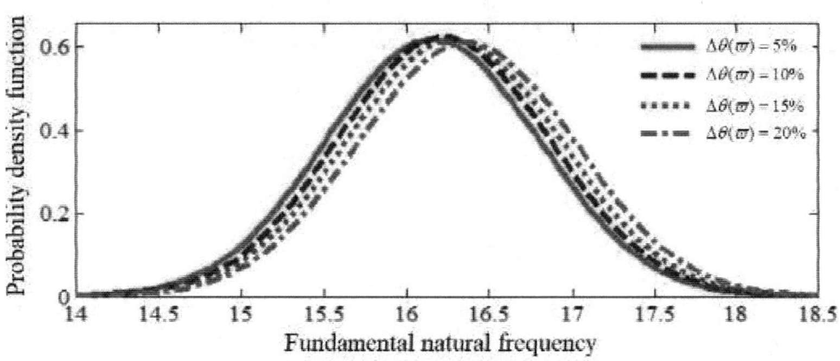

(a)

IOP Conf. Series: Materials Science and Engineering **798** (2020) 012036 doi:10.1088/1757-899X/798/1/012036

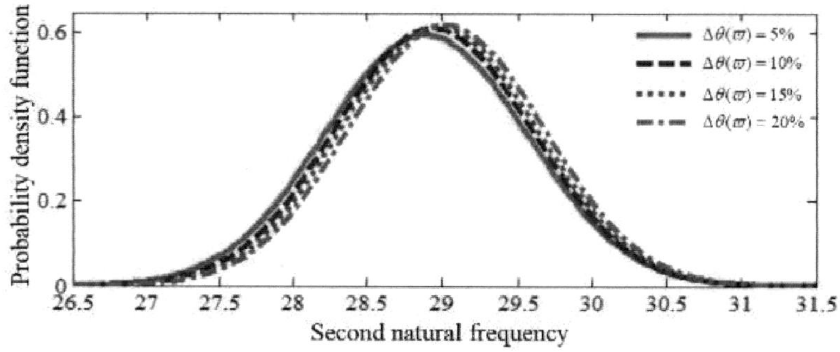

(b)

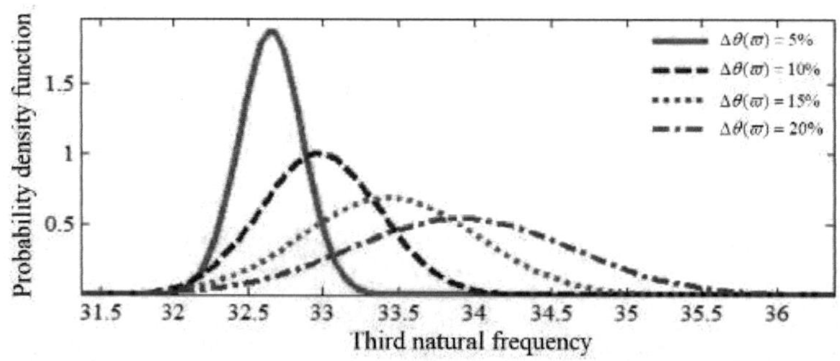

(c)

Figure 4. Stochastic natural frequencies (rad/sec) because of variation in ply-orientation angle with different degree of stochasticity.

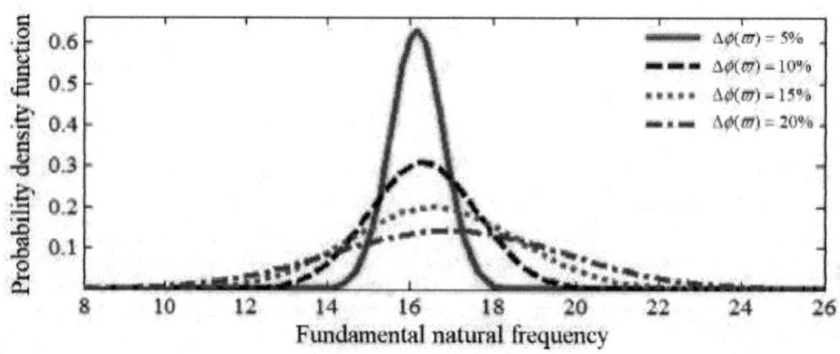

(a)

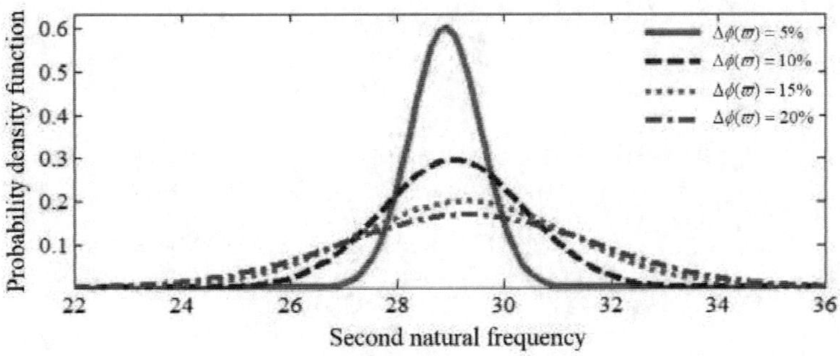

(b)

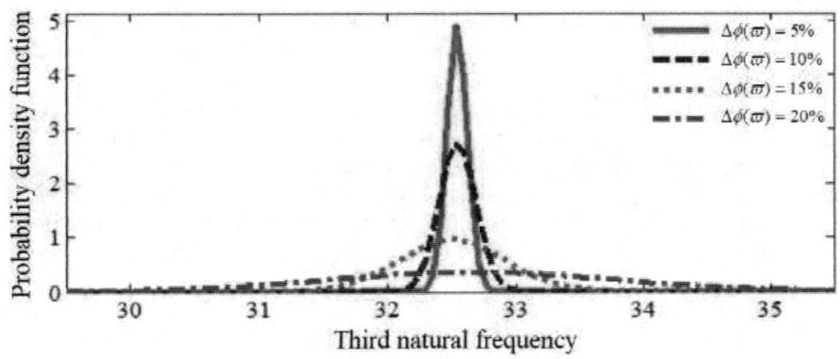

(c)

Figure 5. Stochastic natural frequencies (rad/sec) because of variation in skew angle with different degree of stochasticity.

4. Conclusion

The novelty of this paper lies in uncertainty quantification of natural frequencies by employing PNN approach. PNN is employed in conjunction with conventional FE based Monte Carlo Simulation (MCS) in order to get benefitted from the capability of MCS (probabilistic characterization) and diminishing the lacuna of computational expensiveness. The present algorithm is observed as computationally efficient with adequate accuracy. Here, layer wise bottom up stochastic propagation approach is employed for natural frequency analysis, considering individual variation of random input parameters. The hybrid FE-PNN approach presented in this study is comprehensive in nature i.e. it can be employed to any complex structure. This study can be further extended by incorporating the noise in the system.

References

1. Kumar, R.R., Pandey, K.M., Dey, S., 2019. Probabilistic assessment on buckling behavior of sandwich panel: - A radial basis function approach. Structural Engineering and Mechanics. Techno press, **71 (2)**, pp. 197–210.
2. Karsh, P.K., Kumar, R.R. and Dey, S., 2019. Radial Basis Function Based Stochastic Natural Frequencies Anaysis of Functionally Graded Plates. International Journal of Computational Methods. doi.org/10.1142/S0219876219500610
3. Mukhopadhyay, T., 2018. A multivariate adaptive regression splines based damage identification methodology for web core composite bridges including the effect of noise. Journal of Sandwich Structures & Materials, **20(7)**, pp.885-903.
4. Karsh, P.K., Mukhopadhyay, T., Chakraborty, S., Naskar, S. and Dey, S., 2019. A hybrid stochastic sensitivity analysis for low-frequency vibration and low-velocity impact of functionally graded plates. Composites Part B: Engineering, **176**, p.107221.
5. Karsh, P.K., Mukhopadhyay, T. and Dey, S., 2020. A Stochastic Investigation of Effect of Temperature on Natural Frequencies of Functionally Graded Plates. In Advances in Structural Engineering and Rehabilitation (pp. 41-53). Springer, Singapore.
6. Mukhopadhyay, T., Chakraborty, S., Dey, S., Adhikari, S. and Chowdhury, R., 2017. A critical assessment of Kriging model variants for high-fidelity uncertainty quantification in dynamics of composite shells. Archives of Computational Methods in Engineering, **24(3)**, pp.495-518.
7. Naskar, S., Mukhopadhyay, T. and Sriramula, S., 2018. Probabilistic micromechanical spatial variability quantification in laminated composites. Composites Part B: Engineering, **151**, pp.291-325.
8. Kumar, R.R., Karsh, P.K., , V., Pandey, K.M., Dey, S., (2019). Stochastic natural frequency analysis of skewed sandwich plates. Engineering Computations, doi.org/10.1108/EC-01-2019-0034.
9. Wang, C. M., Ang, K. K., Yang, L., and Watanbe, E. 2000. Free vibration of skew sandwich plates with laminated facings. *Journal of sound and vibration* **235(2)** pp. 317-340.
10. Chalak, H. D., Chakrabarti, A., Iqbal, M. A., and Sheikh, A. H. 2013. Free vibration analysis of laminated soft core sandwich plates. *Journal of Vibration and Acoustics* **135(1)** pp. 011-013.
11. Sayyad, A. S., and Ghugal, Y. M. 2015. On the free vibration analysis of laminated composite and sandwich plates: a review of recent literature with some numerical results. *Composite Structures* **129** pp. 177-201.
12. Kulkarni, S. D., and Kapuria, S. 2008. Free vibration analysis of composite and sandwich plates using an improved discrete Kirchhoff quadrilateral element based on third-order zigzag theory. *Computational mechanics* **(42.6)** pp. 803-824.
13. Kumar R. R., Pandey K. M., 2017. Static structural and modal analysis of gas turbine blade. InIOP Conference Series: Materials Science and Engineering **225**(1) pp. 012102). IOP Publishing.
14. Kumar RR, Mukhopadhyay T, Pandey KM, Dey S. 2018. Stochastic buckling analysis of sandwich plates: The importance of higher order modes. International Journal of Mechanical Sciences. **152** pp. 630-643
15. Karsh PK, Mukhopadhyay T, Dey S. 2019. Stochastic low-velocity impact on functionally graded plates: Probabilistic and non-probabilistic uncertainty quantification. Composites Part B: Engineering. **159** pp. 461-80.

16. Mukhopadhyay T, Naskar S, Karsh PK, Dey S, You Z. 2018. Effect of delamination on the stochastic natural frequencies of composite laminates. Composites Part B: Engineering. **154** pp. 242-56.
17. Karsh PK, Mukhopadhyay T, Dey S. 2018. Stochastic dynamic analysis of twisted functionally graded plates. Composites Part B: Engineering. **147** pp. 259-78.
18. Karsh PK, Mukhopadhyay T, Dey S. 2018. Spatial vulnerability analysis for the first ply failure strength of composite laminates including effect of delamination. Composite Structures. **184** pp. 554-67.
19. Karsh PK, Mukhopadhyay T, Dey S. 2018. Stochastic Investigation of Natural Frequency for Functionally Graded Plates. In IOP Conference Series: Materials Science and Engineering. **326 (1)**, p. 012003. IOP Publishing.
20. Kumar, R.R., Mukhopadhyay, T., Naskar, S., Pandey, K.M. and Dey, S., 2019. Stochastic low-velocity impact analysis of sandwich plates including the effects of obliqueness and twist. Thin-Walled Structures, **145**, pp.106411.
21. Karsh, P.K., Kumar, R.R. and Dey, S., 2019. Stochastic impact responses analysis of functionally graded plates. Journal of the Brazilian Society of Mechanical Sciences and Engineering, **41(11),** p.501.
22. Kumar, R.R., Mukhopadhya, T., Pandey, K.M. and Dey, S., 2020. Prediction capability of polynomial neural network for uncertain buckling behavior of sandwich plates. In Handbook of Probabilistic Models, Butterworth-Heinemann pp. 131-140.
23. Kumar, R.R., Pandey, K.M. and Dey, S., 2020. Stochastic Free Vibration Analysis of Sandwich Plates: A Radial Basis Function Approach. In Reliability, Safety and Hazard Assessment for Risk-Based Technologies, Springer, Singapore pp. 449-458
24. Dey S, Mukhopadhyay T, Adhikari S. 2017. Metamodel based high-fidelity stochastic analysis of composite laminates: A concise review with critical comparative assessment. Composite Structures. **(171)** 227-50.

IOP Conf. Series: Materials Science and Engineering **798** (2020) 012037 doi:10.1088/1757-899X/798/1/012037

Three dimensional natural transform and its applications

Arvind Kumar Sinha
National institute of technology Raipur (C.G), 492010
E-mail: aksinha.maths@nitrr.ac.in

Srikumar Panda
National institute of technology Raipur (C.G), 492010
E-mail: srikumarpanda79@gmail.com

Abstract. We have discussed the concept of three dimensional natural transform and its applications in material science. We have solved the particular type of third order partial differential equations (known as Mboctara equation) using the three dimensional natural transform operator.

1. Introduction

Integral transforms have been found wide applications for solving the problems in many fields of Mathematics as well as other sciences and technologies [1-4]. So many generalization have been done in the field of integral transform [5-21]. The natural transform was introduced by Khan and Khan [22] in the year 2008 then its detailed properties was investigated by Belgacem [23] in the year 2012. Firstly it was used to solve the problems of Maxwell's equations [24,25,26]. In the year 2017 the double natural transform was established by Kilicman and omran [27] to solve the problems of general telegraph, wave and partial integro-differential equations. Mboctara equation is a special type of non-homogeneous partial differential equation (PDE) which is helpful to analysis the nature of collective motion of micro particles in the material. Specally, in material science, we can analyse any material with Mboctara equation with the study of its geometrical and structural design. Using this technique, we can solve PDE with singularities of given type and we can present exact solution for material science problems.

2. Definition of the three-dimensional natural transform:(3-D natural transform)

Definition 2.1.
Let $h(\alpha, \beta, \gamma)$ be a function and $\alpha, \beta, \gamma \in \Re_+$, then 3-D natural transform is stated as

$$N_+^3\{h(\alpha,\beta,\gamma)\} :=: R_+^3[(\xi,\eta,\vartheta);(\lambda,\nu,\varpi)] = \int_0^\infty \int_0^\infty \int_0^\infty e^{-(\xi\alpha+\eta\beta+\vartheta\gamma)}\, h(\alpha\lambda, \beta\nu, \gamma\varpi)\, d\alpha\, d\beta\, d\gamma. \quad (1)$$

We can write (1) in other form as

$$N_+^3\{h(\alpha,\beta,\gamma)\} = \frac{1}{\lambda\,\nu\,\varpi}\int_0^\infty\int_0^\infty\int_0^\infty e^{-\left(\frac{\xi\alpha}{\lambda}+\frac{\eta\beta}{\nu}+\frac{\vartheta\gamma}{\varpi}\right)} h(\alpha,\beta,\gamma)\,d\alpha\,d\beta\,d\gamma \tag{2}$$

Definition 2.2.
3-D inverse natural transform can be written as the formula

$$N_+^{-3}[N_+^3\,h(\alpha,\beta,\gamma)] = h(\alpha,\beta,\gamma) = \frac{1}{2\pi i}\int_{\kappa-i\infty}^{\kappa+i\infty} e^{\frac{\xi\alpha}{\lambda}}\,d\xi\,\frac{1}{2\pi i}\int_{\rho-i\infty}^{\rho+i\infty} e^{\frac{\eta\beta}{\nu}}\,d\eta$$
$$\frac{1}{2\pi i}\int_{\Xi-i\infty}^{\Xi+i\infty} e^{\frac{\vartheta\gamma}{\varpi}}\,N_+^3\{h(\alpha,\beta,\gamma)\}\,d\vartheta \tag{3}$$

3. 3-D natural transform of some special functions:

(i) Let $h(\alpha,\beta,\gamma) = 1$ for $\alpha, \beta\; and\; \gamma > 0$, then

$$N_+^3\{h(\alpha,\beta,\gamma)\} = \frac{1}{\lambda\nu\varpi}\int_0^\infty\int_0^\infty\int_0^\infty e^{-\left(\frac{\xi\alpha}{\lambda}+\frac{\eta\beta}{\nu}+\frac{\vartheta\gamma}{\varpi}\right)}\,d\alpha\,d\beta\,d\gamma = \frac{1}{\xi\eta\vartheta}.$$

(ii) Let $h(\alpha,\beta,\gamma) = e^{\kappa\alpha+\rho\beta+\Xi\gamma}$, where κ, ρ, Ξ are any constants, then

$$N_+^3\{h(\alpha,\beta,\gamma)\} = \frac{1}{\lambda\,\nu\,\varpi}\int_0^\infty\int_0^\infty\int_0^\infty e^{-\left(\frac{\xi\alpha}{\lambda}+\frac{\eta\beta}{\nu}+\frac{\vartheta\gamma}{\varpi}\right)} e^{\kappa\alpha+\rho\beta+\Xi\gamma}\,d\alpha\,d\beta\,d\gamma$$
$$= \left(\frac{1}{\xi-\kappa\lambda}\right)\left(\frac{1}{\eta-\rho\nu}\right)\left(\frac{1}{\vartheta-\varpi\Xi}\right).$$

(iii) Let $h(\alpha,\beta,\gamma) = e^{i[(\kappa\alpha+\rho\beta+\Xi\gamma)]}$, where, $\kappa, \rho\; and\; \Xi$ are constants, then

$$N_+^3\{h(\alpha,\beta,\gamma)\} = N_+^3\{e^{i[(\kappa\alpha+\rho\beta+\Xi\gamma)]}\}$$
$$= \frac{1}{(\xi-i\kappa\lambda)\,(\eta-i\rho\nu)\,(\vartheta-i\Xi\Omega)}$$
$$= \frac{(\xi+i\kappa\lambda)\,(\eta+i\rho\nu)\,(\vartheta+i\Xi\Omega)}{(\xi^2+\kappa^2\lambda^2)\,(\eta-i\rho\nu)\,(\vartheta-i\Xi\Omega)}$$
$$= \frac{(\xi\eta-\kappa\rho\lambda\nu)\varphi-(\kappa\lambda\eta\varpi+\xi\rho\nu\varpi)\Xi+(\xi\eta\varpi-\kappa\rho\lambda\nu)\Xi+(\kappa\lambda\eta+\rho\nu\xi)\varphi}{(\xi^2+\kappa^2\lambda^2)(\eta^2+\rho^2\nu^2)(\vartheta^2+\Xi^2\varpi^2)}$$

Consequently,

$$N_+^3\{\cos(\kappa\alpha+\rho\beta+\Xi\gamma)\} = \frac{(\xi\eta-\kappa\rho\lambda\nu)\varphi-(\kappa\lambda\eta\varpi+\xi\rho\nu\varpi)\Xi}{(\xi^2+\kappa^2\lambda^2)(\eta^2+\rho^2\nu^2)(\vartheta^2+\Xi^2\varpi^2)}$$
$$N_+^3\{\sin(\kappa\alpha+\rho\beta+\Xi\gamma)\} = \frac{(\xi\eta\varpi-\kappa\rho\lambda\nu)\Xi+(\kappa\lambda\eta+\rho\nu\xi)\varphi}{(\xi^2+\kappa^2\lambda^2)(\eta^2+\rho^2\nu^2)(\vartheta^2+\Xi^2\varpi^2)}.$$

(iv) Let $h(\alpha,\beta,\gamma) = \{\cosh(\kappa\alpha+\rho\beta+\Xi\gamma)\} = \frac{1}{2}\Big[N_+^3\{e^{(\kappa\alpha+\rho\beta+\Xi\gamma)}\} + N_+^3\{e^{-(\kappa\alpha+\rho\beta+\Xi\gamma)}\}\Big]$

$$= \frac{1}{2}\Big[\frac{1}{(\xi-\kappa\lambda)(\eta-\rho\nu)(\vartheta-\Xi\varpi)} + \frac{1}{(\xi+\kappa\lambda)(\eta+\rho\nu)(\vartheta+\Xi\varpi)}\Big].$$

Similarly, we can obtain

$$N_+^3\{\sinh(\kappa\alpha+\rho\beta+\Xi\gamma)\} = \frac{1}{2}\Big[\frac{1}{(\xi-\kappa\lambda)(\eta-\rho\nu)(\vartheta-\Xi\varpi)} - \frac{1}{(\xi+\kappa\lambda)(\eta+\rho\nu)(\vartheta+\Xi\varpi)}\Big].$$

4. Properties

(i) Linearity property is achieved, for any constants κ, ρ, Ξ such that

$$N_+^3\{\kappa\, h(\alpha,\beta,\gamma) + \rho\, g(\alpha,\beta,\gamma) + \Xi\, s(\alpha,\beta,\gamma)\}$$

$$= \kappa N_+^3\{h(\alpha,\beta,\gamma)\} + \rho N_+^3\{g(\alpha,\beta,\gamma)\} + \Xi N_+^3\{s(\alpha,\beta,\gamma)\}.$$

(ii) Let $N_+^3\{h(\alpha,\beta,\gamma)\} = R_+^3\ [(\xi,\eta,\vartheta);(\lambda,\nu,\varpi)]$ then

$$N_+^3\{e^{-\kappa\alpha-\rho\beta-\Xi\gamma}\, h(\alpha,\beta,\gamma)\} = R_+^3\ [h(\alpha,\beta,\gamma);(\xi+\kappa,\eta+\rho,\vartheta+\Xi);(\lambda,\nu,\varpi)].$$

(iii) If $R_+[h(\alpha);\xi,\lambda]$ denoted to natural transform $h(\alpha)$, $R_+[g(\alpha);\eta,\nu]$ and $R_+[s(\alpha);\vartheta,\varpi]$ is natural transform of a function $g(\alpha)$ and $s(\alpha)$, then

$$N_+^3\{h(\kappa\alpha)\, g(\rho\beta)\, s(\Xi\gamma)\} = \frac{1}{\kappa\lambda\ \rho\nu\ \Xi\varpi} R_+\Big[h(\alpha);\frac{\xi}{\kappa},\lambda\Big] R_+\Big[g(\alpha);\frac{\eta}{\rho},\nu\Big] R_+\Big[s(\alpha);\frac{\vartheta}{\Xi},\varpi\Big], \kappa,\rho,\Xi > 0.$$

(iv) $N_+^3\{\frac{\partial h(\alpha,\beta,\gamma)}{\partial\alpha}\} = \frac{\xi}{\lambda}N_+^3\{h(\alpha,\beta,\gamma)\} - \frac{N_+^3\{h(0,\beta,\gamma)\}}{\lambda}$

(v) $N_+^3\{\frac{\partial^2 h(\alpha,\beta,\gamma)}{\partial\alpha^2}\} = \frac{\xi^2}{\lambda^2}N_+^3\{h(\alpha,\beta,\gamma)\} - \frac{\xi}{\lambda^2}N_+^3\{h(\alpha,\beta,\gamma)\} - \frac{1}{\lambda}N_+^3\frac{\partial h(0,\beta,\gamma)}{\partial\alpha}$

(vi) $N_+^3\{\frac{\partial h(\alpha,\beta,\gamma)}{\partial\beta}\} = \frac{\eta}{\nu}N_+^3\{h(\alpha,\beta,\gamma)\} - \frac{N_+^3\{h(\alpha,0,\gamma)\}}{\nu}$

(vii) $N_+^3\{\frac{\partial^2 h(\alpha,\beta,\gamma)}{\partial\beta^2}\} = \frac{\eta^2}{\nu^2}N_+^3\{h(\alpha,\beta,\gamma)\} - \frac{\eta}{\nu^2}N_+^3\{h(\alpha,0,\gamma)\} - \frac{1}{\nu}N_+^3\frac{\partial h(\alpha,0,\gamma)}{\partial\gamma}$

(viii) $N_+^3\frac{\partial^3 h(\alpha,\beta,\gamma)}{\partial\alpha\,\partial\beta\,\partial\gamma} = \frac{\xi\eta\vartheta}{\lambda\nu\varpi}N_+^3\{h(\alpha,\beta,\gamma)\} - \frac{\xi\eta}{\lambda\nu}N_+^3\{h(\alpha,\beta,0)\} - \frac{\xi\vartheta}{\lambda\varpi}N_+^3\{h(\alpha,0,\gamma)\}$
$- \frac{\eta\vartheta}{\nu\varpi}N_+^3\{h(o,\beta,\gamma)\} + \frac{\xi}{\lambda}N_+^3\{h(\alpha,0,0)\} + \frac{\eta}{\nu}N_+^3\{h(0,\beta,0)\} + \frac{\vartheta}{\varpi}N_+^3(0,0,\gamma)\} - \frac{h(0,0,0)}{\lambda\nu\varpi}$

(ix) Convolution property: Suppose that $h(\alpha,\beta,\gamma)$ and $g(\alpha,\beta,\gamma)$ have the 3-D natural transforms, thus

$$N_+^3\{(h***g)(\lambda,\nu,\varpi);(\xi,\eta,\vartheta)\} = \lambda\nu\varpi N_+^3\{h(\lambda,\nu,\varpi);(\xi,\eta,\vartheta)\}N_+^3\{g(\lambda,\nu,\varpi);(\xi,\eta,\vartheta)\}.$$

5. Applications

5.1 If 3-D natural transform of the function $u(\alpha,\beta,\gamma)$ is denoted by

$$U[(\xi,\eta,\vartheta);(\lambda,\nu,\varpi)]$$

and if we consider the following third-order partial differential equation (known as the Mboctara equation) with the following conditions

$$\partial_{\alpha,\beta,\gamma}u(\alpha,\beta,\gamma) + u(\alpha,\beta,\gamma) = 0, \tag{4}$$

$$u(\alpha,\beta,0) = e^{\alpha+\beta},\ u(\alpha,0,\gamma) = e^{\alpha-\gamma},\ u(0,\beta,\gamma) = e^{\beta-\gamma},\ u(\alpha,\beta,1) = e^{\alpha+\beta-1},$$

then we are interested to find its exact solution in geometrical form.

Solution: We apply 3-D natural transform (4), we get

$$\frac{\xi\eta\vartheta}{\lambda\nu\varpi}N_+^3\{U(\xi,\eta,\vartheta);(\lambda,\nu,\varpi)\}+N_+^3\{U(\xi,\eta,\vartheta);(\lambda,\nu,\varpi)\}=N_+^3\{G(\xi,\eta,\vartheta);(\lambda,\nu,\varpi)\}.$$

Here,

$$\begin{aligned}&N_+^3\{G(\xi,\eta,\vartheta);(\lambda,\nu,\varpi)\}\\&=\frac{\xi\eta}{\lambda\nu}N_+^3\{U(\xi,\eta,0);(\lambda,\nu,0)\}+\frac{\xi\vartheta}{\lambda\varpi}N_+^3\{U(\xi,0,\vartheta);(\lambda,0,\varpi)\}+\frac{\eta\vartheta}{\nu\varpi}N_+^3\{U(0,\eta,\vartheta);(0,\nu,\varpi)\}\\&-\frac{\xi}{\lambda}N_+^3\{U(\xi,0,0);(\lambda,0,0)\}-\frac{\eta}{\nu}N_+^3\{U(0,\eta,0);(0,\nu,0)\}-\frac{\vartheta}{\varpi}N_+^3\{U(0,0,\vartheta);(0,0,\varpi)\}+N_+^3U(0,0,0)\end{aligned} \tag{5}$$

Factorising the right hand side of the equation (5)

$$U(\xi,\eta,\vartheta;\lambda,\nu,\varpi)=\frac{G(\xi,\eta,\vartheta)}{1+\xi\eta\vartheta} \tag{6}$$

Now applying the inverse 3-D natural transform on the previous equation we obtain the following equation

$$u(\alpha,\beta,\gamma)=N_{\alpha\beta\gamma}^{-1}\left[U(\xi,\eta,\vartheta;\lambda,\nu,\varpi)\right]=\frac{G(\xi,\eta,\vartheta)}{1+\xi\eta\vartheta}=e^{\alpha+\beta-\gamma} \tag{7}$$

This is the exact solution of Mboctara equation (4) and its solution in geometrical form is given by the figure (e) with the given conditions of figures (a) to (d).

IOP Conf. Series: Materials Science and Engineering **798** (2020) 012037 doi:10.1088/1757-899X/798/1/012037

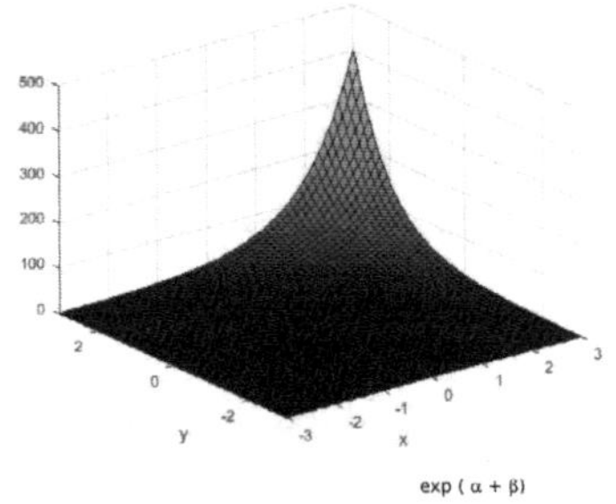

(a) For the condition $u(\alpha, \beta, 0) = e^{\alpha+\beta}$

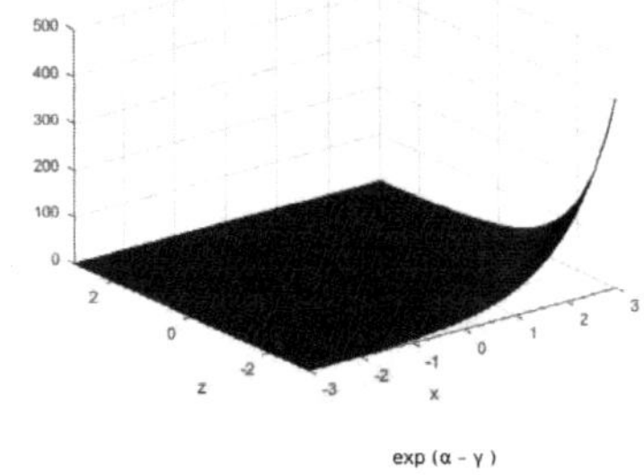

(b) For the condition $u(\alpha, 0, \gamma) = e^{\alpha-\gamma}$

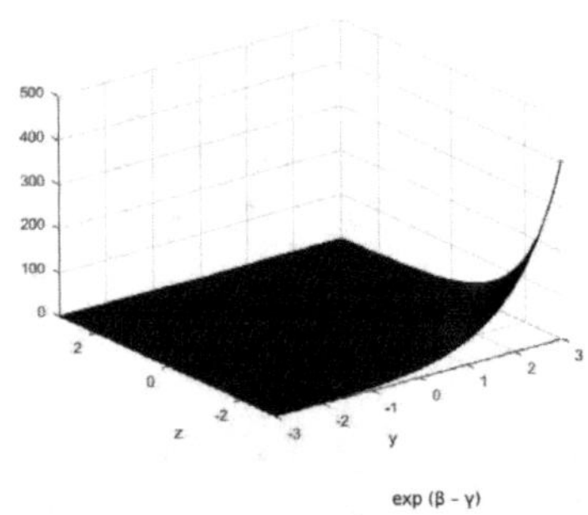

(c) For the condition $u(0, \beta, \gamma) = e^{\beta-\gamma}$

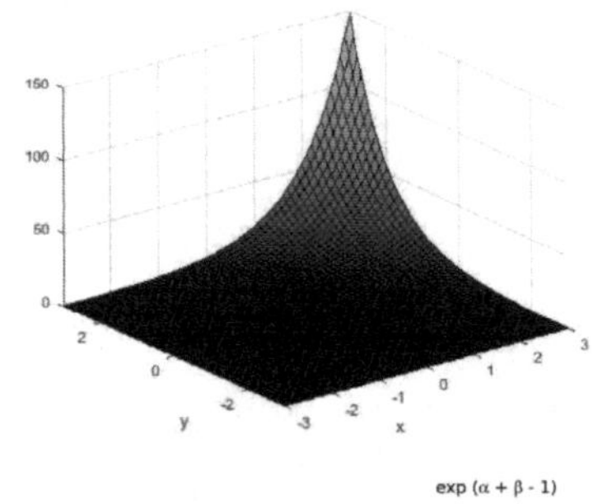

(d) For the condition $u(\alpha, \beta, 1) = e^{\alpha+\beta-1}$

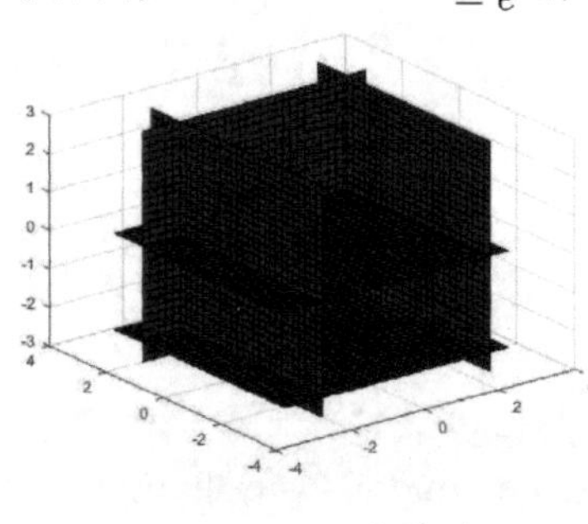

(e) Exact solution of the mboctara equation (4)

5.2 Let us consider the following non homogeneous Mboctara equation

$$\partial_{\alpha\beta\gamma} u(\alpha, \beta, \gamma) + u(\alpha, \beta, \gamma) = cos(\alpha)\cos(\beta)\cos(-\gamma) - \sin(\alpha)\sin(\beta)\sin(-\gamma) \tag{8}$$

subjected to the following conditions:

$$u(\alpha, \beta, 0) = cos(\alpha)cos(\beta), \partial_\gamma u(\alpha, \beta, 0) = \partial_\alpha(0, \beta, \gamma) = \partial_\beta(\alpha, 0\gamma) = 0,$$
$$u(\alpha, \frac{\pi}{2}, \gamma) = u(\alpha, \beta, \frac{\pi}{2}) = u(\frac{\pi}{2}, \beta, \gamma) = 0, \tag{9}$$

then we are interested to find its exact solution.

Solution: We apply 3-D natural transform (8), we get that

$$\frac{\xi\eta\vartheta}{\lambda\nu\varpi}N_+^3\{U(\xi,\eta,\vartheta)\}+N_+^3\{U(\xi,\eta,\vartheta)\}$$
$$=N_+^3\{G(\xi,\eta,\vartheta)\}+\frac{\xi\eta\vartheta}{(\xi^2+\lambda^2)+(\eta^2+\nu^2)+(\vartheta^2-\varpi^2)}-\frac{\lambda\nu\varpi}{(\xi^2+\lambda^2)+(\eta^2+\nu^2)+(\vartheta^2-\varpi^2)} \tag{10}$$

Factorising the right side of (10), we obtain the

$$\{U(\xi,\eta,\vartheta)\}$$
$$=[G(\xi,\eta,\vartheta)]\Big[\frac{\xi\eta\vartheta+\lambda\nu\varpi}{\lambda\nu\varpi}\Big]+\xi\eta\vartheta\Big/\frac{(\xi^2+\lambda^2)+(\eta^2+\nu^2)+(\vartheta^2-\varpi^2)}{1+\frac{\xi\eta\vartheta}{\lambda\nu\varpi}}-$$
$$\lambda\nu\varpi\Big/\frac{(\xi^2+\lambda^2)+(\eta^2+\nu^2)+(\vartheta^2-\varpi^2)}{1+\frac{\xi\eta\vartheta}{\lambda\nu\varpi}} \tag{11}$$

Now applying the inverse 3-D natural transform on the previous equation, we obtain the following solution

$$u(\alpha,\beta,\gamma)=N_{\alpha\beta\gamma}^{-1}\Big\{[G(\xi,\eta,\vartheta)]\Big[\frac{\xi\eta\vartheta+\lambda\nu\varpi}{\lambda\nu\varpi}\Big]\Big\}$$
$$=\xi\eta\vartheta\Big/\frac{(\xi^2+\lambda^2)+(\eta^2+\nu^2)+(\vartheta^2-\varpi^2)}{1+\frac{\xi\eta\vartheta}{\lambda\nu\varpi}}-\lambda\nu\varpi\Big/\frac{(\xi^2+\lambda^2)+(\eta^2+\nu^2)+(\vartheta^2-\varpi^2)}{1+\frac{\xi\eta\vartheta}{\lambda\nu\varpi}} \tag{12}$$

$$=\cos(\alpha)\,cos(\beta)\,cos(-\gamma).$$

This is the exact solution for non homogeneous Mboctara equation.

6. Discussion and Conclusion

This work deals with three dimensional natural transform operator and its inverse. Fundamental properties of three dimensional natural transform are obtained. Further, some applications on three dimensional natural transform are presented. Specially, in material science, we can analyze any material whose crystal phenomena is related with Mboctara equation with the study of its geometrical and structural design. Using this techniques, we can solve PDE with singularity of given type of any crystaline material and we can present its exact solution in geometrical form in material sciences problems.

Acknowledgment

The authors are extremely thankful to Department of Mathematics, National Institute of Technology Raipur C.G., India, for providing facilities, space and opportunity for the work.

References

[1] Debnath L and Bhatta D 2007 *Integral transforms and applications* (CRC Press: London)
[2] Sneddon IN 1972 *The use of integral transform* (Mcgraw-Hill, New-York)
[3] Brychkov A, Glaeske HJ, Prudnikov AP and Tuan VK 1992 *Multidimensional integral transformations* (Gordon and Breach Science Publishers; New York)
[4] Zemanian AH 1968 *Generalized integral transformations* (Dover Publications; New York)
[5] Spiegel MR 1965 *Schaum's outline of theory and problems of Laplace transforms* (McGraw-Hill; New York)
[6] Atangana A 2013 *Abstract and Applied analysis.* **10** 2013
[7] Watugala GK 1993 *Internat. J. Math. Ed. Sci. Tech.* **24**, 35–43
[8] Asiru MA 2001 *Internat. J. Math. Ed. Sci. Tech* **32** 906–910
[9] Eltayeb H and Kilicman A 2010 *Appl. Math. Sci.* **4** 1089–1098
[10] Eltayeb H and Kilic¸man A 2010 *Math. Probl. Eng.* **2010** 13
[11] Kilicman A, Eltayeb H, Agarwal RP 2010 *Abstr. Appl. Anal.* **2010** 11
[12] Belgacem FBM 2006 *Nonlinear Stud.* **13** 23-41
[13] Belgacem FBM *Appl. Math. Sci.***4** 3665–3686
[14] Belgacem FBM and Karaballi AA 2006 *J. Appl. Math. Stoch. Anal.* 23
[15] Belgacem FBM, Karaballi AA and Kalla SL 2003 *Math. Probl. Eng.* 103–118
[16] Hayder A (2009) *J. Kerbala Uni.* **7** 33-46
[17] Eltayeb H and Kilicman A 2010 *Appl.Math.Sci,* **4** 109-118
[18] Kilicman A and Gadain H E 2009 *Lobachevskii J. Math,* **30** 214-223
[19] Eltayeb H and Kilicman A 2010 *Malays. J. Math. Sci.* **4** 17-30
[20] Eltayeb H and Kilicman A 2012 *Appl. Math.* **2012** 13 pages
[21] Eltayeb H and Gadain H.E 2009 *Lobachevskii .J. Math.* **30** 214-223
[22] Khan ZH and Khan WA 2008 *NUST J. Eng. Sci.* **1** 127–133
[23] Belgacem FBM and Silambarasan R 2012 *Int. J. Math. Eng. Sci. Aerosp.* **3** 99–124
[24] Belgacem FBM and Silambarasan R 2012 *AIP Conf. Proc.* **1493** 106–110
[25] Belgacem FBM and Silambarasan R 2011 *Prog. Electromagnetic Research Symposium Proc.* 899-902
[26] Belgacem FBM and Silambarasan R 2012 *Int. J. Math. Eng. Sci. Aerosp.* **3** 313-323
[27] Kilicman A and Omran M 2017 *J.Nonlinear Sci. Appl.* **10** 1744-1754

IOP Conf. Series: Materials Science and Engineering **798** (2020) 012038 doi:10.1088/1757-899X/798/1/012038

Computational approach of tumor growth in human body with a significant technique the rough set

Arvind Kumar Sinha[1], Nishant Namdev[2*]

[1,2]Department of Mathematics, National Institute of Technology Raipur C.G., India.

*Corresponding author's e-mail address: aksinha.maths@nitrr.ac.in, nishantnamdev84@yahoo.in

Abstract. Tumors are the most threatening issue everywhere throughout the world. The development of tumor cells is dubious in the human body because of its unusual phenomena. The Rough set is a rising and the most special mathematical device to manage uncertain circumstances. A scientific model is given for tumor cells population development with carrying capacity and by the Rough set in uncertain circumstances. In this methodology, the mathematical analysis of the nonlinear behavior of tumor cells population is set up via carrying capacity and simulation by using Euler's method. The **accuracy** of the carrying capacity of the number of tumors cells **99.53%** correct according to our model. The paper is an interface between mathematical modeling, numerical computation, simulation, and implementation of application on biomedical systems, which is an oriented idea to biology. **Keywords:** tumor, carrying capacity, tumor growth, rough set, non-linear behavior.

1. Introduction

Tumors are the most threatening issue everywhere throughout the world. It is the fundamental driver of death of human life and puts a substantial burden on the medical system because of the endless characteristics. Tumors are gathering of diseases incorporating isolated cells improvement with the likelihood to develop to a different part of the body [1] [2].These show up distinctively in connection to benign tumors, which do not grow to various parts of the body [2]. Possible signs and responses of the tumor cell development contain a bump, abnormal passing on, and weight loss [3].Tumors are perceived by specific symptoms and indications or screening tests [1].

Over the earlier decades, a consistent advancement related to tumor research has been performed [4]. Numerous scientific models are used to support, appreciate, and treat the tumor. Models are used to perceive how tumors form [5] and shape [6-8]. They are used to streamline or even modify the modern approach [4] [9], prognosticate the suitability of different prescriptions [10], or a mix of different medicines [11-13] and give knowledge into the development of resistance from medication. Numerous examinations are done to improve the capacity and to deliver a result of the immediate medical care for tumors [14-15].

This article concentrates on the ordinary differential equation (ODE) model for the nonlinear behavior of tumor cell enlargement in the human body. Different ODE models are proposed to address tumor growth and are reliably utilized to make assumptions regarding the sufficiency of development medication [10] [16] [17-18]. The uncontrolled development of cells that attack close-by parts of our body causes tumor growth [10]. The behavior of the tumor cells describes by a system of differential equations with time-dependent coefficients [19] [20].

The analytical dynamics of nonlinear behavior of tumor cell growth have not been studied in uncertain situations by the Rough set, which is a useful asset to consider the conduct of tumor cell development in any suspicious circumstance. Along these lines, in this scientific model, the dynamics of nonlinear behavior of tumor cells and carrying capacity is given with simulation by using Euler's method in MATLAB. A relation between the number of tumor cells and carrying capacity is established and also validated by the Rough set. The complete workflow of the mathematical model of tumor cell growth and validation using the Rough set is in Figure 1.

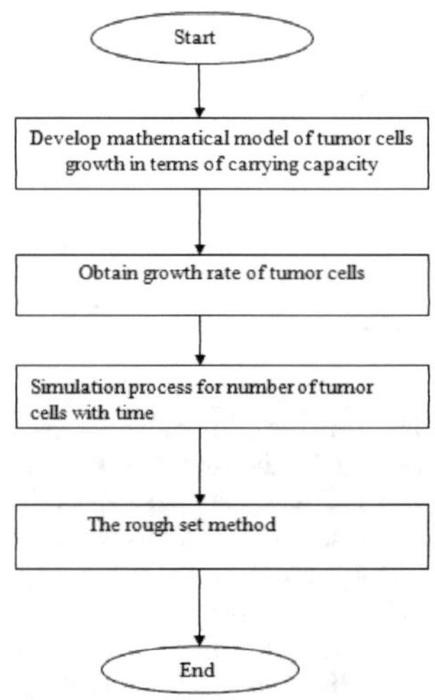

Figure 1. The workflow of the mathematical approach of tumor cells growth with validation using the Rough set.

2. Materials and methods

2.1 Data source and collection

Experimental data for EAT (Ehrilch ascites tumors) in a mouse and the results of approximation using the logistic equation for $K = 150 \times 10^7$ and $\theta = 0.6$which is obtained from (Krug and Taubert, 1985) [21] [22].

2.2Mathematical Approach of Tumors Cells Growth

The measure of disease cells in tumors is difficult to survey in perspective on unsurprising changes in time. The number of tumor cells increases, and the number of cells die liable to the considered time distinction. It allows us to expect the cell cycle length of an individual tumor cell is twenty-four hours. By then, through the range of one day, the probability that the cell isolates are close 100%. We can acknowledge that the likelihood of this cell to partition inside the time length of one hour is 1/24. In reality, the right number of cells in a tumor cell population not known, so the population level can explicitly imply the above case. It is accepted that all cells partition once if dt = twenty-four hours. So also, if dt = one hour, just a small amount of cell in the population (around 1/24) is relied upon to separate [23].

2.3 Mathematical Analysis of nonlinear behaviour of tumors cells in terms of carrying capacity

To find the number of diseased cells in the tumor in a human body is a hard task because of constant changes in tumor growth with time [23]. The number of cells changes according to time,

$$\frac{dS}{dt} = S(t)[e(t) - f(t)] \qquad \ldots(1)$$

$$\frac{dS}{dt} = \theta(t)S(t) \qquad \ldots(2)$$

where S stands the number of tumor cells concerning time t (in days), e and f respectively producing and dying tumor cells, $\frac{dS}{dt}$ is the per capita growth rate of tumor cells population, $e(t) - f(t) = \theta(t)$ and $\theta(t)$ is the function that controls the growing the population of cells and strongly connected with access to nutrient and space availability.

A per capita tumor cell development, subject to tumors size and concerning carrying limit K, is given by the logistic model [24].

$$\frac{dS}{dt} = \theta S(\frac{K-S}{S})$$

Also equivalent to

$$\frac{dS}{dt} = \theta S(1 - \frac{S}{K}) \qquad \ldots(3)$$

at $S(t=0) = S_0, K(t=0) = K_0$

So, we obtain,

$$S = \frac{KS_0}{S_0 + (K - S_0)e^{-\theta t}} \qquad \ldots(4)$$

Let $f(S) = \theta(1 - \frac{S}{K})$ and if the growth of tumor cell will be exponential and in a real-life situation, it is not possible the growth of cells exponentially because after some time growth of the cells will reach to constant position so now from equation (3),

$$\frac{dS}{dt} = Sf(S) \qquad \ldots(5)$$

The carrying limit K is considered as a variable representing the tumor cell growth.

$$\frac{dK}{dt} = \theta S^{2/3} \qquad \ldots(6)$$

The fraction $\frac{2}{3}$ has taken because the carrying capacity is proportional to the tumor surface [25].

2.4 Maximum size of tumor cell

For the solution of the carrying capacity, now from the system (6),

$$\frac{dK}{dt} = \theta S^{2/3}$$

So, putting the value of the S in the above equation, we obtain,

$$\frac{dK}{dt} = \theta \left\{ \frac{KS_0}{S_0 + (K - S_0)e^{-\theta t}} \right\}^{2/3}$$

We simplify this equation, and we obtain

$$K = \left\{ \frac{\gamma t + K_0^{1/3} + \frac{1}{2B_0} K_0^{4/3}}{3(1 - \frac{2}{3} e^{-\gamma t}) \frac{1}{2B_0} e^{-\gamma t}} \right\}^{3/4} \quad \ldots (7)$$

In the above expression, since K is carrying capacity of tumor cells, it shows that K is the maximum number of tumor cells in the human body at any part; it means, it forms a tumor of the maximum size.

3. Result of the model

The equation (7) shows the expression of the carrying capacity. The number of cells increases with time (days), and after some time growth of the number of tumor cells reaches the constant state (Figure 2). If the growth of tumor cells reaches the carrying capacity, then a few cells from the tumor leaves and make another tumor.In the case of the absence of the dead cells, the tumor cells grow at a rate proportional to the current population of the tumor cells.

4. The Rough set

The Rough set is concerned with the characterization and examination of imprecise data information and considered as one of the principal non-statistical methodologies in information investigation [26-27]. The concept of the Rough set has turned into an effective method in different issues, for example, representation of uncertain learning, information investigation, and assessment of data[27-29].

4.1 Validation with the Rough set

The data that we have taken is more appropriate data for explaining to our model because we only need the approximate number of tumor cells growth with time (days) for the validation of the model, and in searching for the estimated number of tumor cells with time, we found this experimental data in the exact form. The data is reliable and most suitable for work; that is why we have used it for the model. This work can be applied to the same type of experimental data.

The data set of the number of tumor cells and its approximation is taken, and by using Rough Set Exploration System (RSES 2.2.2) [30], we found that the number of tumor cells increases with time (days) and after sometime growth of the tumors cells shows the constant behavior (Figure3).

4.2 The mechanism used for the Rough set

A Rough set is a scientific apparatus that can be used to administer with imperfect learning and uncertain situations. The main principle of the Rough set is the indiscernibility association, which is used to estimate to what degree things are like or related.

From (Krug and Taubert, 1985) [21] [22], we have two attributes (i.e., number of tumor cells and time) for one object. The number of tumor cells which are below the carrying limit belongs to the lower approximation, and some of the cells that cross the carrying limit belong to the upper limit. That gives us a Rough set.So, here, the dynamics of growth of tumor cells is dealt with the Rough set. We take the data set (Krug and Taubert, 1985, [21] [22]) of approximate values of the number of tumor cells with time, and we found the relation of the number of tumor cells with time (Figure 3). We found the relationship through the simulation process (Figure 2)that is the same as the relation that generated through the Rough Set Exploration System (RSES 2.2.2) (Figure 3).

4.3 The result from the Rough set

By using the approximate values of the number of tumor cells (Krug and Taubert, 1985, [21] [22]) and the Rough Set Exploration System (RSES 2.2.2), the relationship found between the number of tumor cells and time, that shows the number of tumor cells increases with time, and after some time, it reaches the constant state (Figure 3).

4.4 General statistics for the "number of cells of tumor" attribute

Using the Rough set, the statistics for the "number of cells of tumor" attribute are seen. Here, the mean value of the number of cells of tumors is 94.317, the standard deviation is 58.47, the lower limit of values is 3.33, and the upper limit of values is 149.3 in Table 1. The maximum number of cells growth values (i.e., 8) is in the range between 134.703 and 149.3, and the minimum number of cells growth values (i.e., 0) is in the range between 32.524 and 47.121 and 76.315 and 90.912 for the number of cells of tumor attribute in Figure 4 (A).

4.5 General statistics for the "time" attribute

Using the Rough set, we can see the statistics for the "time" attribute. Here, the mean value of time is 8.5, the standard deviation is 5.339, the lower limit of values is 0, and the upper limit of values is 17 in Table 2. The maximum number of values (i.e., 2) are in the range ([0, 1.7], [1.7, 3.4], [3.4, 5.1], [6.8, 8.5], [8.5, 10.2], [11.9, 13.6], [13.6, 15.3] and [15.3, 17.2]) and the minimum number of values (i.e., 1) are in the range ([5.1, 6.8] and [10.2, 11.9]) for the time attribute Figure 4 (B).

5. Conclusion and Discussion

Tumors are the primary reason for death and put a massive load on the medical system because of the chronic characteristics of the disease and typically undesirable effects produced by many of the treatments. The Rough set is involved with the characterization and examination of imprecise data information and considered as one of the principal non-statistical methodologies in information investigation.

The fundamental assumptions from various references of international reputed research articles are taken and formulated the mathematical model of nonlinear tumor cells growth in the human body, which validated by using the Rough set. We found that the relation between the number of tumor cells and time that the number of tumor cells increases, and after some time, it reaches to constant state (Figure 2). The Rough set for the validation of the model applied and found the result using the Rough Set Exploration System (RSES 2.2.2) that the relationship between the number of tumor cells and time, that the number of tumor cells increases with time, and after some time it reaches the constant state (Figure 3).

The result from the simulation process using MATLAB and the Rough Set Exploration System (RSES 2.2.2) is the same, i.e., the number of tumor cells increases with time, and after some time, it reaches the carrying capacity that is the growth of tumor cells reaches the constant state.

Further, experimentally, the results of approximation using the logistic equation for $K = 150\times10^7$ [21] [22]), and in our model, we have obtained the estimate of the number of tumor cells ($\approx 149.30\times10^7$) in the simulation process. So our model is only differing from $K = 0.7\times10^7$ significant digits, and we have obtained the **accuracy** of carrying capacity of the number of tumor cells **99.53% correct** in the body according to our model.

Therefore, these advances offer novel insights for tumor growth, which further supports research in tumor cell dynamics.

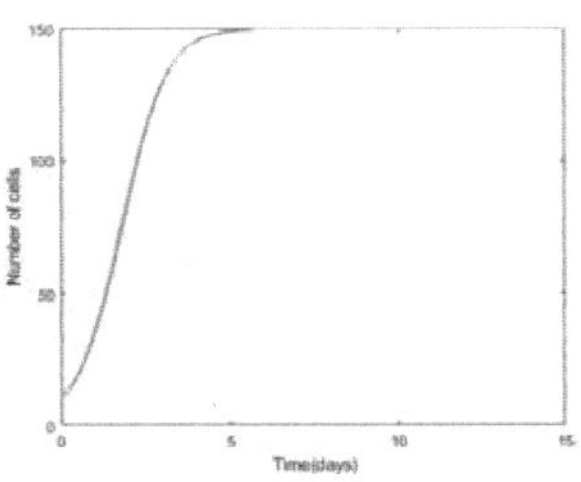

Figure 2. The number of tumor cells increases with time, and after some time, it reaches the constant state.

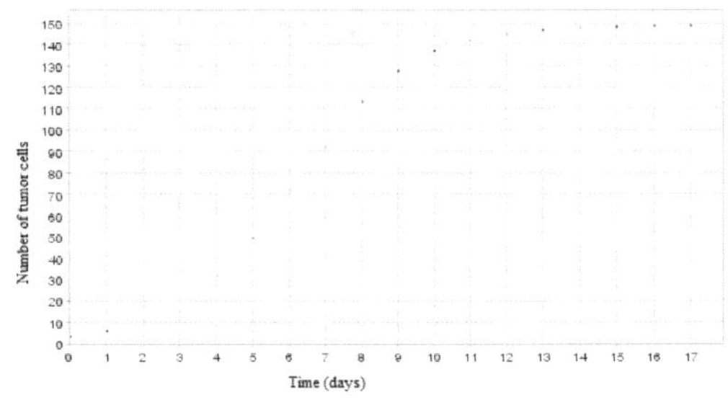

Figure 3. The number of tumor cells increases with time, and after some time, it reaches the constant state constant state by the Rough set exploratory system (2.2.2).

Table 1. General statistics of the "number of tumor cells"attribute

General Statistics

Number of objects: 18
Number of attributes: 2

Attributes: Number of tumor cells
Status: condition
Type: numeric
Precision: 4
Mean: 94.317
Standard deviation: 58.47
Minimum: 3.33
Maximum: 149.3

Attributes	Attribute range
Number of tumor cells	Lower limit of values: 3.33
Time	Upper limit of values: 149.3

Table 2. General statistics of the "time" attribute

General Statistics

Number of objects: 18
Number of attributes: 2

Attributes: Time
Status: condition
Type: numeric
Precision: 4
Mean: 8.5
Standard deviation: 5.339
Minimum: 0
Maximum:17

Attributes	Attribute range
Number of tumor cells	Lower limit of values: 0
Time	Upper limit of values: 17

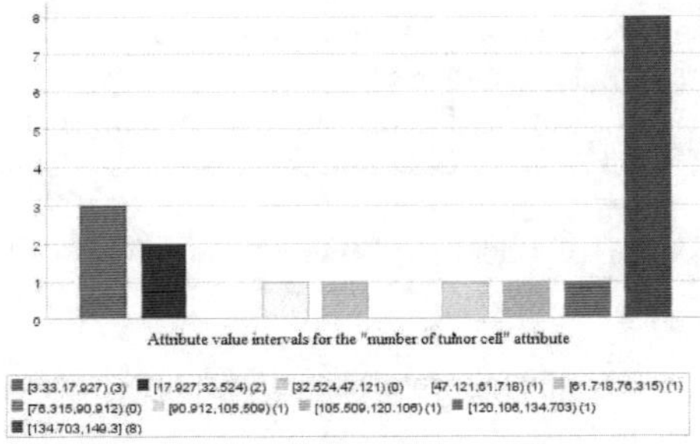

Figure 4 (A). Attribute value interval for the "number of tumor cells" attribute

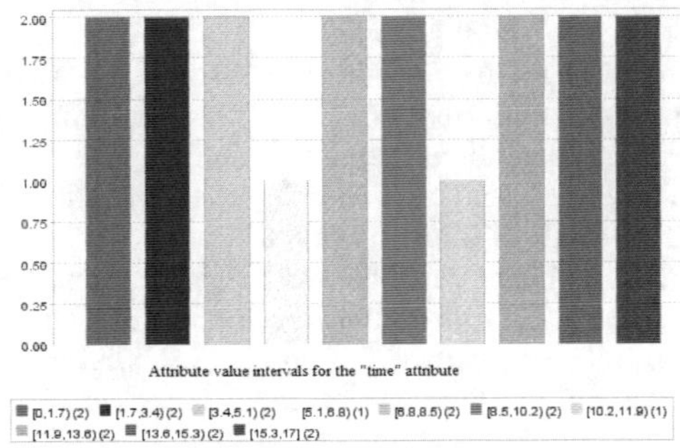

Figure 4 (B). Attribute value interval for the "time" attribute

Figure 4 (A) - (B): The bar diagrams found from the Rough set using the parameters, number of tumor cells and time **(A)** attribute value interval for the "number of tumor cells" attribute which shows that the maximum number of tumor cells growth values (i.e., 8) is in the range between 134.703 and 149.3 and the minimum number of cells growth values (i.e., 0) are in the range between 32.524 and 47.121 and 76.315 and 90.912 for the number of cells of tumor attribute; **(B)** attribute value interval for the "time" attribute which shows that the maximum number of values (i.e., 2) are in the range between [0, 1.7], [1.7, 3.4], [3.4, 5.1], [6.8, 8.5], [8.5, 10.2], [11.9, 13.6], [13.6, 15.3] and [15.3, 17] and the minimum number of values (i.e., 1) are in the range between [5.1, 6.8] and [10.2, 11.9] for the time attribute.

Acknowledgments
The authors are extremely thankful, to the Department of Mathematics, NIT Raipur (C. G.), India for providing facilities, space and an opportunity for the work.

Conflict of Interest
The authors declare that there is no conflict of interest.

References

[1] Clark W H 1991 Tumour progression and the nature of cancer*Br. J. Cancer***64** 631–644. Doi: 10.1038/bjc.1991.375
[2] Elder D E, Rodeck U, Thurin J, Cardillo F, Clark WH, Stewart R and Herlyn M 1989 Antigenic profile of tumor progression stages in human melanocytic nevi and melanomas *Can. Research***49** 5091–5096.
[3] Kerbel R S, Waghorne C, Korczak B, Lagarde A and Breitman M L 1988 Clonal dominance of primary tumours by metastatic cells: genetic analysis and biological implications *Can. Surveys* **7** 597–629.
[4] Batmani Y and Khaloozadeh H 2013 Optimal drug regimens in cancer chemotherapy: A multi-objective approach *Comp. Bio. Medicine***43** 2089–95. Doi: 10.1016/j.compbiomed.2013.09.026.
[5] Huang X, Ning J and Wahed A S 2014 Optimization of individualized dynamic treatment regimes for recurrent diseases *Stat. Medicine***33**2363–78. Doi: 10.1002/sim.6104.
[6] Elias J,Dimitrio L, Clairambault J and Natalini R 2014 The p53 protein and its molecular network: Modelling a missing link between dna damage and cell fate *BiochimBiophys Acta Proteins Proteomics***1844** 232–47. Doi: 10.1016/j.bbapap.2013.09.019.
[7] Laird A K 1965 Dynamics of tumour growth: Comparison of growth rates and extrapolation of growth curve to one cell*Br. J. Cancer***19**278–91. Doi: 10.1038/bjc.1965.32.
[8] Laird A K 1964Dynamics of tumors growth *Br.J. Cancer.***13**490–502. Doi:

10.1038/bjc.1964.55.

[9] Brodin N P, Vogelius I R, Eriksson T B, AfRosenschold P M, Maraldo M V, Aznar M C, Specht L and Bentzen S M 2014 Optimizing the radiation therapy dose prescription for pediatric medulloblastoma: Minimizing the life years lost attributable to failure to control the disease and late complication risk*ActaOncologica* **53**462–70. Doi: 10.3109/0284186X.2013.858824.

[10] Ercan B, İlhanÖ andSenolK 2019Dynamical behaviour of fractional order tumor model with Caputo and conformable fractional derivative*Chaos, Solitons & Fractals***123** 43-51. Doi: https://doi.org/10.1016/j.chaos.2019.03.032.

[11] Panetta J C 1998 A mathematical model of drug resistance: Heterogeneous tumors *Math. Biosciences***147** 41–61. Doi: 10.1016/S0025-5564(97)00080-1.

[12] Sakode C M, Padhi R, Kapoor S, Rallabandi V P S and Roy P K 2014 Multimodal therapy for complete regression of malignant melanoma using constrained nonlinear optimal dynamic inversion *Biomed. Sig. Pro. Control***13** 198–211. Doi: 10.1016/j.bspc.2014.04.010.

[13] De Pillis L G, Gu W and Radunskaya A E 2006 Mixed immunotherapy and chemotherapy of tumors: modeling, applications and biological interpretations *J. Theo. Biology* 238841–62. Doi: 10.1016/j.jtbi.2005.06.037.

[14] Rihan F A, Lakshmanan S and Maurer H 2019Optimal control of tumor-immune model with time-delay and immuno-chemotherapy*Appl. Math. Computation***353**147-165. Doi: https://doi.org/10.1016/j.amc.2019.02.002.

[15] Huang Y, Zhang Z andHu B 2017Bifurcation for a free-boundary tumor model with angiogenesis*Nonli. Analy.: Re. Wor. Applications***35**483-502. Doi: https://doi.org/10.1016/j.nonrwa.2016.12.003.

[16] Jin Y, Yuanshun T andRobert A C2019Modelling effects of a chemotherapeutic dose response on a stochastic tumor-immune model*Chaos, Solitons & Fractals***123** 1-13. Doi: https://doi.org/10.1016/j.chaos.2019.03.029.

[17] Parajdi L 2014 Modeling the treatment of tumor cells in a solid tumor*J. Nonli. Sci. Applications***7** 188-195.

[18] Alavi S A, Norabadi J and Arjmand M 2012 Optimal Control Brain Tumor System with Drog and Its Stability*J. Math. Comp. Science***4** 473 – 486.

[19] Zheng J andCui S 2019Analysis of a tumor-model free boundary problem with a nonlinear boundary condition*J. Math. Analy. Applications***478** 806-824. Doi: https://doi.org/10.1016/j.jmaa.2019.05.056.

[20] Kassem M A E H, Hemeda A A and Abdeen M A 2020 Solution of the tumor-immune system by differential transform method *J. Nonli. Sci. Applications*13 9–21. Doi: 10.22436/jnsa.013.01.02.

[21] Krug H and Taubert G 1985Zur praxis der anpassung der logistischen function an das wachstum experimenteller tumoren*Arch. Geschwulstforsch***55** 235–244.

[22] Forys U and Czochra A M 2003 Logistic equations in tumour growth modelling, *Int. J. Appl. Math. Comput. Science* **13** 317–325.

[23] Enderling H and Chaplain A J M 2014Mathematical modeling of tumor growth and treatment*Curr Pharm Design***20**4934-40. Doi: 10.2174/13816128196661311251150434.

[24] Hahnfeldt P, Panigrahy D, Folkman J and Hlatky L 1999 Tumors development under angiogenic signaling: a dynamical theory of tumors growth, treatment response, and postvascular dormancy*Can. Research***59** 4770-5.

[25] Iwata K, Kawasaki K and Shigesada N 2000 A dynamical model for the growth and size distribution of multiple metastatic tumors *J. Theo. Biology***203** 177–86.

[26] Jain M and Bhatia M P S 2014A rough set based approach to classify node behavior in mobile Ad hoc networks*J. Math. Comp. Science***11** 64-78.

[27] Polkowski L 2002*Advance in soft computing: Rough sets mathematical foundations*(Physical-Verlag A Springer-Verlag Company).Doi: 10.1007/978-3-7908-1776-8.

[28] Peters G, Lingras P and Slezak D 2012*Rough sets selected methods and applications in management and engineering*(Springer).Doi: 10.1007/978-1-4471-2760-4.

[29] Lin T Y and Cercone N1997 *Rough sets and data mining analysis of imprecise data*(Springer).Doi: 10.1007/978-1-4613-1461-5

[30] *RSES 2.2 User's Guide* 2005*Warsaw University*. http://logic.mimuw.edu.pl/»rses.

An efficient way of material classification using uncertainty assessment

Arvind Kumar Sinha[1*] **and Pradeep Shende**

Department of Mathematics, National Institute of Technology, Raipur, Chhattisgarh India

*Corresponding author's e-mail address: aksinha.maths@nitrr.ac.in

Abstract. Analysis of material properties using the available data is a compulsory part of the material classification and is considered as a challenging task in industrial processes. Classification refers to the segregation of objects using known data or knowledge. The uncertainties in the data increase the complexities of material classification. In this paper, we present a methodology to measure uncertainties in the attributes-values. An example illustrates the efficiency, validity, and effectiveness of the proposed method. We find the most significant features for the classification of glass materials using uncertainty evaluation and optimization.

1. Introduction

Knowledge-based systems use available data to find useful information out of it. It is a very challenging task, as incomplete or inexact data may yield misleading information. Such data leading to wrong information is termed as uncertainties in the data set [1] . In today's time, knowledge is accumulating at the exponential pace, and with this hoard of information, data evaluation complexities surge intensely. Finding useful patterns out of the available information becomes obscured due to the presence of uncertainties in it. Many approaches are studied in the literature to deal with uncertain values in the data-set. One of the very effective ways to deal with vagueness is fuzzy sets introduced by Zadeh [2]. In 1982, Pawlak [3] introduced the rough set theory (RST) as a new approach to deal with uncertainties. Both the Fuzzy set theory and RST finds vast scientific applications in almost all fields. RST requires only available information and is very useful in reasoning about data and to find helpful information out of it. The material information stored using various attributes of the material is called an information system. We use the data of elements in the form of the decision table. The collection of elements or instances or examples whose information is stored in the information system using various attributes is the universe of discourse or universal set. The set of all features is called the feature space or the set of attributes. These features lead to the labeling of objects using the decision feature. The values of the decision attributes are termed as decision class, and each decision class denotes the concept in the information system. The information system whose instances are labeled using the decision attribute is term as a decision system. The learning from the decision system is supervised learning, and learning from examples without the decision parameter is termed as unsupervised learning. The feature values add knowledge to the information system and are called as granules of knowledge [4]. Material classification is the segregation of distinct elements using the various attributes. Due to missing information, different objects seem indistinguishable. In such cases, the classification of elements using the most significant parameters becomes mandatory. Elements

classification using RST is studied in [5]. Uncertainty assessment is useful in finding the significance of the attribute [6]. Also, it helps to find the most suitable combination of features that keeps the classification of the elements intact. Finding features subset out of the given set of features is termed as feature subset selection (FSS).
[7]proposed a wrapper method for FSS based on the evaluation of subsets of the feature set for classification. FSS using a genetic algorithm is studied in [8]. Optimization of the feature space is studied in [9]. Decision-rule is another way of representing data. These rules help tremendously in recognition of new instances. These rules also find useful in the categorization of new elements [10]. The decision rules generation in case of missing information is studied in [11,12]. Statistical inference based decision rules generation is studied in [13]. The variable precision rough set model is discussed in [14], which is the generalization of RST using optimization of uncertainties allowing small errors in the rule set. The data reduction and feature optimization processes mainly evaluate uncertainties in the information system. We assess the uncertainties in the feature values by using the range of decision classes.

2. Data pre-processing

Data pre-processing is the process of cleaning the raw data employing the removal of missing values, if any, reducing noise and non-linearity, discretization processes, etc. We remove those elements whose attribute values outdistance from the remaining values. In the case of continuous attribute values, we get a vast number of rule-set. Discretization is the representation of continuous attribute values to its categorical counterparts. The complete data preprocessing method is represented in Figure 1.

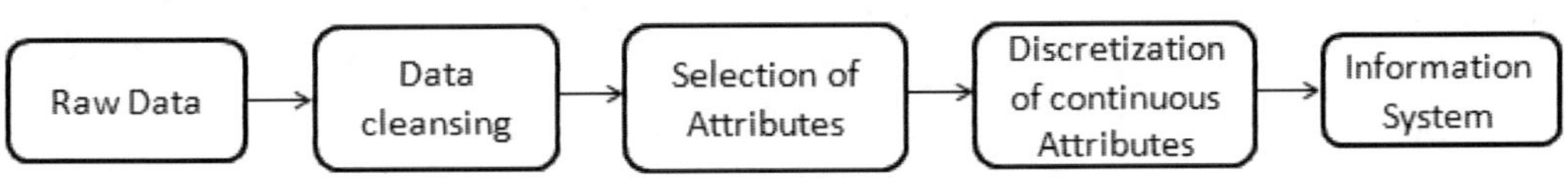

Figure 1. Material data pre-processing

2.1. *Rough set reducts*

The uncertain features severely escalate the noise in the information system. Finding the combination of features among the given feature set, which keeping the classification of objects intact, is termed as feature subset selection (FSS). The rough set provides skillful algorithms for FSS. Let $I=(U, A \cup \{d\})$ is the information system where U is the universal set of elements, A is the features space and d is the decision feature. Let any $\alpha \in A$ is the feature whose uncertain region is to be determine, $\alpha(o)$ is the feature value of object o for feature α and V_α is the set of feature values of feature α then V_α is given by the function $f: \alpha \rightarrow V_\alpha$ defined as $f(o) = V_{\alpha(o)}, \quad o \in U$.

Let $B \subseteq A$, define an equivalence relation $R(B)$ on U as follows:

$$for\ o_i, o_j \in U, \quad o_i\ R(B)\ o_j \ \ iff\ \ V_{\alpha(o_i)} = V_{\alpha(o_j)}, \ \forall \alpha \in B \qquad (1)$$

The cells of the relation $R(B)$ are

$$[o_i]_{R(B)} = \{o_j : o_i\ R(B)\ o_j\} \qquad (2)$$

Each feature-value represents a concept in the decision system I. Let δ be a concept in I, approximation of δ using feature-values of $B \subseteq A$ by creating the $B-lower$ approximation and $B-upper$ approximation of δ denoted by $B_*(\delta)$ and $B^*(\delta)$ respectively where,

$$B_*(\delta) = \{o : [o]_{R(B)} \subseteq \delta\} \quad and \quad B^*(\delta) = \{o : [o]_{R(B)} \cap \delta \neq \phi\} \tag{3}$$

$B-Boundary$ Region of δ $BN_R(\delta) = B^*(\delta) - B_*(\delta)$ amount to the elements whose precise categorization into the concept δ is not possible using the feature B. The rough set incorporates the concepts with an imprecise boundary region. This boundary region embodies the uncertain-region.We call the feature δ as dispensable in $B \subseteq A$ if $[o_i]_{R(B)} = [o_i]_{R(B-\{a\})}$, $\forall o_i \in U$. The non-dispensable features are termed as indispensable. The indispensable elements set are described as independent set. If $B \subseteq A$ is independent and $[o_i]_{R(B)} = [o_i]_{R(A)}$, $\forall o_i \in U$ then B is described as reduct of A.

3. Uncertainties assessment methodology

The equivalence relation $R(\alpha)$ on U using the feature $\alpha \in A$ is defined as

$$o_1\ R(\alpha)\ o_2 \quad iff \quad V_{\alpha(o_1)} = V_{\alpha(o_2)} \tag{4}$$

The cells of the relation $R(\alpha)$ are obtained as

$$[o]_{R(\alpha)} = \{\xi : o\ R(\alpha)\ \xi\}, \quad o, \xi \in \alpha \tag{5}$$

For the concepts δ_i, $i = 1, 2, ..., k$ in the decision system, we designate the membership of any $o \in U$ in the concept δ_i as

$$\mu_{\delta_i}^{\alpha}(o) = \frac{|\delta_i \cap [o]_{R(\alpha)}|}{|[o]_{R(\alpha)}|} \tag{6}$$

where, $|\cdot|$ denotes the cardinality. We designate the feature-range with corresponding to decision-class δ_i as

$$\alpha_{\delta_i} = \{o \in U : \mu_{\delta_i}^{\alpha}(o) > 0\} \tag{7}$$

If $0 < \mu_{\delta_i}^{\alpha}(o) < 1$ then o is the imprecise case and its categorization into the concept δ_i is not possible. The uncertain-region $Unc(\alpha)$ of the feature α is

$$Unc(\alpha) = \bigcup_{i \neq j} \{\alpha_{\delta_i} \cap \alpha_{\delta_j}\}, \qquad i, j = 1, 2, ..., k \tag{8}$$

The certain-space $Cer(\alpha)$ of the feature α is

$$Cer(\alpha) = U \setminus Unc(\alpha) \tag{9}$$

Let $R(\alpha)_1$, $R(\alpha)_2$,..., $R(\alpha)_\eta$ be the classes of the relation $R(\alpha)$ on the feature α, then we prescribe the significance-factor σ_α of the feature α as:

$$\sigma_\alpha = \sum_{i=1}^{\eta_\alpha} \frac{|Cer(\alpha) \cap R(\alpha)_i|}{|R(\alpha)_i|} \tag{10}$$

Figure 2 is the flowchart of the feature subset selection (FSS) using the uncertain-region optimization of the features.

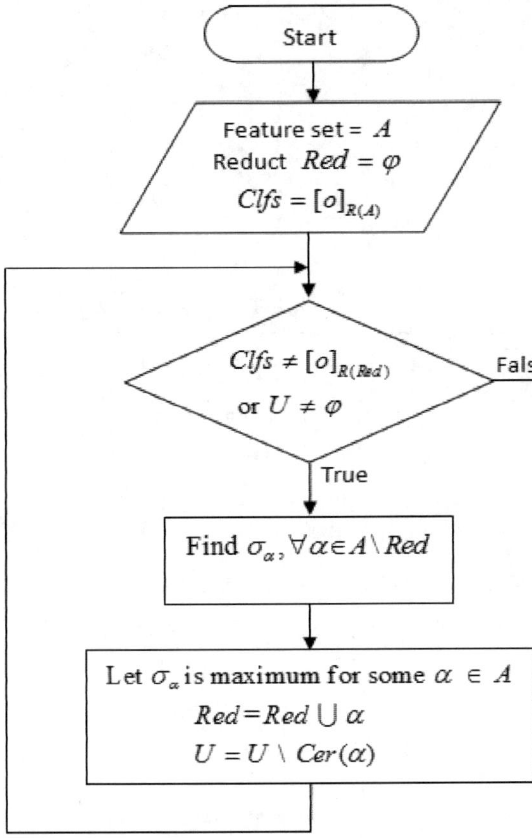

Figure 2. Flowchart of FSS using uncertainty optimisation

Following example demonstrate the methodology of FSS using certainty region of features. The Table 1 is the decision table consisting of information of seven elements using four features α_1, α_2, α_3 and α_4. Using the information given in Table 1, the classification of the examples is given by $Clfs = \{\{E1, E3, E6\}, \{E2\}, \{E5\}\}$. Examples $E4$ and $E7$ are conflicting cases, and its precise classification is not possible.

Table 1. Example of the Decision table.

Example	α_1	α_2	α_3	α_4	Decision
$E1$	2	1	0	1	1
$E2$	0	2	0	1	0
$E3$	0	1	1	2	1
$E4$	1	2	0	1	1
$E5$	0	0	1	2	2
$E6$	1	0	2	2	1
$E7$	1	2	0	1	2

For Table 1, feature space $A = \{\alpha_1, \alpha_2, \alpha_3, \alpha_4\}$ and universal set of examples $U = \{E1, E2, E3, E4, E5, E6, E7\}$. There are three decision classes $\delta_1 = 0, \delta_2 = 1, \delta_3 = 2$. Using the equivalence relation defined in equation (4), for feature α_1 we get the classes as $[E1]_{\alpha_1} = \{E1\}, [E2]_{\alpha_1} = [E3]_{\alpha_1} = [E5]_{\alpha_1} = \{E2, E3, E5\}, [E4]_{\alpha_1} = [E6]_{\alpha_1} = [E7]_{\alpha_1} = \{E4, E6, E7\}$ For feature α_2, we get the classes as $[E1]_{\alpha_2} = [E3]_{\alpha_2} = \{E1, E3\}$, $[E5]_{\alpha_2} = [E6]_{\alpha_2}$, $[E2]_{\alpha_2} = [E4]_{\alpha_2} = [E7]_{\alpha_2} = \{E2, E4, E7\}$. For feature α_3, the classes are $[E1]_{\alpha 3} = [E2]_{\alpha_3} = [E4]_{\alpha_3} = [E7]_{\alpha_3} = \{E1, E2, E4, E7\}$, $[E3]_{\alpha_3} = [E5]_{\alpha_3} = \{E3, E5\}$, $[E4]_{\alpha_3} = [E7]_{\alpha_3} = \{E4, E7\}$. Also the classes for the feature α_4 are given by $[E1]_{\alpha_4} = [E2]_{\alpha_4} = [E4]_{\alpha_4} = [E7]_{\alpha_4} = \{E1, E2, E4, E7\}$ and $[E3]_{\alpha_4} = [E5]_{\alpha_4} = [E6]_{\alpha_4} = \{E3, E5, E6\}$. The certaintyregion for the features is obtained as $Cer(\alpha_1) = \{E1\}, Cer(\alpha_2) = \{E1, E3\}, Cer(\alpha_3) = \{E6\}, Cer(\alpha_4) = \phi$. Using equation (10), we get the significance of the features as $\sigma_{\alpha_1} = 1, \sigma_{\alpha_2} = 2, \sigma_{\alpha_3} = 1, \sigma_{\alpha_4} = 0$. The feature with maximum significance factor is α_2. Hence adding feature α_2 in the set $\mathrm{Re}d$ and removing the elements which belongs to $Cer(\alpha_2)$, the reduced decision table obtained is specified in

Table 2. Reduced decision table.

Example	α_1	α_3	α_4	Decision
$E2$	0	0	1	0
$E4$	1	0	1	1
$E5$	0	1	2	2
$E6$	1	2	2	1
$E7$	1	0	1	2

We have $\text{Red} = \{\alpha_2\}$. Note that $[o]_{R(A)} \neq [o]_{R(\text{Red})}$. Proceeding in the same manner, here we get the classes as $[E2]_{\alpha_1} = [E5]_{\alpha_1} = \{E2, E5\}$, $[E4]_{\alpha_1} = [E6]_{\alpha_1} = [E7]_{\alpha_1} = \{E4, E6, E7\}$, $[E2]_{\alpha_3} = [E4]_{\alpha_3} = [E7]_{\alpha_3} = \{E2, E4, E7\}$, $[E5]_{\alpha_3} = \{E5\}$, $[E6]_{\alpha_3} = \{E6\}$, and for feature we get $[E2]_{\alpha_4} = [E4]_{\alpha_4} = [E7]_{\alpha_4} = \{E2, E4, E7\}$, $[E5]_{\alpha_4} = [E6]_{\alpha_4} = \{E5, E6\}$. The certainty region is given by $Cer(\alpha_1) = \phi$, $Cer(\alpha_3) = \{E5, E6\}$, $Cer(\alpha_4) = \phi$. The significance factor is $\sigma_{\alpha_1} = 0$, $\sigma_{\alpha_3} = 2$, $\sigma_{\alpha_4} = 0$. Therefore the feature added to set Red is α_3. After eliminating the examples belonging to $Cer(\alpha_3)$ we get the reduced decision table shown in Table 3.

Table 3. Reduced decision table.

Example	α_1	α_4	Decision
$E2$	0	1	0
$E4$	1	1	1
$E7$	1	1	2

Here $[o]_{R(A)} \neq [o]_{R(\text{Red})}$. Here we get $[E2]_{\alpha_1} = \{E2\}$, $[E4]_{\alpha_1} = [E7]_{\alpha_1} = \{E4, E7\}$., $[E2]_{\alpha_4} = [E4]_{\alpha_4} = [E7]_{\alpha_4} = \{E2, E4, E7\}$. The certainty region is given by $Cer(\alpha_1) = \{E2\}$, and $Cer(\alpha_4) = \phi$. The significance factor obtained as $\sigma_{\alpha_1} = 1$ and $\sigma_{\alpha_4} = 0$. Thus the feature added to the set Red is α_1. At this stage we get $[o]_{R(A)} = [o]_{R(\text{Red})}$.. Therefore the reduct set obtained is $\text{Red} = \{\alpha_1, \alpha_2, \alpha_3\}$. The feature α_4 is dispensable.

4. Research Data

In this section, we present the data used in the study. We use the data of materials concentration in the glass given in the UCI machine learning repository [15]. The glass classification utilizing the information of its nine constituent elements is given in the data. The data set comprises information of a total of 214 objects. The elements are classified into six classes using the nine features. All the features are continuous-valued. Table 4 gives detailed of the data-set used in the study. The feature noted as RI shows the refractive index of the glass. Other features notation is the standard notation of elements. Among the 214 objects, 163 objects are the example of windows glass, and 51 objects are the example of the non-windows glass. Among the 163 examples of the glass of the windows, 87 objects are float-processed, and 56 objects are non-float processed. We applied the maximum discernibility discretization algorithm for feature-values discretization. The algorithm is implemented in ROSETTA toolkit [16] which provides a platform for rough set methods. Using the discretized features, we applied the FSS using uncertainty assessment; we found the reduct of the feature space as $\{RI, Na, Mg, Si, Ba\}$. To compare the classification accuracy we used the SVM classification algorithm [17] with 10-fold validation.The comparison of the classification accuracy using SVM to the feature set and its reduct is given in Table 5.

Table 4. Description of the data set used in the study.

Sr.No.	Feature	Min	Max	Average	SD	Correlation With Class
1.	*RI*	1.5112	1.5339	1.5184	0.0030	-0.1642
2.	*Na*	10.73	17.38	13.407	0.8166	0.5030
3.	Mg	0	4.49	2.684	1.4424	-0.745
4.	Al	0.29	3.5	1.445	0.4993	0.5988
5.	*Si*	69.81	75.41	72.6509	0.7745	0.1515
6.	*K*	0	6.21	0.4971	0.6522	-0.0100
7.	*Ca*	5.43	16.19	8.9570	1.4232	0.0007
8.	*Ba*	0	3.15	0.1750	0.4972	0.5751
9.	*Fe*	0	0.51	0.0570	0.0974	-0.1879

There are 70 examples of float-processed windows building window glass and denoted by class 1. Decision class 3 comprises of Seventeen examples are the float processed vehicle windows glass. Seventy-six examples of the non-float glass of building windows are denoted by class 2. Decision class 5 are examples of container glass. Decision classes 6 and 7 represent the examples of tableware glass and headlamps, respectively.

Table 5. Accuracy obtained using feature space and its reduct.

Set	TP	FP	Precision	Recall	F-measure	Accuracy
Feature Space	0.67	0.15	0.63	0.67	0.65	67.29%
Reduct	0.72	0.11	0.71	0.73	0.7	72.89%

Table 5 shows that the accuracy obtained in the case of reduct is significantly higher than the accuracy of full feature space. The features RI and Na found the most significant for the classification of the glass of class 1. Feature Barium Ba found most notable for the categorization of headlamp glass. Silicon Si found most significant for the classification of tableware glass. Calcium Ca obtained as a most significant feature for the classification of non-float processed building window glass.

5. Conclusion

Imprecise or inexact data may yield misleading information. In this work, we presented a novel method for assessing the uncertainties in the feature space. A flowchart for finding the combination of features among the feature space, which keeps the entire information about the dataset as the feature space is specified. We introduced an innovative methodology for feature selection using uncertainty optimization. Using the proposed method, we evaluated the glass material dataset and obtained the accuracy of classification significantly more than the accuracy of the entire feature space.

References

[1] Alchian A A 1950 Uncertainty, evolution, and economic theory *J. Polit. Econ.* **58** 211–21
[2] Zadeh L A 1965 Fuzzy sets *Inf. Control* **8** 338–53
[3] Pawlak Z 1982 Rough sets *Int. J. Comput. Inf. Sci.* **11** 341–56
[4] Pawlak Z 1998 Granularity of knowledge, indiscernibility and rough sets *1998 IEEE International Conference on Fuzzy Systems Proceedings. IEEE World Congress on Computational Intelligence (Cat. No. 98CH36228)* vol 1 pp 106–10
[5] Pawlak Z 1984 Rough classification *Int. J. Man. Mach. Stud.* **20** 469–83
[6] Jurado K, Ludvigson S C and Ng S 2015 Measuring uncertainty *Am. Econ. Rev.* **105** 1177–216
[7] Kohavi R and John G H 1997 Wrappers for feature subset selection *Artif. Intell.* **97** 273–324
[8] Yang J and Honavar V 1998 Feature subset selection using a genetic algorithm *Feature extraction, construction and selection* (Springer) pp 117–36
[9] Koller D and Sahami M 1996 *Toward optimal feature selection*
[10] Apté C, Damerau F and Weiss S M 1994 Automated learning of decision rules for text categorization *ACM Trans. Inf. Syst.* **12** 233–51
[11] Grzymala-Busse J W and Werbrouck P 1998 On the best search method in the LEM1 and LEM2 algorithms *Incomplete Information: Rough Set Analysis* (Springer) pp 75–91
[12] Holmström B and Myerson R B 1983 Efficient and durable decision rules with incomplete information *Econom. J. Econom. Soc.* 1799–819
[13] Cencov N N 2000 *Statistical decision rules and optimal inference* (American Mathematical Soc.)
[14] Ziarko W 1993 Variable precision rough set model *J. Comput. Syst. Sci.* **46** 39–59
[15] Dua D and Graff C 2017 {UCI} Machine Learning Repository
[16] Nguyen H S 1998 Discretization problem for rough sets methods *International Conference on Rough Sets and Current Trends in Computing* pp 545–52
[17] Cortes C and Vapnik V 1995 Support-vector networks *Mach. Learn.* **20** 273–97

IOP Conf. Series: Materials Science and Engineering **798** (2020) 012040 doi:10.1088/1757-899X/798/1/012040

Second-formulation adaptive wavelet optimization technique applicable in material sciences

Arvind Kumar Sinha
Department of Mathematics, National Institute of Technology Raipur, India
E-mail: aksinha.maths@nitrr.ac.in

Radhakrushna Sahoo
Department of Mathematics, National Institute of Technology Raipur, India
E-mail: radhakrushnasahoo92@gmail.com

Abstract. We developed a second-formulation adaptive wavelet optimize finite element (SFAWOFE) method on a general manifold with the help of approximation theory. The SFAWOFE method used to solve Burger's equation with periodic and Dirichlet boundary conditions. The beauty of the SFAWOFE process is the numerical solution is optimized for the finite element method and does not affect the computational algorithm. We construct an algorithm for numerical results that have been optimized finite element method on a diffusion wavelet, which formed employing the adaptive wavelet plot. The algorithms and trial problems used to calculate magnetostatic networking range dipoles in material science. For the trial problem, the processing time checked out the SFAWOFE method has calculated and analyzed the analogous processing time checked out the finite element method on a general manifold. Finally, we verify the SFAWOFE method convergence for each trial problem and high efficient.

Key words: Adaptive wavelet; approximation theory; finite element method; trial problem.

1. Introduction

Wavelet first interpolated by Grossmann and Morlet in 1980 [1] since then they maintain expressively ingrowing a great deal of intellect and metallurgy [2, 3]. Wavelets are construction wedges for familiar data sheets or challenges. In general, we shall wavelet forms an orthonormal basis for multiresolution analysis with the climbing function of a wavelet frame [4]. Wavelets are exploited, in the wavelet frame assignment of an imperious numerical rank on an adaptive plot to discover the network has to be clarified or coarcted to characterize the solution optimally are faster to calculate. Alternatively, of developing the results in form of wavelets, the adaptive transform is utilized to dictate the wavelet lattice [5, 6, 7]. The significant arrogance of second-formulation wavelets is the same as always wavelets are concocted straight real-time domain while one-hit-wonder to solve your problems; therefore, the invite formulated is boundary conditions disparate.
Excellent wavelet compaction features[8, 9] short fulcrum, departure moment [10], and the capacity to determine confined framework formed the wavelets a useful appliance future exploited meant for the numerical solution of the partial differential equation(PDEs) [11, 12]. Wavelet optimized finite element (WOFE) interpolated by Jameson [4]. In WOFE, wavelets exploited

IOP Conf. Series: Materials Science and Engineering **798** (2020) 012040 doi:10.1088/1757-899X/798/1/012040

to obtain an adaptive plot. Alternatively, we developing the climbing function, and climbing transformation of the wavelet transform is utilized for the solution, while the finite element grid has to be clarified or coarcted [13]. In this framework, we are going to solve Burger's equation with periodic and Dirichlet boundary conditions with the help of an adaptive wavelet optimized finite element technique. The least-squares method exploited for the approach of the partial differential equation implicated in that manner Burger's equation, and such as second-formulation wavelet exploited to dictate wheres limited dissimilarity plot needs to be clarified or coarcted to express the solution precisely.

2. Second-formulation wavelet transform

This explanation of the second-formulation multiresolution analysis convinces expeditious second-formulation wavelet transform. Accustomed procedure in the vicinity of the framework for wavelets with rare up heave is aware of ensure:

1. Delineation the diffusion wavelet transform with out an allusion to thaw climbing functions ϕ, $\overline{\phi}$ and wavelet function ψ, $\overline{\psi}$.
2. The climbing function ϕ, $\overline{\phi}$ and wavelet function ψ, $\overline{\psi}$ a retail compile adopting diffusion wavelet transform.

Consequently, the most significant gait is the framework of diffusion wavelet transform and inverse diffusion wavelet transform, which is deliberate follow.
Conclave function $h \in l_2(R)$, the climbing function co-active $\eta_{k,m}$ and wavelet co-active $\kappa_{k,m}$ are defined as ensue:
$\eta_{k,m} = \langle h, \overline{\phi}_{k,m}\rangle$, $\kappa_{k,m} = \langle h, \overline{\psi}_{k,m}\rangle$, where $\overline{\phi}_{k,m} = 2^{k/2}\overline{\phi}(2^k r - n)$. Let us postulate $\eta_{s,m}$ is given for stationary s and we want to calculate $\eta_{k,m}$ and $\kappa_{k,m}$ for $k < s$. This can be done concluded the iterative application of the elaboration relation

$$\phi(r) = 2\sum_n f_n \phi(2r - n)$$

can be written as

$$\begin{aligned}\overline{\psi}_{k,m} &= 2^{k/2}\overline{\phi}(2^k r - m),\\ &= 2^{k/2} \times 2\sum_n \overline{f}_n \overline{\phi}(2(2^k r - m) - n),\\ &= \sqrt{2}\sum_n \overline{f}_{n-2m}\overline{\psi}_{k+1,n}.\end{aligned}$$

$$\begin{aligned}\eta_{k,m} &= \langle h, \overline{\phi}_{k,m}\rangle,\\ &= \langle h, \sqrt{2}\sum_n \overline{f}_{n-2m}\overline{\phi}_{k+1,n}\rangle,\\ &= \sqrt{2}\sum_n \overline{f}_{n-2m}\eta_{k+1,n}.\end{aligned}$$

similarly,

$$\kappa_{k,m} = \sqrt{2}\sum_n \overline{g}_{n-2m}\eta_{k+1,n}.$$

The inverse transform iterative benediction the formula

$$\eta_{k+1,n} = \sqrt{2}\sum_m f_{n-2m}\eta_{k,m} + \sqrt{2}\sum_m g_{n-2m}\kappa_{k,m}. \tag{1}$$

The consequence beeline programmer is certified as the adaptive wavelet transform. Now that the fundamental scoop method, we hoop against $\{f, \overline{f}^0, g^0, \overline{g}\}$ to place $\{f, \overline{f}, g, \overline{g}\}$ and hence from the set $\{\phi^0, \overline{\phi^0}, \psi^0, \overline{\psi^0}\}$ to place $\{\phi, \overline{\phi}, \psi, \overline{\psi}\}$. Thither diminish the quantity of operators as an analogy to thaw garden-variety process. The matters emulate this

$$\begin{aligned}\overline{\phi}_{k,m} &= 2^{k/2}\overline{\phi}(2^k r - m)\\ &= 2^{k/2}(2\sum_n \overline{f}^0_n \overline{\phi}(2(2^k r - m) - n) + \sum_n p_{-n}\overline{\psi}(2^k r - m - n))\\ &= \sqrt{2}\sum_n \overline{f}^0_{n-2m}\overline{k+1,n} + \sum_n p_{m-n}\overline{\psi}_{k,n}.\end{aligned}$$

Therefore,

$$\eta_{k,m} = \sqrt{2}\sum_n \overline{f}^0_{n-2m}\eta_{k+1,n} + \sum_n p_{m-n}\psi_{k,n}.$$

We know

$$\overline{\psi}(r) = 2\sum_n \overline{g}_n\overline{\phi}(2r - n),$$

then we have

$$\kappa_{k,m} = \sqrt{2}\sum_n \overline{g}_{n-2m}\eta_{k+1,n}.$$

They acquire the inverse transform with Eq-1 like

$$\eta_{k+1,n} = \sqrt{2}\sum_m f_{n-2m}(\eta_{k,m} - \sum_t p_{m-t}\kappa_{k,t}) + \sqrt{2}\sum_m g^0_{n-2m}\kappa_{k,m}.$$

Consequently, we hoop the following procedure for adaptive scoop wavelet transform (Diffusion wavelet transform DWT) and more polar (polar diffusion wavelet transform PDWT).

DWT

Step-1: (Computational of unliganded co-active)

$$\eta_{k,m} = \sqrt{2}\sum_n \overline{f}^0_{n-2m}\eta_{k+1,n}$$

$$\kappa_{k,m} = \sqrt{2}\sum_n \overline{g}^0_{n-2m}\eta_{k+1,n}.$$

Step-2 (Computation of uprear co-active)

$$\eta_{k,m} = \eta_{k,m} + \sum p_{m-n}\kappa_{k,m}.$$

PDWT

Step-1

$$\eta_{k,m} = \eta_{k,m} - \sum_n p_{m-n}\kappa_{k,m},$$

Step-2

$$\eta_{k+1,m} = \sqrt{2}\sum_m f_{n-2m}\eta_{k,m} + \sqrt{2}\sum_m g^0_{n-2m}\kappa_{k,m}.$$

3. Second-formulation adaptive wavelet optimized finite element method

The many rear applications of Burger's equation in mathematics and material sciences. For example, it observed as an equation that models convulsion, influx flows, audio communication in a haze, etc. It can also see as an agitation often-used ugly Laplace equation, what type of crook investigated are discussed in this framework. They evaluate a bar built arise of temperature fetching material. Let E be the elongation of the bar, and B be the fixed sliced region of the bar. We suppose that the only endpoint of the bar can replace the temperature, and the edgewise rectus of the bar is desolated. Again, it is considered that the temperature of the bar is stable all over any layer given the sliced region.

Let illustrate the heat of location r and time t by $v(t, r)$ and subsequent arete bounds often quilted substance tufted bar: n, the thermal energy; $\rho(r)$, the solidity often-used content at point r; $c(r)$, the particular temperature oftentimes bar at that position $r, h_1(t, r)$, the ratio of heat-producing respectively time. With the significance bounds, the equation controlling the heat within the bar is inclined

$$\rho(r)c(r)\frac{\partial v(t,r)}{\partial t} = \frac{\partial}{\partial r}(k(r)\frac{\partial}{\partial r}v(t,r)) + h_1(t,r). \tag{2}$$

We farther suppose that the solidity, the explicit temperature, and the thermal energy are isolated of the location; with this taking up Eq-2 gets

$$\frac{\partial v(t,r)}{\partial t} = \frac{m}{\rho c}\frac{\partial^2 u(t,r)}{\partial r^2} + \frac{h_1(t,r)}{\rho c}.$$

Granting $\nu = \frac{m}{\rho c}$ and $h(t,r) = \frac{h_1(t,m)}{\rho c}$, the above equation grows

$$\frac{\partial v(t,r)}{\partial t} = \nu\frac{\partial^2 v(t,r)}{\partial r^2} + h(t,r). \tag{3}$$

Now we examine a confusion Eq-3 in such a way

$$\frac{\partial v(t,r)}{\partial t} + v(t,r)\frac{\partial v(t,r)}{\partial r} = \nu\frac{\partial^2 v(t,r)}{\partial r^2} + h(t,r). \tag{4}$$

Equation-4 viscosity even constant ν . Because the partial equation involves in time and area respectively, the necessity of this type equation contains initial condition $v(0, r) = v_0(r)$ to solve the problem. The real thing two standard categories of two issues:

1. Periodic boundary condition

$$v(t,r) = v(t,r+L)$$

2. Dirichlet boundary condition

$$v(t,0) = v_L(t), v(t,L) = v_R(t).$$

3.1. Finite element method for two boundary conditions

Introduce the region isthmian interruption $[a, b]$ with point discretion the region be $a = r_1 < r_2 < r_3 < \cdots r_I = b$. Applying the Riesz representation theory, the first and second derivatives for good approximation behavior of the finite-dimensional domain is inclined

$$\begin{aligned} &\frac{dv(r_1)}{dr} = \frac{v(r_2)-v(r_0)}{r_2-r_0}, \frac{dv(r_2)}{dr} = \frac{v(r_3)-v(r_1)}{r_3-r_1}, \\ &\cdots\cdots\cdots \\ &\frac{dv(r_{I-1})}{dr} = \frac{v(r_I)-v(r_{I-2})}{r_I-r_{I-2}}, \frac{dv(r_I)}{dr} = \frac{v(r_{I+1})-v(r_{I-1})}{r_{I+1}-r_{I-1}}, \end{aligned} \tag{5}$$

$$\frac{d^2v(r_1)}{dr^2} = \frac{2v(r_0)}{(r_0 - r_1)(r_0 - r_2)} + \frac{2v(r_1)}{(r_1 - r_0)(r_1 - r_2)} + \frac{2v(r_2)}{(r_2 - r_0)(r_2 - r_1)},$$
$$\frac{d^2v(r_2)}{dr^2} = \frac{2v(r_1)}{(r_1 - r_2)(r_1 - r_3)} + \frac{2v(r_2)}{(r_2 - r_1)(r_2 - r_3)} + \frac{2v(r_3)}{(r_3 - r_1)(r_3 - r_2)},$$
$$\cdots\cdots\cdots$$
$$\frac{d^2v(r_{I-1})}{dr^2} = \frac{2v(r_{I-2})}{(r_{I-2} - r_{I-1})(r_{I-2} - r_I)} + \frac{2v(r_{I-1})}{(r_{I-1} - r_{I-2})(r_{I-1} - r_I)} + \frac{2v(r_I)}{(r_I - r_{I-1})(r_I - r_{I-2})},$$
$$\frac{d^2v(r_I)}{dr^2} = \frac{2v(r_{I-1})}{(r_{I-1} - r_I)(r_{I-1} - r_{I+1})} + \frac{2v(r_I)}{(r_I - r_{I-1})(r_I - r_{I+1})} + \frac{2v(r_{I+1})}{(r_{I+1} - r_{I-1})(r_{I+1} - r_I)}. \quad (6)$$

Not even r_0 and r_{I+1} are not inclined to in our discretion region. Conditional on the different kinds of the properties elaborate we-group disparate finite element method as clarified following.

3.2. Periodic boundary conditions

For fear that of periodic boundary conditions, every indecision is expected to be cogent modulo $I - 1$ (as $v(a) = v(b)$). Consequently, we handle r_0 are r_{I+1} as r_2. Supplementary, we suppose that $r_1 - r_0 = r_2 - r_1$ and $r_{I+1} - r_I = r_I - r_{I-1}$. Hence, in this compact, $E^{(1)}$ and $E^{(2)}$ are first and second finite element matrix in such a manner $[v'(r_1), v'(r_2), \cdots v'(r_I)]' = E^{(1)}[v(r_1), v(r_2), v(r_3), \cdots v(r_I)]'$ and $[v''(r_1), v''(r_2), \cdots v''(r_I)]' = E^{(2)}[v(r_1), v(r_2), v(r_3), \cdots v(r_I)]'$ are specified

$$E^{(1)}(1,2) = -E^{(1)}(1, I-1) = \frac{1}{2(r_2 - r_1)},$$
$$E^{(1)}(I,2) = -E^{(1)}(I, I-1) = \frac{1}{2(r_I - r_{I-1})},$$
$$E^{(1)}(j,j-1) = E^{(1)}(j,j+1) = -\frac{1}{r_{j+1} - r_{j-1}},$$
$$E'(j,k) = 0, \forall otherjandk$$
$$-\frac{1}{2}E^{(2)}(1,1) = E^{(2)}(1,2) = E^{(2)}(1, I-1) = \frac{1}{(r_2 - r_1)^2}$$
$$E^{(2)}(I,2) = E^{(2)}(I, I-1) = -\frac{1}{2}E^{(2)}(I,I) = \frac{1}{(r_I - r_{I-1})^2}$$
$$E^{(2)}(j,j-1) = \frac{2}{(r_{j-1} - r_j)(r_{j-1} - r_{j+1})}, 2 \le j \le I-1$$
$$E^{(2)}(j,j) = \frac{2}{(r_j - r_{j-1})(r_j - r_{j+1})}, 2 \le j \le I-1$$
$$E^{(2)}(j,j+1) = \frac{2}{(r_{j+1} - r_{j-1})(r_{j+1} - r_j)}, 2 \le j \le I-1$$
$$E^{(2)}(j,k) = 0 \forall otherjandk$$

3.3. Dirichlet boundary conditions.

In anticipation of Dirichlet boundary condition the behavior is identified at the bound points, i.e. $v(a) = v_L$ and $v(b) = v_R$ that indicates $v(r_1) = v_L$ and $v(r_I) = v_R$. While solution is familiar at that bound points, we have lone one possible the $I - 2$ strangers especially $v(r_2), v(r_3), \cdots v(r_{I-1})$. Consequently, we examine the each one but first and last equations of 5 and 6 and write $v(r_1) = v_L, v(r_I) = v_R$. We acquire the succession

$$\begin{pmatrix} v'(r_2) \\ v'(r_3) \\ \vdots \\ v'(r_{I-2}) \\ v'(r_{I-1}) \end{pmatrix} = E^{(1)} \begin{pmatrix} v(r_2) \\ v(r_3) \\ \vdots \\ v(r_{I-2}) \\ v(r_{I-1}) \end{pmatrix} + \begin{pmatrix} -\dfrac{v_L}{r_3 - r_1} \\ 0 \\ \vdots \\ 0 \\ \dfrac{v_R}{r_I - r_{I-2}} \end{pmatrix}$$

$$\begin{pmatrix} v''(r_2) \\ v''(r_3) \\ \vdots \\ v''(r_{I-2}) \\ v''(r_{I-1}) \end{pmatrix} = E^{(2)} \begin{pmatrix} v(r_2) \\ v(r_3) \\ \vdots \\ v(r_{I-2}) \\ v(r_{I-1}) \end{pmatrix} + \begin{pmatrix} -\dfrac{2v_L}{(r_1 - r_2)(r_1 - r_3)} \\ 0 \\ \vdots \\ 0 \\ \dfrac{2v_R}{(r_I - r_{I-1})(r_I - r_{I-2})} \end{pmatrix}$$

where $E^{(1)}$ and $E^{(2)}$ are the matrices of order $(I-2) \times (I-2)$ with subsequent field

$$E^{(1)}(1,2) = \frac{1}{(r_3 - r_1)}, E^{(1)}(I-2, I-3) = -\frac{1}{Z_I - r_{I-2}},$$

$$E^{(1)}(j, j-1) = -\frac{1}{Z_{j+2} - r_j}, E^{(1)}(j, j+1) = -\frac{1}{Z_{j+2} - r_j}$$

$$2 \leq j \leq I-3$$

$$E^{(1)}(j,k) = 0, \forall other j and k$$

$$E^{(2)}(1,1) = \frac{2}{(r_2 - r_1)(r_2 - r_3)}, E^{(2)}(1,2) = \frac{2}{(r_3 - r_1)(r_3 - r_2)},$$

$$E^{(2)}(I-2, I-3) = \frac{2}{(r_{I-2} - r_{I-1})(r_{I-2} - r_I)},$$

$$E^{(2)}(I-2, I-2) = \frac{2}{(r_{I-1} - r_{I-2})(r_{I-1} - r_I)}$$

$$E^{(2)}(j, j-1) = \frac{2}{(r_j - r_{j+1})(r_j - rj + 2)}, 2 \leq j \leq I-3$$

$$E^{(2)}(j, j) = \frac{2}{(r_{j+1} - r_j)(r_{j+1} - r_{j+2})}, 2 \leq j \leq I-3$$

$$E^{(2)}(j, j+1) = \frac{2}{(r_{j+2} - r_j)(r_{j+2} - r_{j+1})}, 2 \leq j \leq I-3$$

$$E^{(2)}(j,k) = 0 \forall other j and k$$

4. Finite element algorithm for solving PDEs

Coming to have found all the necessary materials for SFAWOFE, we describe the algorithm for SFAWOFE

1. Discretionary the domain of the PDEs.
2. Application Riesz representation theory for finite difference matrices disposed of in sec() to discretionary the function intricate in the PDE approaching.
3. Assume the time $t_1, t_2, \cdots$ are determined towards mesh transformation(note-taking the option of those times is dispute conditional, providing the problem is such a manner for a solution is quickly fluctuating then t_j's must be favored nearer). Inception with the initial

condition, which takes an application of a time combination strategy (Least-Squares method in that matter), we acquire results at the time t_1, i.e., $v(t_1)$.

4. We apply $v(t_1$ and the correlated mesh points T^{t_1} to acquire $T^{t_1+\triangle t}$ applying the method explained in the behind section.
5. Computing the least-square approximation elaborate in the PDE on $T^{t_1+\triangle t}$which is convert PDEs to ordinary differential equations(ODEs).
6. Integrate the concluding mesh of ODEs in time using the Least-Squares method to acquire the results at $t = t_1 + \triangle t$.
7. Repetition the step-6 as late as $v(t_2)$ is achieved. Coming by found $v(t_2)$ proceed to step-4.

4.1. Numerical outcomes

Investigation of the admitted SFAWOFE method weighs against deliberate Burger's equation with two boundary conditions as periodic and Dirichlet s. The equation deliberate is specified as ensues

$$\frac{\partial v(t,r)}{\partial t} + v(t,r)\frac{\partial v(t,r)}{\partial r} = \nu\frac{\partial^2 v(t,r)}{\partial r^2}$$

with initial condition given by $v(r,0) = f(r)$ and appropriate boundary conditions. Using the above algorithm on SFAWOFE, as clarified in Section-3. Calculating in operation boundary conditions exploited coming trial problems are deliberated.

4.2. Trial problem

In the paper we contemplate Burger's equation on the region $[0, 1]$

$$\frac{\partial v(t,r)}{\partial t} + v(t,r)\frac{\partial v(t,r)}{\partial r} = \nu\frac{\partial^2 v(t,r)}{\partial r^2}$$

with initial condition $v(r,0) = v_0(r) = sin2\pi r$ and boundary condition is periodic and Dirichlet i.e. $v(r,t) = v(r+1,t)$ and $v(t,0) = v_1(t)$, $v(t,1) = v_1(t)$. Analytic solution for the Burger,s equation is

$$v(r,t) = \frac{\int_{-\infty}^{\infty} \frac{r-\tau}{t} exp\left(\frac{-(r-\tau)^2}{4\nu t}\right) exp\left(\frac{cos(2\pi\tau)}{4\pi r}\right) d\tau}{\int_{-\infty}^{\infty} exp\left(\frac{-(r-\tau)^2}{4\nu t}\right) exp\left(\frac{cos(2\pi\tau)}{4\pi r}\right) d\tau}.$$

The analytic solution of an equation is a changeless wave, which is modified a sharp Grenadian at $r = 0.5$. For $\epsilon = 10^{-5}$ and $\nu = 0.004$. Fig-1(a) exhibitions the comparison between a numerical and analytic solution of Burger,s equation. Fig-1(b) shows the relation between compaction error and the number of expressive coefficients. Fig-1(c) compares the PT time checked out the AWOFE method with the PT time checked out the finite element method for calculating the solution at the time $t = 2^n\triangle t$. We can observe that from Fig-1(e) at time $t = 2^5\triangle t = 0.3601$, the PT time is checked out the AWOFE method is 5% less than the time checked out finite element method. Fig-1(d) and Fig-1(f) exhibitions the convergence of the process corresponding to different boundary conditions and adaptive grid with time. Fig-1(g) exhibitions the point to point error at various times, at it is detected that error is the highest value for the diversion as anticipated. Therefore a large number of mesh grid will be taken in the taken interval, and titular lead to generating a diffusion plot. The error can be minimized while the decreases values of ϵ, an error is directly pro personal to ϵ. Therefore, ϵ is to be taken a tiny number; the relation between error and mathematical cost stability properly handles. Now, weddings

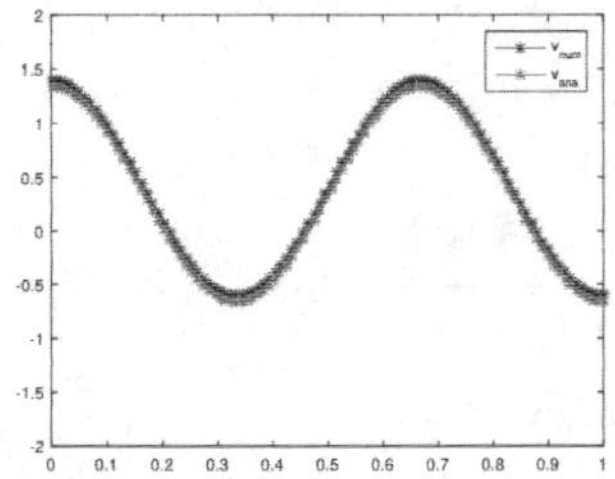

(a) Numerical vs analytic solution

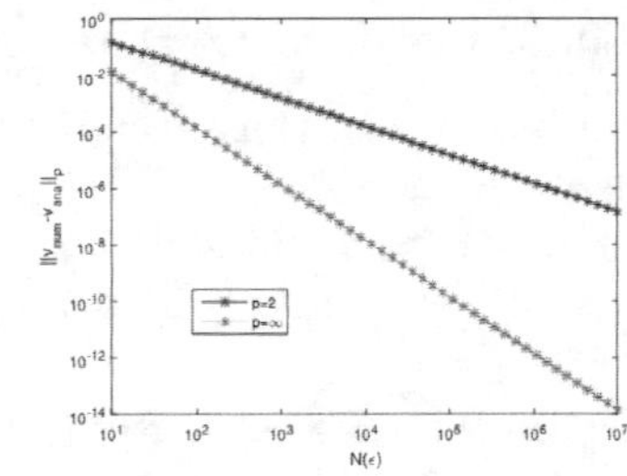

(b) compaction error vs expressive coefficient.

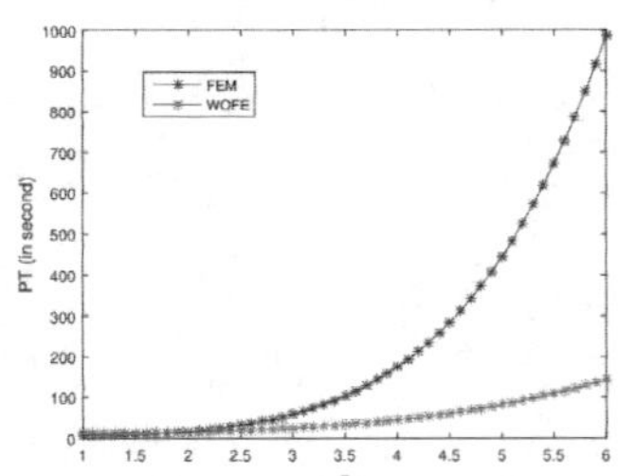

(c) PT for calculating the solution at time $t = 2^n \triangle t$.

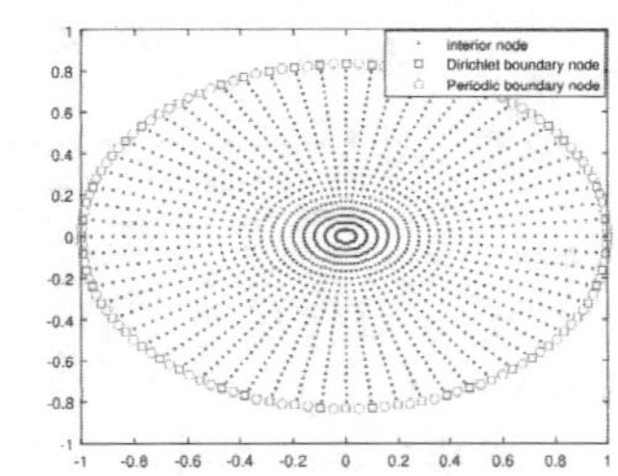

(d) comparison in boundary condition for Bourger,s equation .

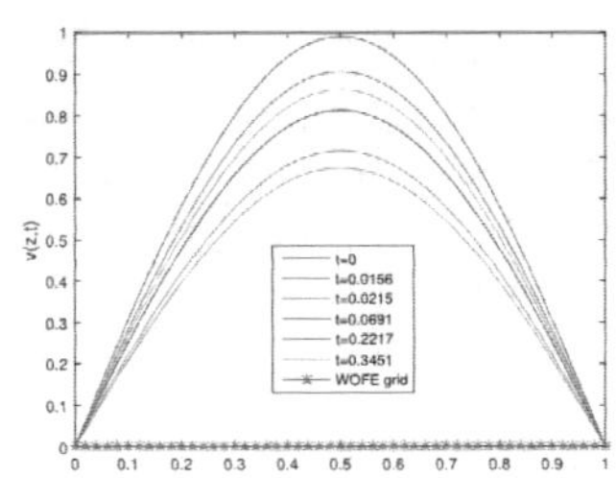

(e) Solution with respect to FEM grid at unlike times.

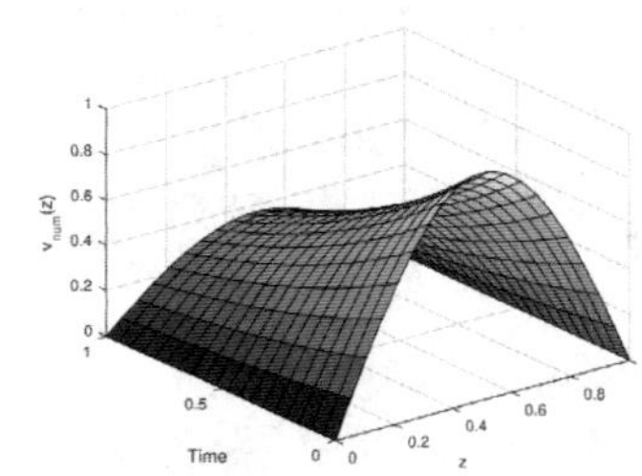

(f) Adaptive wavelet with time.

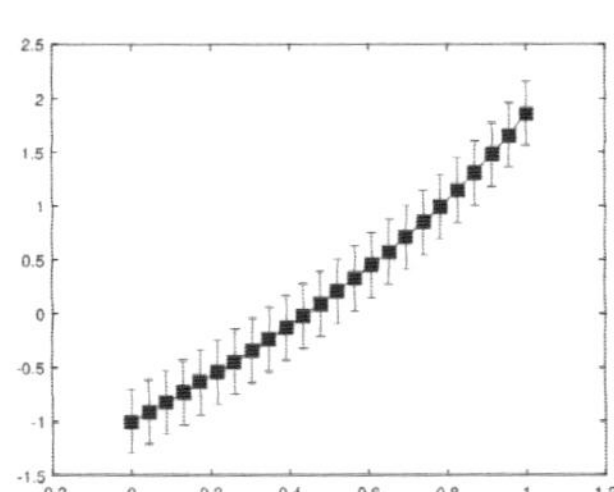

(g) Point to point error at different times.

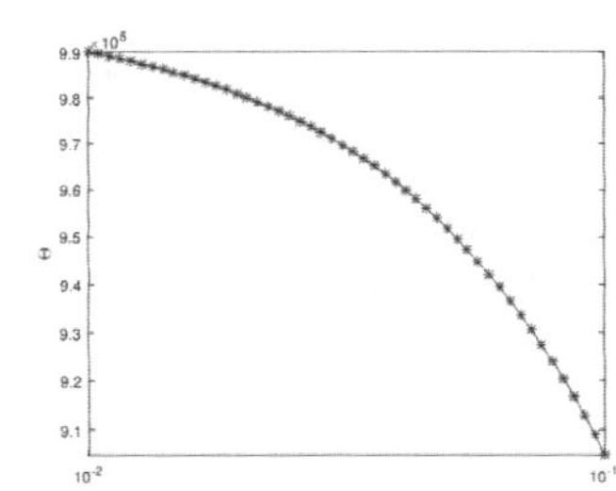

(h) ν vs Θ.

Figure 1: Results for trial problem.

the time compaction diversion as $\Theta = \dfrac{PT(\epsilon = 0)}{PT(\epsilon)}$. Fig-1(h) shows the diversion of Θbeing a purpose of ν; it can be checked that decrease isthmian value of ν, the increased cost of , Θ, i.e., the technique executes superior for minimum costs of ν. Table-1 provides diversion of $PT(\epsilon)$ with ϵ for $\nu = 0.0002$. We hold checked thatch ϵ lower Θ also lower. The cost of Θ increases the adaptive wavelet technique is more efficient as compare finite element method.

Table 1: The execution of SFAWOFE for trial problem.

ϵ	10^{-3}	10^{-4}	10^{-5}	10^{-6}
$PT(\epsilon)$	0.0892	0.2361	0.2386	0.3034
Θ	3.7116	2.7353	2.6133	1.9840

5. Discussion and Conclusion

In this framework, the second-formulation adaptive wavelet optimized finite element method established, individually considered as SFAWOFE for the solution of Burger's equation with periodic and Dirichlet boundary conditions. We have developed a finite element algorithm for the resolution of PDEs and calculate numerical results of Burger's equation for the time integration trial problem. The trial problem used for computing magnetostatic networking rang dipole in material science, which used for fast computation of process time checked out the SFAWOFE method are analyzed with process time checked out the finite element method. We are getting verified the performance of the purposed way for Burger,s equation. We have comprehensive this method for disturbance designing and problem-solving for fluid analysis in our future work.

Acknowledgments

The authors are incredibly thankful to NIT Raipur, India, for giving an opportunity for the work.'

References

[1] Grossmann A. and Morlet J., (1984) *SIAM. J. Math. Anal.*, **15** 723–736.
[2] Daubechies I., (1992) *SIAM, Philadelphia.*
[3] Maday Y., Perrier V. and Ravel J.C. (1991), *C. R. Acad. Sci.*, **312**, 405–410.
[4] Jameson L., (1998) *SIAM. J. Sci. Comput.*, **19**, 1980–2013.
[5] Dahmen W., (1997) *Act. Num.*, **6**, 55–228.
[6] Dahmen W., Kunoth A. and Urban K. A. (1996) *Act. Num.*, **56**, 259–301.
[7] Dahmen W., Urban K. and Vorloeper J., (2002) *World Scientific, Singapore*, pp. 39–80.
[8] Frohlich J. and Schneider K. (1994) *Eur. J. Mech. B Fluids*, **13**, 439–471.
[9] Frohlich J. and Schneider K., (1997) *J. Comput. Phys.*, **130**, 90–174.
[10] Gagnon A. and Lina J. M., (1994) *J. Phys. A Math. Gen.*, **27**, 8207–8230.
[11] Liandrat J. and Tchamitchian P., (1990) *NASA CR-187480: ICASE*, Report No. 90–83.
[12] Jameson L., (1994) *Technical Report NASA CR-191601: ICASE* , Report No. 94–9.
[13] Mehra M. and Rathish B. V. K., (2005) *Comm. Numer. Meth. Meth. Eng.*, **21**, 313-326.

Theoretical investigation on absorption spectral profile of tetrazole liquid crystals: Hexyl, nonyl end groups and solvent effects

Punyatoya Das and P. Lakshmi Praveen
Department of Physics, Veer Surendra Sai University of Technology, Burla-768018, Sambalpur, Odisha, India, Email: daspunyatoya012@gmail.com
Email: plpraveen_phy@vssut.ac.in

Abstract. In this article, a theoretical investigation on absorption spectral profile of tetrazole liquid crystals 4-[(2-alkyl)-2H-tetrazol-5-yl] phenyl 4-alkyloxybenzoates (*n*TP*m*B; with *n*=m=6, 9) has been carried out. The hexyl, nonyl, and solvent (DMSO) effects on absorption spectral profiles have been investigated using the CNDO/S, and INDO/S methods. Population analysis of these molecules in the gaseous state has been performed by generalized atomic polar tensor (GAPT) charge distribution scheme for AM1, PM3, MNDO, CNDO/S, and INDO/S methods. The observed $\pi \rightarrow \pi^*$, and $n \rightarrow \pi^*$ electronic transitions have been reported. The hexyl, nonyl end groups and solvent effects on absorption profiles have been discussed to analyse the photo stability of the molecules. Further, a through comparative analysis of reactivity descriptors has been made. The reported data may offer an insight in determining the end use of compounds.

1. Introduction

Long-range orientation ordered liquid crystals (LCs) are superior over normal crystals due to some of its special properties from application point of view [1]. LCs may contain various kinds of functional molecular systems which are electronegative [2, 3]in nature. LCs are structurally flexible enough so that it can be mixed with other compatible substances in any proportion by varying molecular structure, reorientation and improving physical properties like photo sensitivity and stability, optical behaviour, absorption of radiation. Qualitative character in the field of optics multiply its application broadly in the area of photonics, manufacture of photo sensitive devices, optical switching and signal processor. Remarkable changes will appear in properties of LCs due to several molecular arrangements with complex stratified structure.

Understanding of spectroscopic properties is compulsory for checking the viability of LCs for photonic applications [4, 5]. Spectroscopy plays vital role for observation of properties of number of anisotropic molecular compounds. It can characterize LC molecules and find their structure on the basis of absorption spectrum as a function of specific wavelength and frequency. Significance of absorption spectrum demonstrates the radiation available for absorption for the interacting sample. The resultant spectral profile acts as a fingerprint from which data are matched with reference parameters if available in the literature. Absorption spectrum gives more accurate results in a simpler way. Qualitatively it can measure the concentration of a molecule in a solution because molecules absorb energy at a specific wavelength [6]. Besides that, polarization of the electronic transition in UV and visible radiation and dipole-dipole interaction in intermolecular level will be observed. From the absorption spectra stable region can be analysed at particular wavelength, frequency and corresponding energy can be known. As the absorbed radiation transferred into heat and gradually warm up the LC molecules [7], parameters like order parameter, dielectric constant, optical anisotropy will be affected due to the absorption of certain frequency of radiation.

This article has been intended to offer a comparative picture of UV-Visible absorption spectral parameters of *n*TP*m*B (*n*=*m*=6, 9) molecules using the CNDO/S [8, 9], and INDO/S [10] methods along with CIS method [11] using the semiempirical ZINDO [12-14] Hamiltonian for excited states calculation. Charge transfer process of the molecules in the gaseous state has been performed by generalized atomic polar tensor (GAPT) charge distribution scheme for AM1, PM3, MNDO,

CNDO/S, and INDO/S methods. The shifts in absorption bands have been discussed. The spectral parameters have been reported to explain the photo sensitivity of the compounds. The LC behaviour and the transition temperatures of the compounds have been reported [15].

2. Methodology

The CNDO/S and INDO/S are versions of the ordinary CNDO, INDO procedures reparametarized to reproduce experimental excitation energies. The SCF step is frequently followed by limited configuration interaction (CI) calculations over a truncated set of singly excited configurations. In both the methods, the wave functions of the ground, and electronically excited states of $\pi\pi^*$-type have been once investigated in the one-configurationally approximation. The UV-Visible absorption spectrum has been calculated using the CI method on the basis of the obtained multi electron wave functions.

The CNDO/S and INDO/S methods including all valence electrons along with configuration interaction analysis have been used to calculate the UV-Visible absorption spectra of the molecules. A revised version of QCPE program 174 by Jeff Reimers, University of Sydney, and co-workers have been used for these calculations [16]. The general structural parameters of the systems such as bond lengths, and bond angles have been taken from the published crystallographic data [15]. The electrochemical properties have been estimated using the formulae reported in Table 4.

3. Results and Discussion

The molecular structures of 6TP6B and 9TP9Bthat have been constructed based on published crystallographic data [15] have been shown in Figure 1. Atomic net charges have been calculated using AM1, PM3, MNDO, CNDO/S, INDO/S schemes have been reported in Table 1 and 2 for 6TP6B and 9TP9B respectively. The AM1, PM3, and MNDO charges have been obtained by MOPAC calculations available via VEGA ZZ package [17]. The atomic partial charge analysis and intermolecular charge transfer of the compounds have been analyzed as given below.

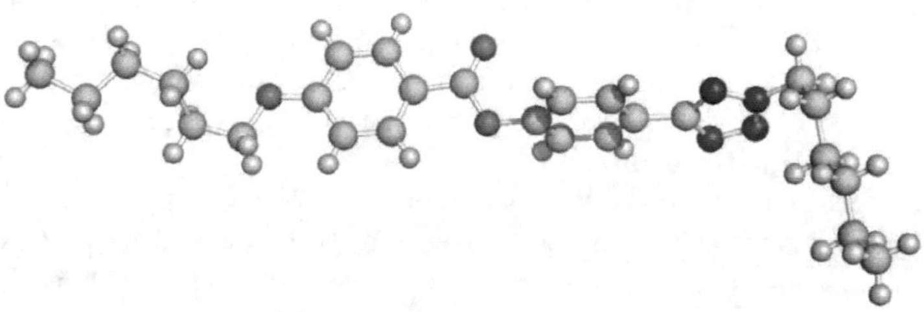

A.6TP6B

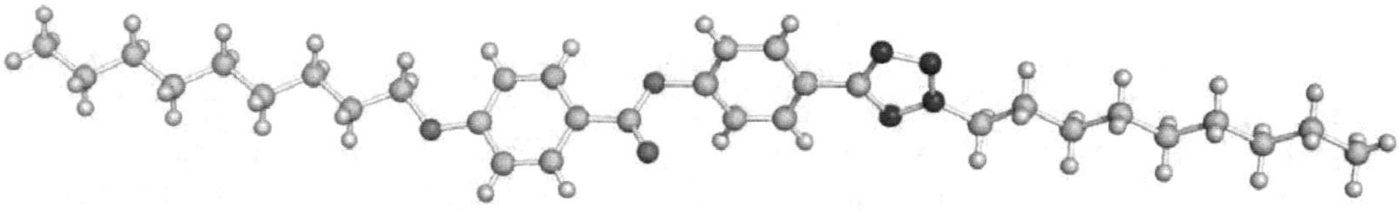

B. 9TP9B

Figure 1 Molecular structures of (A) 6TP6B and (B) 9TP9B compounds

3.1. Atomic partial charge analysis

The atomic charges of *n*TP*m*B (*n*=*m*=6, 9)molecules have been reported in Table 1 and 2. From the tables, discrepancies may be observed in terms of individual atomic partial charges. However, much agreement has found out in general in case of group charges. Carbon and nitrogen atoms carry both positive and negative charges. Most of the carbon atoms in the end chain carry negative charge for 6TP6B. Most negative charged atom is O_{32}. The next maximum negative charged atoms are O_{33} and O_{31} respectively. The maximum positive charge acquired by C_2 followed by C_1 and C_{12}. Similarly,C_2 contain maximum positive charge for 9TP9B molecule. Next two maximum positive charged atoms are C_1 and C_{12}. O_{38} is the most negative charged atom followed by O_{39} and N_{33}. All the five different schemes are in agreement in estimating the atomic partial charges. Hence, the intermolecular charge transfer is stronger in both the molecules between side to aromatic groups and end to aromatic groups. From Table 1 and 2, it may be observed that the strongest positive and negative charges in both molecules are nearly same. Hence, appreciable difference in terms of phase transition temperature may not be found. This is in accord with literature details [15].

Table 1 Interatomic charge distribution for 6TP6B molecule (heavy atoms) using CNDO/S, INDO/S, AM1, PM3, and MNDO methods.

S.No.	Atom	Atomic Charges				
		CNDO/S	INDO/S	AM1	PM3	MNDO
1	C_1	0.199	0.187	-0.092	-0.056	-0.019
2	C_2	0.406	0.452	0.363	0.433	0.417
3	C_3	0.004	0.008	-0.022	-0.029	-0.020
4	C_4	-0.017	-0.016	-0.083	-0.052	-0.006
5	C_5	-0.040	-0.024	-0.144	-0.124	-0.073
6	C_6	0.146	0.147	0.082	0.096	0.124
7	C_7	-0.044	-0.028	-0.144	-0.120	-0.073
8	C_8	-0.015	-0.013	-0.080	-0.050	-0.001
9	C_9	-0.017	-0.005	-0.160	-0.176	-0.132
10	C_{10}	0.003	0.009	-0.039	-0.009	0.020
11	C_{11}	-0.060	-0.037	-0.177	-0.162	-0.098
12	C_{12}	0.160	0.162	0.123	0.132	0.168
13	C_{13}	-0.072	-0.048	-0.229	-0.208	-0.148
14	C_{14}	-0.003	0.001	-0.027	-0.001	0.032
15	C_{15}	0.040	0.068	0.009	-0.035	0.202
16	C_{16}	-0.044	-0.019	-0.178	-0.119	-0.032
17	C_{17}	-0.054	-0.027	-0.162	-0.107	-0.009
18	C_{18}	-0.057	-0.029	-0.157	-0.101	-0.007
19	C_{19}	-0.057	-0.030	-0.158	-0.098	-0.022
20	C_{20}	-0.097	-0.052	-0.210	-0.109	0.029
21	C_{21}	0.054	0.102	-0.024	0.055	0.180
22	C_{22}	-0.048	-0.021	-0.161	-0.121	-0.010
23	C_{23}	-0.058	-0.030	-0.158	-0.100	-0.006
24	C_{24}	-0.055	-0.026	-0.157	-0.102	-0.007
25	C_{25}	-0.056	-0.028	-0.159	-0.106	-0.021
26	C_{26}	-0.099	-0.055	-0.211	-0.107	0.029
27	N_{27}	-0.251	-0.251	-0.057	-0.210	-0.095
28	N_{28}	0.068	0.148	-0.157	0.221	-0.174
29	N_{29}	-0.077	-0.100	0.018	-0.121	0.019
30	N_{30}	-0.218	-0.225	-0.106	-0.091	-0.137
31	O_{31}	-0.242	-0.272	-0.236	-0.222	-0.283
32	O_{32}	-0.448	-0.510	-0.336	-0.364	-0.339
33	O_{33}	-0.260	-0.309	-0.208	-0.185	-0.277

IOP Conf. Series: Materials Science and Engineering **798** (2020) 012041 doi:10.1088/1757-899X/798/1/012041

Table 2 Interatomic charge distribution for 9TP9B molecule (heavy atoms) using CNDO/S, INDO/S, AM1, PM3, and MNDO methods.

S.No.	Atom	Atomic Charges				
		CNDO/S	INDO/S	AM1	PM3	MNDO
1	C_1	0.202	0.188	-0.090	-0.057	-0.010
2	C_2	0.407	0.453	0.363	0.433	0.417
3	C_3	0.005	0.009	-0.021	-0.027	-0.022
4	C_4	-0.013	-0.011	-0.081	-0.050	-0.006
5	C_5	-0.043	-0.028	-0.144	-0.121	-0.075
6	C_6	0.145	0.146	0.081	0.094	0.124
7	C_7	-0.036	-0.020	-0.144	-0.117	-0.072
8	C_8	-0.017	-0.016	-0.083	-0.054	-0.005
9	C_9	-0.015	-0.003	-0.159	-0.172	-0.132
10	C_{10}	-0.005	-0.001	-0.027	-0.003	0.031
11	C_{11}	-0.065	-0.042	-0.229	-0.205	-0.148
12	C_{12}	0.156	0.160	0.123	0.130	0.168
13	C_{13}	-0.058	- 0.036	-0.176	-0.158	-0.098
14	C_{14}	0.001	0.008	-0.040	-0.011	0.021
15	C_{15}	0.038	0.064	0.011	-0.039	0.203
16	C_{16}	-0.046	0.022	-0.174	-0.122	-0.036
17	C_{17}	-0.058	-0.032	-0.158	-0.098	-0.008
18	C_{18}	-0.058	- 0.031	-0.157	-0.101	-0.008
19	C_{19}	-0.059	- 0.032	-0.157	-0.101	-0.008
20	C_{20}	-0.059	-0.032	-0.157	-0.101	-0.008
21	C_{21}	-0.058	-0.030	-0.157	-0.102	-0.007
22	C_{22}	-0.058	- 0.031	-0.158	-0.097	-0.021
23	C_{23}	-0.098	- 0.053	-0.210	-0.109	0.029
24	C_{24}	0.052	0.100	-0.023	0.053	0.180
25	C_{25}	-0.049	-0.023	-0.165	-0.115	-0.017
26	C_{26}	-0.060	-0.031	-0.156	-0.098	-0.011
27	C_{27}	-0.059	-0.031	-0.157	-0.101	-0.008
28	C_{28}	-0.059	-0.032	-0.157	-0.101	-0.008
29	C_{29}	-0.059	-0.031	-0.157	-0.101	-0.008
30	C_{30}	-0.058	-0.029	-0.157	-0.102	-0.007
31	C_{31}	-0.057	-0.030	-0.158	-0.097	-0.021
32	C_{32}	-0.097	-0.053	-0.210	-0.109	0.029
33	N_{33}	-0.256	-0.255	-0.060	-0.213	-0.096
34	N_{34}	0.062	0.139	-0.155	0.230	-0.175
35	N_{35}	-0.067	-0.091	0.017	-0.124	0.019
36	N_{36}	-0.224	-0.231	-0.108	-0.090	-0.139
37	O_{37}	-0.242	-0.272	-0.236	-0.220	-0.283
38	O_{38}	-0.451	-0.513	-0.336	-0.363	-0.339
39	O_{39}	-0.264	-0.314	-0.208	-0.187	-0.278

3.2. Absorption spectral profile

(a) Pure compounds of 6TP6B and 9TP9B:

The absorption spectra of *n*TP*m*B (*n*=*m*=6, 9) pure molecules using CNDO/S, and INDO/S methods have been shown in Figure 2 and 3 respectively. CNDO/S method estimates two band spectra whereas; INDO/S method estimates one band spectra. The 6TP6B and 9TP9B molecules exhibits absorption maxima at 240.42nm, 238.67nm (CNDO/S), and 249.22nm, 248.63nm (INDO/S) respectively. However, the extinction coefficient has been decreased from 0.80 to 0.79 (CNDO/S) and 1.11 to 1.01 (INDO/S) respectively. These bands arises from the HOMO→LUMO transition, and is assigned as $\pi \rightarrow \pi^*$ transition. The other absorption bands corresponding to the remaining wavelengths

IOP Conf. Series: Materials Science and Engineering **798** (2020) 012041 doi:10.1088/1757-899X/798/1/012041

in UV region also indicate the more numbers of $\pi \rightarrow \pi^*$ transitions at other wavelengths. Therefore, both methods are in agreement with each other by exhibiting blue-shift of wavelength, and hypochromic effect when hexyl end group is replaced by nonyl end group.

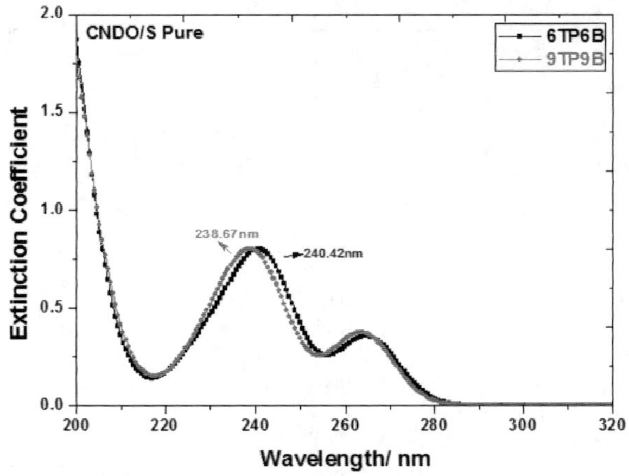

Figure 2 Absorption spectra of pure 6TP6B and 9TP9B molecules using CNDO/S Method

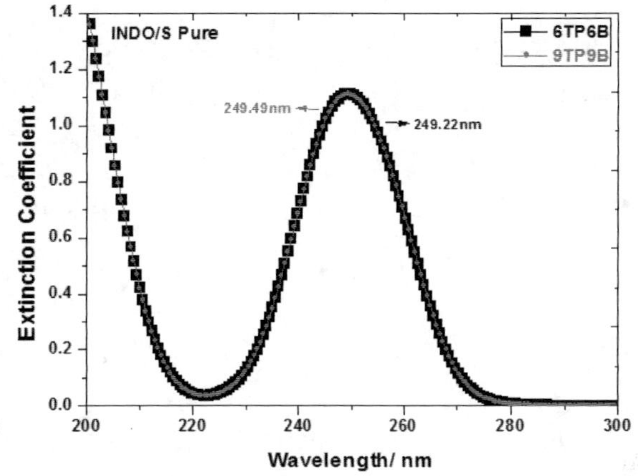

Figure 3 Absorption spectra of pure 6TP6B and 9TP9B molecules using INDO/S method

The spectral parameters of *n*TP*m*B (*n*=*m*=6, 9) pure molecules have been reported in Table 3. The above mentioned blue-shift is possible due to the excitation of the electron that causes an increment in transition energies (Table 3). Therefore, this makes the absorption band to poor interaction electrostatically. From the literature, it has been observed that the 9TP9B exhibits little higher transition temperature [15] compared to 6TP6B. So, the high phase stability of 9TP9B plays a vital role for increment in the transition energies. This leads to differences in charge distribution of the compound, and a reduced delocalization of electrons. Hence, it may be concluded that the ground, and excited $n \rightarrow \pi^*$ transitions are not visualized due to the less molecular rigidity offered by the ring system.

b) Solvent (DMSO) effect

The spectral parameters of *n*TP*m*B (*n*=*m*=6, 9) molecules in DMSO are reported in Table 3 using CNDO/S, and INDO/S methods. CNDO/S method estimates two band spectra of 6TP6B in DMSO with 257.42nm as absorption wavelength, and INDO/S method estimates four band spectra with 321.87nm as absorption wavelength in DMSO (fig 4 and fig 5). Therefore, the solvent causes a red-shift of wavelength compared to pure molecules (Table 3). Further, hypochromic effect has been observed for 6TP6B molecule in DMSO. Therefore, the solvent enhances the UV stability of 6TP6B molecule in DMSO compared to its pure form.

The CNDO/S method estimates four band spectra of 9TP9B in DMSO with 205.27nm as absorption wavelength, and INDO/S method estimates five band spectra in DMSO with 301.95nm as absorption wavelength. Hence, INDO/S method predicts a red-shift of wavelength compared to pure molecules (Table 3). Hypochromic effect has been observed for 9TP9B molecule in DMSO.

Hence, it may be understood from the spectral parameters reported in Table 3 that the UV stability of 6TP6B is more compared to 9TP9B in pure form. The DMSO effect does not alter the UV stability of 6TP6B as it is evident from Table 3. Moreover, increment of end chains from hexyl to nonyl groups does not alter the nature of shift of wavelength in DMSO. But the solvent enhances the UV stability of individual compounds from its pure form to solvent. Further, both methods predict a lesser band gap for 6TP6B molecule which concludes the conductivity of 6TP6B.

IOP Conf. Series: Materials Science and Engineering **798** (2020) 012041 doi:10.1088/1757-899X/798/1/012041

Table 3.The absorption bands (AB), extinction coefficients (EC), Oscillator strength (*f*), Vertical transition energy (E_V), dipole moment (μ),HOMO (H), LUMO (L) energies, and the band gap (E_g=E_{LUMO}-E_{HOMO}) values of *n*TP*m*B (*n*=*m*=6, 9) pure molecules and in DMSO using CNDO/S, and INDO/S methods.

Molecule	Method		AB/ nm	EC*	*f*	E_V/ eV	μ/D
6TP6B	CNDO/S	Pure	**240.42**	0.80	0.51	5.14	6.81
			264.45	0.35	0.35	4.68	11.93
		DMSO	**257.42**	0.53	0.53	4.81	5.09
			286.13	0.12	0.12	4.33	7.06
			H= -9.13	L= -1.53	E_g= 7.60		
	INDO/S	Pure	**49.22**	1.11	0.12	4.97	7.80
		DMSO	209.37	0.09	0.09	5.90	4.83
			321.87	0.55	0.49	3.86	5.65
			468.94	0.02	0.02	2.64	5.99
			583.78	0.16	0.16	2.12	6.09
			H = -7.87	L = -0.41	E_g = 7.46		
9TP9B	CNDO/S	Pure	**238.67**	0.79	0.68	5.16	5.94
			263.28	0.37	0.37	4.70	11.66
		DMSO	**205.27**	0.62	0.55	6.02	4.13
			238.08	0.30	0.30	5.21	9.13
			683.22	0.11	0.11	1.82	5.80
			789.07	0.20	0.21	1.58	6.04
			H= -9.19	L= -1.56	E_g= 7.63		
	INDO/S	Pure	**248.63**	1.01	0.30	4.95	7.68
		DMSO	213.47	0.05	0.05	5.81	4.64
			301.95	0.26	0.26	4.11	8.78
			396.87	0.02	0.02	3.12	6.10
			454.29	0.34	0.31	2.72	5.99
			478.32	0.14	0.14	2.59	5.68
			H = -7.92	L = -0.39	E_g=7.53		

*EC unit: 10^4 dm^3 mol^{-1} cm^{-1} **Bold**value represents UV absorption maxima

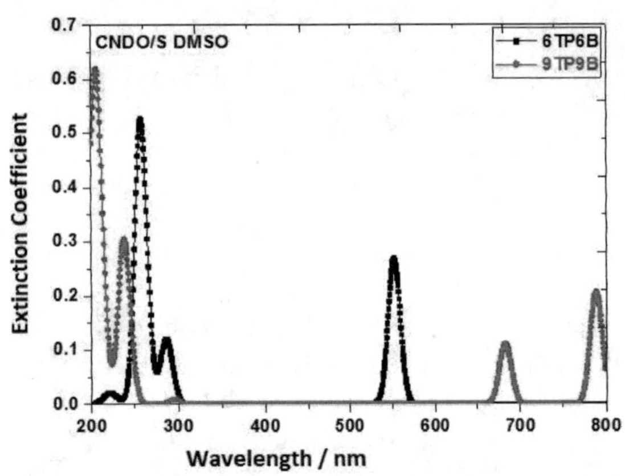

Figure 4 Absorption spectra of 6TP6B and 9TP9B molecules in DMSO using CNDO/S Method

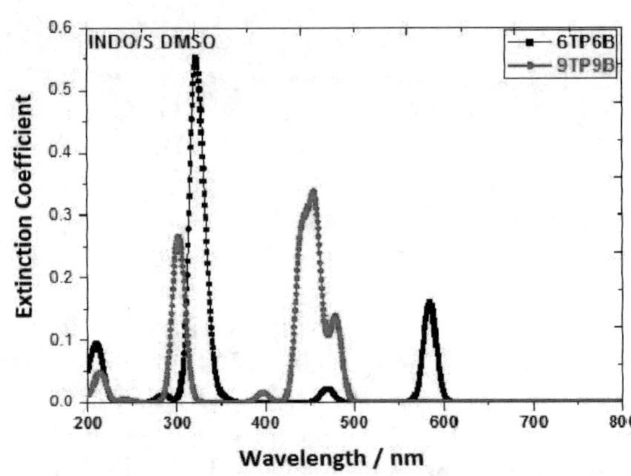

Figure 5 Absorption spectra of 6TP6B and 9TP9B molecules in DMSO using INDO/S Method

(c) Electrochemical properties

The electrochemical properties of the *n*TP*m*B (*n*=*m*=6, 9) molecules using CNDO/S, INDO/S methods have been reported in Table 4. The estimated energy gap and chemical hardness values offer an understanding of the chemical stability of the compounds. The larger energy gap indicates the larger kinetic stability in accordance with the principle of softness-hardness. In general, the more energy gap of the molecule offers more hardness, and also provides higher thermal and kinetic stability. The smaller gap lowers the kinetic stability and offers higher chemical reactivity. The parameter ΔN_{max} reveals the propensity of the molecule to get additional electronic charge from the environment. It has been observed from the Table that both molecules exhibit same values of maximum charge transfer. This indicates the almost similar charge distribution behaviour of the molecules which is in agreement with the partial charge analysis (section 3.1). Further, it may be observed from Table 4 that almost all parameters increase from hexyl to nonyl group, and both methods are exhibits good agreement in estimating all the parameters.

Table 4 Calculated values of ionization energy I=($-E_H$), electron affinity A=($-E_L$), electro negativity χ= (I+ A)/2, chemical hardness η=(I-A)/2, electronic chemical potential μ= -(I+A)/2, electrophilic index $\omega = \mu^2/\eta$, and softness $S=1/\eta$, maximum charge transfer ΔN_{max} = - (μ/η), of *n*TP*m*B (*n*=*m*=6, 9)molecules using CNDO/S, and INDO/S methods.

Molecule	Parameter	CNDO/S method	INDO/S method
6TP6B	I/eV	9.13	7.87
	A/eV	1.53	0.41
	χ/eV	5.33	4.14
	η/eV	3.80	3.73
	μ/eV	-5.33	-4.14
	ω/eV	7.47	4.60
	S/eV^{-1}	0.26	0.27
	ΔN_{max}	1.40	1.11
9TP9B	I/eV	9.19	7.92
	A/eV	1.56	0.39
	χ/eV	5.37	4.15
	η/eV	3.81	3.76
	μ/eV	-5.37	-4.15
	ω/eV	7.57	4.58
	S/eV^{-1}	0.26	0.26
	ΔN_{max}	1.41	1.10

4. Conclusions

1. The strongest positive and negative charges in both molecules are nearly same. Hence, appreciable difference in terms of phase transition temperature may not be found. This is in accord with literature details.
2. Both methods are in agreement with each other by exhibiting blue-shift of wavelength, and hypochromic effect when hexyl end group is replaced by nonyl end group in pure form.
3. The UV stability of 6TP6B is more compared to 9TP9B in pure form. The DMSO effect does not alter the UV stability of 6TP6B. But the solvent enhances the UV stability of individual compounds from its pure form to solvent.

References

[1] Andrienko D 2018 Introduction to liquid crystals *J. Mol. Liq.* **267** 520-541
[2] Kirsch P 2015 Fluorine in liquid crystal design for display applications *J. Fluor. Chem.* **177** 29-36
[3] Saccone M, Pfletscher M, Dautzenberg E, Dong YR, Michal CA, Geise M 2019 Hydrogen-bonded liquid crystals with broad-range blue phases*J Mat. Chem C* **7** 3150-3153

[4] Praveen PL, Ojha DP 2014 Photo sensitivity, substituent and solvent induced shifts in UV-Visible absorption bands of naphthyl-ester liquid crystals: a comparative theoretical approach *Liq. Cryst.* **41** 872-882

[5] Chang CM, Lin YH, Srivastava AK, Chigrinov VG 2018 An optical system via liquid crystal photonic devices for photobiomodulation*Sci. Reports.* **8** 4251 1-10

[6] Praveen PL 2018 7-Hexyloxy-3-[4'-(3-methylbutyloxy) phenyl]-4H-1-benzopyran-4-one: Study of Smectic behaviour and UV absorption profile*Mol. Cryst. Liq. Cryst.* **667** 44-53

[7] Misra AK, Tripathi PK, Manohar R 2015 Fluorescence, UV absorbance and dielectric studies of fluorescent dye doped ferroelectric liquid crystal*J. Non-Crystalline Solids.***412** 1-4

[8] Bouman T, Lightner DA 1979The octant rule. 5. Nature of the third nodal surface. Anunderstanding of "anti-octant" and front octant effects by a CNDO/S study of rotatory strengths of the carbonyl n .fwdarw. .pi.* transition*J.Am. Chem. Soc.***98** 3145-.3154

[9] Volosov AP, Zublov VA 1977 Role of the configuration interaction in the CNDO/S calculation of optical rotatory strengths*Theor. Chim. Acta.***44** 375-384

[10] Komori H, Inai Y 2006 Electronic CD Study of a Helical Peptide Incorporating *Z*-Dehydrophenylalanine Residues: Conformation Dependence of the Simulated CD Spectra *J. Phys. Chem. A.***110** 9099-9107

[11] Hanemann T, Bohm MC, Haase W, Wu ST 1992 Calculation of UV/VIS absorption spectra of liquid crystals and dye molecules *Liq. Cryst.* **11** 917-927

[12] Markovic Z, Manojlovic N, Zlatanovic S 2008 Electronic absorption spectra of substituted anthraquinones and their simulation using ZINDO/S method *J. Ser. Soc. Comp. Mech.*273-79

[13] Quotschalla U, Hanemann T, Bohm MC, Haase W 1991 Conformation analysis and absorption properties of anthraquinone dyes – a quantum chemical approach *Mol. Cryst. Liq. Cryst.* **207** 103-116

[14] Szabo A,Ostlund NS 1989 Modern Quantum Chemistry: Introduction to AdvancedElectronic Structure Theory, McGraw-Hill: New York

[15] Tariq M, Hameed S, Bechtold IH, Bortoluzzi AJ, Merlo AA 2013 Synthesis andcharacterization of some novel tetrazoleliquid crystals*J. Mat. Chem. C.***1** 5583-5593

[16] Praveen PL, Ojha DP 2012 Calculation of spectral shifts in UV–visible region and photo stability of thermotropic liquid crystals: Solvent and alkyl chain length effects*J. Mol. Liq.***169** 110-116

[17] Pedretti A, Villa L, VistoliG 2002VEGA: a versatile program to convert, handle and Visualize molecular structure on windows-based PCs*J. Mol. Graph Model.***21** 47-49

IOP Conf. Series: Materials Science and Engineering **798** (2020) 012042 doi:10.1088/1757-899X/798/1/012042

Study of the Radiotherapy Treatment Margins in Prostate Cancer with Fuzzy Logic Model

S K Patnaikuni[1,2], S M Saini[1], R M Chandola[2], V Choudhary[2]

[1]Department of Physics,National Institute of Technology Raipur, Chhattisgarh, India.
[2] Department of Radiotherapy, Pt.JNM Medical college, Raipur, Chhattisgarh, India
e-mail:smsaini.phy@nitrr.ac.in

Abstract. Start your abstract here...External beam radiation therapy (EBRT) is a process in which therapeutic radiation dose is delivered to targettissues whilst seeking the risk of healthynormal and critical organs is minimal. In case of Complex radiation therapy technique such as volumetric modulated arc radiotherapy (VMAT),there is requirement of precise and optimal treatment margin selection so that not only the resulting dose distributions gives high tumor control probability(TCP) but also dose close to critical organs will get minimal. A fuzzy logic application was used here to derive the optimal radiotherapy treatment margins to be used in radiation therapy treatment planning. Here major all possible radiotherapy uncertainties accounted such as translational set-up, organdelineation and organ motion-induced errors for calculating the quantitative variations in target volume radiobiological parameters for treatment plans using fixed step sized increments of treatment margins up to maximum possible variations of tumor volumes.Therules and membership functions were adopted from dosimetricresults of treatment planning for the fuzzy inference system. The imprecision and smoothness of the original fuzzy output was corrected with application of a convolution technique. The results demonstrate that performance of the applied fuzzy model margins against current used radiotherapy margins was consistent inlower range of error magnitude but beyond that a significant non-linearly margin performance was found when considering the normal tissue complication probability (NTCP) and constraint factors into account in the margin recipe formulation. With the new margins obtained then applied for the study of prostate cancer treatment planning and results were compared well with current volumetric modulated arc radiotherapy (VMAT)technique.

Key words: PTV (Planning target volume), TCP, NTCP, volumetric modulated arc radiotherapy (VMAT).

1. Introduction

In the external beam radiotherapy, the therapeutic radiation dose is delivered to targettissues whilst seeking the risk of healthy normal and critical organs is minimal. Before the treatment, the area of interest to be irradiated, all target volumes and normal structures should be delineated clearly .The inability of current imaging modalities may lead to the inter and intra-observer variability when considering the microscopic tumor spread for particular type of cancer. In the external beam radiotherapy treatment planning the energy, geometry of radiation beam,dose prescription and optimization are considered to achieve the therapeutic gain of treatment. During the fractionated radiotherapy treatment course, many variations such as translational set-up,organdelineation and organ motion errors affect the planned treatment volume with actual treatment volume so that as a result the delivered dose to planned volumes which may differs from the actual received dose to planned volumes. To avoid this significant dose variation, treatment volumes must be defined with precise margins so that to achieve the therapeutic gain of treatment by maintain the geometrical accuracy. Many standard recommendations for the definition of margins applied in radiotherapy treatment planning are available via ICRU reports 50 and 62 [1-2]and formulations are based on probabilistic dose distributions. The gross tumor volume (GTV) which is the identifiable or palpable tumor and a clinical target volume (CTV) is defined with extra margin to GTV to considering the effects of microscopic growth of tumor, whilst the inclusion of additional margin to CTV for geometrical accuracy results is the planning target volume (PTV). The radiobiological concepts for tumor volumes and normal structures are control probability (TCP) which is the probability of killing all tumor cells in a volume.Thenormal tissue complication probability (NTCP) which is the damage that occurs to normal tissues and critical organs respectively.These biological parameters are calculated to get the possible radiation therapy rationaleor probability of cure without complications which is obtained by estimating the Uncomplicated Tumor Control Probability using the margin recipe formulation.

In the present study we considered uncertainties i.e. translational set-up, organ contouring and organ motion errors to estimate the required margins in radiotherapy with fuzzy inference system. To quantify mathematically relationship between the above said radiobiological parameters, radiotherapy margins as well as radiotherapy uncertaintiesis difficult, however the distinct advantage of fuzzy logic is thatallow the linkage of these parameters through the use of fuzzy rules and membershipfunctions to have possible output [5,6-8].Advanced radiotherapy treatment delivery techniques such as intensity modulated radiotherapy (IMRT), volumetric modulated arc therapy (VMAT) and Tomotherapy required precise treatment margins so that treatment volumes must be defined with precise margins to achievehigh dose close to critical organs and dose escalation. The application of the derived fuzzy margin was applied this study and compared to currently usedmargins to assesssuch a clinical situation.

2. Materials and methods

The TheSugeno type fuzzy inference system was used considering knowledge based rules and Gaussian membership functions in the PTV margin modeling as it gives results which would be consistent with the expected output. From variousstudies [9-10], the possible variations of prostate tumormotion displacement was observed to be 0-15 mm. With help of earlier sources and our institutional image guiding techniques the procedure first started with defining variable PTV margins (0mm to 16 mm) symmetrically with treatment planning system (TPS) using typical incremental error magnitudes as obtained during radiotherapy treatment. For each PTV, the treatment plans were generated for prostate cancer treatment. The dosimetric statistics of each plan were used to calculate baseline values of TCP and NTCP as starting point used for comparisonsinMatlab program and also subsequent calculations of new TCP and NTCP values corresponding to each plan which has new

PTV margin with stepped symmetricalincrement. The crisp output was evaluated with final defuzzification and taken care for smoothness of the fuzzy output processing with Gaussian convolution kernel. Then using final PTV margin estimated from the fuzzy model, the VMAT treatment planningwas performed using the commercial Eclipse 15.6(Varian Medical Systems) treatment planning system (TPS) in which The AAA algorithm is configured in the Eclipse Beam Configuration task to provide the increased speed and accuracy required for modern treatment techniquesthen final plan was compared with currently applied van Herk margins for prostate cancer. All treatment plans planned with different PTV marginswhich were varing symmetrically from 0 to 16 mm in 2 mm step sized increment from minimum to maximum possible variation of PTV. The dose prescription was set as conventional dose fractionation with 70Gy.Theplan passing criteria so that 95% isodose line covered the PTV on all slices to achieve the high conformit y of dose distribution. For each plan the dosimetric results ofthe prostate volume (CTV) and major normal structures (rectum and bladder) were calculated. The Equivalent Uniform Dose (EUD) modeling[11] , which was recommended for the true radiobiological outcomes and here was used to calculate the radiobiological parameters TCP and NTCP using a Matlab-based simulation tool based on the concept EUD which is defined as the uniform dose distribution giving an equivalent survival fraction to that of a heterogeneous dose distribution. In calculation of radiobiological parameters i.e loss of prostate TCP which corresponding to the difference between new TCP and baseline value(TCP_0) and the increase in rectal NTCP which corresponding to the difference between new NTCP and baseline value($NTCP_0$) for each plan. The loss of prostate TCP and increase in rectal NTCP were denoted as ΔTCP, ΔNTCP respectively.Foreach of the treatment plans using the different PTV margin with stepped increment size, ΔTCP and ΔNTCPwere calculated. The fuzzyrules were defined based mainly on expert knowledge and clinically assessable so that the balance which allows the risk between increase in NTCP and reductionthe PTV margin also the loss in and increasing the PTV margin size respectively. Here the theloss of prostate TCP and increase in rectal NTCP were chosen as two inputs forFuzzy system and one output, i.e. PTV margin for each normal tissue nearby prostate PTV. Six membership functions i.e. Very High, High, Medium, Small, Very Small, Almost Zero and were chosen for the input and output values considering knowledge based andexisting normal tissue toxicity limitations. Preselected increase in rectal NTCPas well as the irradiated volume of the anterior wall of rectum werechosen so as to allow the algorithm to select margins forrectal complications. The optimization of fuzzy rules[13-14]derived from the input data and using the conditions imposed on the margin limits. The weighted average technique was used to calculate the final outputof the SugenoFIS.

3. Results and discussion

The resulted output function of the Sugeno FIS obtained from the defuzzification as 3D surface view in which each PTV margin point corresponds to a specific value of two inputs. i.e ΔTCP and ΔNTCP from a matrix representation of the output function from 3D surface of particular OAR and target sub volumes of OAR combination. Also it was observed when ΔNTCP increasing, the results in a PTV margin decreases. Also the loss in TCP increasing the resultant PTV margin would be increasing as satisfying the preselected margin requirements. The plan with resultant fuzzy model PTV margin was compared to the plan with PTV commonly used margin recipe van Herk*et al* [4]. Considering the modeling uncertainty into consideration, it was observed that good agreement but between the conventional PTV margin based on van Herk*et al*'s formulation and fuzzy model PTV margin up to 5.5 mm s.d. equivalent errors but on average 0.5 mm bigger margin of fuzzy model than the conventional margin in the considerable clinical range. Due to the conditions imposed in our margin formulation, the fuzzy PTV margin apparently not exceeding approximately 12.5mm for errors above 4.5-5mm s.d. The fuzzy PTV margin observation was linear up to 11 mm range which shows insignificant variation from van Herk margin but variation was observed non-linearly greater than

5mm s.d. as shown in Figure-1. This was due to introducing of knowledge based fuzzy rules, the normal tissue complication rates, rectal tolerance constraints and related biological factors were incorporated into the margin formulation.

To assess the performance of fuzzy derived PTV margin against current margins, The VMAT treatment planning using appropriate plan geometry was performed with new derived fuzzy margin and compared with VMAT in which conventional PTV was sat as target volume based on van Herk formulation. As the fuzzy margin on an average 0.5mm bigger than the conventional margin in the considerable range of clinical acceptance . But higher the error magnitude greater than 5-5.5mm s.d. clinically fuzzy margin would be significant depending on clinical imposed and existing patient related factors. The dose-volume results of the VMAT plans for the prostate PTV, the dose volume histogram (DVH)was shown in Figure- 2(a).which demonstrates more or less similarity was observed between the plans as they were considered in lower s.d. variation depends on tumor staging and growth characteristics of current prostate case. In case of critical organs, rectum was observed as organ at risk to be spared. The dose volume histogram (DVH) of the rectum shown in Figure-2(b),in which intermediate volumes may significant but overall insignificant differences were observed. But as posterior margins increases, biological based fuzzy margins may have significant ΔNTCP results of OARs.

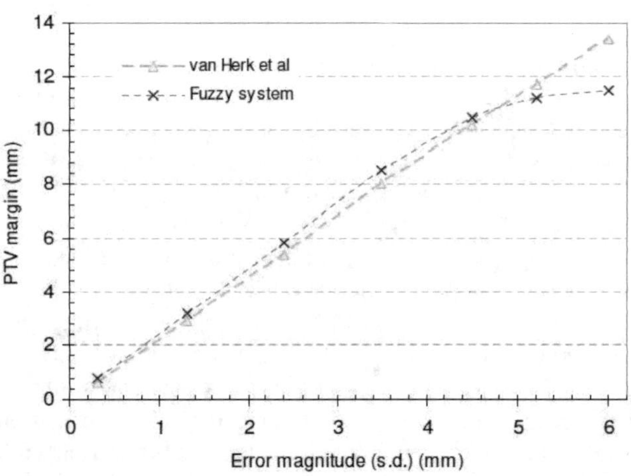

Figure-1.Comparison of van herk*et al* and fuzzy model based ptv margins, in terms of the standard deviation of total displacement errors, using the same corresponding uncertainties as inputs

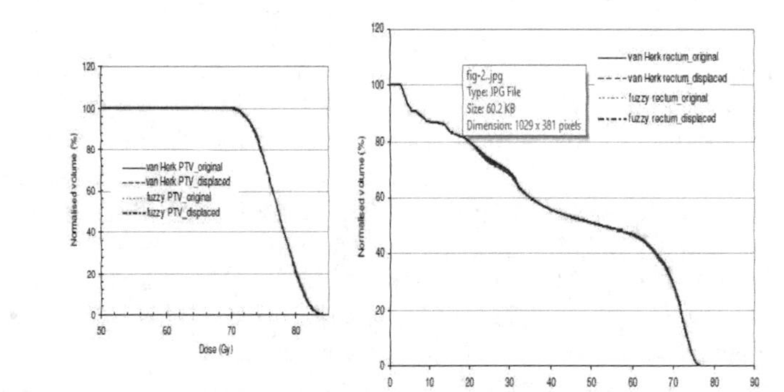

Figure-2(a) Figure-2(b)

Figure-2(a).DV(dose-volume) Histograms for the prostate PTV and Figure-2 (b) rectum for plans using the conventional(van herk) and expected fuzzy margins.

4. Results and discussion

The application of fuzzy logic based PTV margin in the treatment plans was dictated to produce more or less similar results compared to plans using current treatment margins but might be favorable clinical endpoints for organ motion uncertainties when higher the error magnitudes. Here the practical usefulness and advantage of using fuzzy logic was addressed under practical limitation on the margin size as proposed fuzzy approach using not only physical dose constraints but also radiobiological data to optimize the required margin to estimate true outcomes whereas the most existing margin models considering mostly physical dose-volume data alone. Apart of that the fuzzy model is not only relatively simple to implement also gives accurate margin sizes and so applicable to all other volumes of interest to be treated and have organ motion related erros and risk estimation of sub volumes. Consistency of the fuzzy margin compared with the classical van Herkbased margin recipe showed nearly similar. The new fuzzy margin was applied in VMAT treatment planning. The results were compared to classical plans showing good consistency. The practical applications of this fuzzy margin in biological based radiation therapy treatment planning to compare true outcomes are still under deep learning.

References

[1] ICRU Report 50, “Prescribing, Recording and Reporting Photon Beam Therapy,” *International Commission on Radiation Units andMeasurements*, 1993.

[2] ICRU Report 62, “Prescribing, Recording and Reporting Photon Beam Therapy (Supplement to ICRU Report 50),” *International Commission on Radiation Units and Measurements*, 1999.

[3] J. C. Stroom, H. C. J. de Boer, H. Huizenga and A. G. Vissier, “Inclusion of geometrical uncertainties in radiotherapy treatment by means of coverage probability,”*Int J RadiatOncolBiolPhys,* vol. 43,pp. 905-919, 1999.

[4] M. van Herk, P. Remeijer, C. Rasch and J. V. Lebesque, “The probability of correct target dose: Dose-population histograms for deriving treatment margins in radiotherapy,” *Int J RadiatOncolBiolPhys,* vol. 47, pp.1121–1135, 2000.

[5] T. Waschek, S. Levegrun, M. van Kampen, M. Glesner, R. Engenhart-Cabillic and W. Schlegel, “ Determination of target volumes for three-dimensional radiotherapy of cancer patients with a fuzzy system,” Fuzzy Sets Systems, vol. 89, pp. 361-370, 1997.

[6] R. Alcala, J. Cano, O. Cordon, F. Herrera, P. Villar, I. Zwir, “Linguistic modeling with hierarchical systems of weighted linguistic rules,” *International Journal of Approximate Reasoning* ,vol. 32, pp. 187-215, 2003.

[7] N. Baboshin, D. Naryshkin, “On identification of multidimensional fuzzy systems,” *Fuzzy Sets and Systems*, vol. 35, pp. 325-331, 1990.

[8] J. De Oliveira and F. Gomide, “Formal methods for fuzzy modeling and control,” *Fuzzy Sets and Systems*, vol. 121, pp. 1-2, 2001

[9] R. Miralbell, P. Nouet, M. Rouzaud, A. Bardina, N. Hejira and D.Schneider, “Radiotherapy of bladder cancer: Relevance of bladder volume changes in planning boost treatment,” *Int J RadiatOncolBiolPhys*, vol. 41,pp.741-746, 1998.

[10] R. K. Sur, J. Clinkard, W. G. Jones, R. E. Taylor, H. J. Close, A. Chaturvedi and B. Carey, “Changes in target volume during radiotherapy treatment of invasive bladder carcinoma,” *ClinicaOncology,* vol. 5, pp. 30-33, 1993

[11] H. A. Gay and A. Niermierko, “A free program for calculating EUDbased NTCP and TCP in external beam radiotherapy,” *PhysicaMedica,* vol. 23, pp. 115-125, 2007

IOP Conf. Series: Materials Science and Engineering **798** (2020) 012042 doi:10.1088/1757-899X/798/1/012042

[12] B. Mzenda, M. E. Hosseini-Ashrafi, A. Palmer, Liu and D.J.Brown, "Quantification of the effects of random and systematic errors in external beam radiotherapy using a Monte Carlo technique," IPEMBiennial Radiotherapy Meeting, Bath, UK, 2008.
[13] Combs W, Andrews J, "Combinatorial rule explosion eliminated by a fuzzy rule configuration.," IEEE Transactions on Fuzzy Systems, vol.6, pp. 1-11, 1998.
[14] Dubois D, Prade H, Ughetto L, "Checking the coherence and redundancy of fuzzy knowledge rule bases," IEEE Transactions on Fuzzy Systems , vol. 5, pp. 398-417, 1997.
[15] F. Fernández and J. Gutiérrez, "Specification and optimization of fuzzy systems using convolution techniques," *Signal Processing,* vol. 80, pp. 935-949, 2000

AUTHOR INDEX